CHRONIC LYMPHOCYTIC LEUKEMIA RESEARCH FOCUS

CHRONIC LYMPHOCYTIC LEUKEMIA RESEARCH FOCUS

CHADI NABHAN
EDITOR

Nova Biomedical Books
New York

For permission to use material from this book please contact us:
Telephone 631-231-7269; Fax 631-231-8175
Web Site: http://www.novapublishers.com

Library of Congress Cataloging-in-Publication Data

Chronic lymphocytic leukemia research focus / Chadi Nabhan (editor).
p. ; cm.
Includes bibliographical references and index.
ISBN-13: 978-160021-526-1 (hardcover)
ISBN-10: 1-60021-526-2 (hardcover)
1. Lymphocylic leukemia. 2.Chronic leukemia. I.Nabhan, Chadi.
[DNLM: 1. Leukemia, Lymphocytic, Chronic--genetics. 2.Leukemia, Lymphocytic, Chronic- -therapy. 3.Age Factors. 4.Cytogenetics- -methods. 5.Leukemia, B-Cell, Chronic- -genetics. 6.Leukemia, B-Cell, Chronic- -therapy. WH 250 C55673 2006]
RC643.C4855 2006
616.99'419- -dc22 2007008428*s*

Published by Nova Science Publishers, Inc. ✣ New York

Contents

Preface

Chronic lymphocytic leukemia (or "chronic lymphoid leukemia"), known for short as CLL, is a type of leukemia in which too many lymphocytes are produced. Although the malignant lymphocytes in CLL may look normal and mature, they are not and these cells may not cope effectively with infection. CLL is the most common form of leukemia in adults. Men are twice as likely to develop CLL as women. However, the key risk factor is age. Over 75% of new cases are diagnosed in patients over age 50.This new book presents important research from around the world in this field.

Chapter I - For years, chronic lymphocytic leukemia (CLL) has been viewed as an indolent disease that is rarely lethal. Therapeutic options have included purine analogues, alkylating agents, and steroids. Since 1999, there have been a number of publications on the biologic aspects of CLL and various prognostic factors that influence survival. The challenge that hematologists face lies in trying to integrate this information in the management of patients with CLL. Genomic aberrations, mutational status of the immunoglobulin variable heavy chain gene (*IgVH*), CD 38 expression, and overexpression of a tyrosine kinase, ZAP-70, have all been reported to be potential prognostic markers that may identify a subset of patients whose disease course is likely to be aggressive and potentially fatal. CLL patients with poor prognosis usually have the unmutated *IgVH*, chromosomal abnormalities such as 17p or 11q, and overexpress ZAP-70 and CD 38. The important question is what is the best therapeutic strategy for such patients.

Data suggest that patients with 17p deletion and/or p53 mutations fail traditional therapy with purine analogues and rituximab. Several retrospective studies suggested that the humanized anti-CD 52 antibody alemtuzumab may be an attractive option in these patients. Furthermore, additional studies have suggested that patients who achieve molecular disease negativity could attain prolonged duration of remission. In this review the novel prognostic factors in CLL will be highlighted and suggestion on how they should be integrated in the therapeutic armamentarium of CLL will be explored. Finally, we will review some data on minimal residual disease and how this concept is being applied in future clinical trial design.

Chapter II - Chronic lymphocytic leukemia (CLL) is relatively common leukemia manifests by the uncontrolled accumulation of mature B-cells that express CD5. The most intriguing aspect of CLL disease is the heterogeneity in its clinical course and outcome. Studying the mutation status of the immunoglobulin heavy-chain variable region (IgV_H) in patients with CLL reveiled that patients with CLL could have leukemic cells originating from either pre-somatic hypermutation stage and have aggressive disease or from post-somatic hypermutation stage and have more indolent course. These findings have stimulated active research to better understand the biology of CLL and helped clinicians to better manage their patients. Here we review the most recent data accumulated on the biology of CLL and discuss the clinical relevance of this data on the prediction of clinical behavior and outcome as well as therapy. We focus on the how basic science discoveries can be used in the clinical management and stratification of CLL patients for various therapeutic approaches.

Chapter III - Novel biologic and molecular markers have been recently described in chronic lymphocytic leukemia that are likely to challenge the traditional approach of "watchful waiting" in asymptomatic patients and poor-risk prognostic factors. Molecular techniques such as microarray DNA, gene expression profile, polymerase chain reaction, and others, are capable of predicting differences in survival even when patients are stratified to the same clinical stage at diagnosis. Several ongoing clinical trials are evaluating whether early therapeutic intervention for chronic lymphocytic leukemia (CLL), compared to later treatment at onset of disease progression has an effect on survival. This paradigm becomes more complex when patients are diagnosed at a younger age. Also, integrating these markers into early therapeutic decision-making could be uniquely challenging, particularly when selecting the optimal therapeutic approach. Substantial progress has been made in developing new targeted therapies that are not only capable of inducing responses by non-specific cytotoxicity, but by activating different apoptotic mechanisms and other immunomodulatory effects as well. High dose chemotherapy (HDCT) followed by autologous hematopoietic cell transplantation (HCT) induces clinical and molecular responses in patients with CLL; however, it has not yet translated into cure. Conversely, allogeneic HCT appears to be a reasonable alternative option, especially when targeted strategies are incorporated into modern reduced-intensity preparative regimens. In this chapter, we discuss a variety of transplantation strategies that have been explored in the treatment of chronic lymphocytic leukemia, and evaluate the impact of novel prognostic factors in that perspective.

Chapter IV - Chronic lymphocytic leukemia of B-cell lineage (CLL) is the most common leukemia in the Western world and is characterized by an accumulation of clonal B-cells with a unique immunophenotype (including CD5+, CD23+, CD10-, FMC 7-). The presence or absence of somatic hypermutations identifies two subgroups with different clinical behaviors. These two subgroups also have different gene expression signatures as identified by DNA microarray.

Prolonged cell survival is one of the fundamental attributes of CLL cells. Although many abnormalities in apoptosis regulation in CLL have been reported, the exact defects and underlying mechanisms are far from being completely delineated. Further, the clinical relevance of apoptosis regulatory molecules as prognostic or predictive variables is not established.

The authors discuss the relevant literature on apoptosis regulation in CLL, extrinsic and intrinsic pathway abnormalities in CLL, our work on the Bcl-2 family regulators, and research directions in this area. After identifying aberrant expression of Bcl-2, Bax, and Mcl-1 in CLL the authors sequenced BAX and MCL-1 genes. A novel single nucleotide polymorphism (SNP), G125A, in the BAX promoter was found to be associated with reduced

Bax protein expression and poor clinical outcome. Subsequently, we have provided evidence of direct effect of this SNP on gene expression by showing its association with reduced Bax mRNA and by Luciferase gene reporter assay. They also identified a short sequence insertion in the MCL-1 promoter and demonstrated its association with gene expression and clinical end-points and using a functional assay a cause-effect relationship between these promoter changes and gene expression. Using DNA microarray focused on 112 apoptosis regulatory genes, we have identified consistent aberrant expression of 10 molecules (from the Bcl-2 family, IAP proteins, PI3 kinase pathway and death receptors) in CLL compared to normal flow-sorted age and sex-matched B-cells and confirmed this data by RT-PCR.

Target-based treatment, as opposed to "carpet-bombing" chemotherapy requires strong justification from basic cancer biology. In order to improve response rates and develop effective treatments that might be able to replace combination chemotherapy, identification of genes/proteins that play a crucial role in the pathogenesis of CLL is required. Some of these genes/proteins may also help identify patients at risk of aggressive disease. In this article the authors discuss the biologic and clinical relevance of apoptosis regulation in CLL.

Chapter V - Tis11b/Berg36 is a member of a family of proteins involved in post-transcriptional gene regulation. Members of this family bind to AU rich elements in certain mRNA species and increase mRNA stability. mRNA that contain AU rich elements and can be regulated in this way include molecules involved in the regulation of apoptosis eg bcl-2 and several cytokines eg TNF, IL-2. The authors have recently shown that Tis11b/Berg36 is involved in the regulation of apoptosis following Rituximab treatment of CLL cells suggesting a pro-apoptotic role of this gene in CLL cells. Thus B-CLL cells were stimulated with IL-4, anti-CD40, anti-CD40+IL-4 and PMA. It was found that IL-4, CD40 and their combination significantly inhibited spontaneous apoptosis while PMA was able to inhibit spontaneous apoptosis in some but not all patients tested and the overall effect did not reached statistical significance.. Unexpectedly Tis11b/Berg36 mRNA remained unchanged following IL-4 treatment, but expression was induced following anti-CD40 or PMA treatment. Because it was found that regulation of Tis11b/Berg36 mRNA following the latter two treatments was under the control of NF-κB pathway and not p38 (as found for Rituximab treatment) it was hypothesized that this gene may be involved in the regulation of cell cycle or differentiation of B-CLL cells. Indeed it was found that stimulation of B-CLL cells with either PMA or anti-CD40 resulted in increase in sIgM as a marker of B-CLL differentiation. CLL cells were also found to express high basal levels of two other members of this family, Tis11 and Tis11d. From these stimuli, IL-4 or anti-CD40 were found to downregulate the basal expression of Tis11 mRNA. These data suggest that members of the Tis11b/Berg36 family are involved in the regulation of apoptosis and differentiation in B-CLL cells

Chapter VI - B-cell chronic lymphocytic leukaemia results from clonal expansion of cells expressing CD19, CD5 and CD23 with reduced levels of membrane IgM, IgD, and CD79b.A primary genetic abnormality characteristic of CLL has not been identified.Yet the monoclonal nature of the B-lymphocytes imply that an inducing lesion must exist in the progenitor clone. The clinical course in CLL is extremely diverse.Despite the heterogeneity the most consistent feature in CLL is the low membrane B cell receptor (BCR) expression. Three areas of genetic factors influencing the biology and prognosis of CLL will be reviewed.

VH mutations. CLL with unmutated immunoglobulin variable heavy chain (VH) gene displays worse prognosis compared to VH mutated CLL. V1-69 is the most common unmutated gene in CLL while the rearrangement of the V3-21 gene has been associated with poor clinical outcome irrespective of the VH mutation status. Biased use of VH genes may reflect antigenic selection in the development of CLL.

The nature of the cell of origin is discussed.

Genetic imbalances. Losses and gains of DNA have been identified in CLL by fluorescence in situ hybridisation (FISH) and are associated with disease progression..Three consistent recurring abnormalities studied by FISH have been used extensively as prognostic markers. Deletion of 13q14 found in 50% of CLL and when present on its own is considered a good prognostic marker. Deletions of 11q22-23 (20%) is associated with disease progression while 17p deletion (10-15%) identifies the poorest prognostic group. The use of comparative genomic hybridisation (CGH) and array based CGH has allowed identification of other recurrent genetic imbalances.Further epigenetic mechanisms such as methylation can silence genes in the absence of a deletion and inactivating mutation in CLL. The genes of relevance in these regions will be discussed.

Genetic predisposition . The study of inherited susceptibility and identification of susceptibility genes could provide insights into pathogenic mechanism and potential targets for therapeutic intervention. Although familial susceptibility to CLL exists, with the age of onset of CLL being younger in CLL-families than in the general population., to date no gene has been shown to unambiguously confer susceptibility to CLL when mutated in the germ line. The investigation of apparently healthy relatives within CLL families with sublinical levels of CLL clones is an additional approach to increase the numbers of affected individuals to investigate genetic predisposion to CLL.

Chapter VII - The objective of the study was to evaluate if prognostic factors as stage, lymphocyte morphology, lymphocyte doubling time, CD38 expression, genomic aberrations, Ig V_H mutational status and ZAP-70 expression correlate with the in vitro apoptosis of CLL lymphocytes. We also wondered if lymphocytes from patients who were treatment naive or were exposed in the past to different treatment regimens would show a different apoptosis profile. The authors were interested too if this in vitro assay could predict treatment outcome.

In vitro apoptosis of the mononuclear fraction was measured after 24h culturing using the flow cytometric method with Annexin V binding combined with 7-amino actinomycin D (7-AAD). Cells were cultured in cell medium alone, were exposed to pharmacological doses of chlorambucil (Chl), fludarabine (FdA) or methylprednisolone (MP) and were irradiated with 2 Gy.

A total of 187 samples of 110 patients with CLL were successfully analysed.

Mean spontaneous apoptosis of lymphocytes was 27.42 %, with a broad range. 2 Gy, Chl, FdA and MP increased apoptosis respectively with an average of 17.9%, 13.25%, 14.99% and 22.04%.

A higher spontaneous and lower Chl and FdA induced apoptosis was found in high risk stage patients, who were almost all treated in the past. A rapid lymphocyte doubling time was associated in untreated and treated patients with a higher MP sensitivity. Less irradiation and more MP induced apoptosis was seen in samples with unmutated V_H genes. The higher MP sensitivity applied to the treated subgroup only. 11q deletion sensitized lymphocytes to MP apoptosis while 17p deletion made lymphocytes more resistant to FdA without suspected differences between treatment naïve and treated patients. More MP apoptosis was also observed in ZAP-70 positive lymphocytes.

53 patients were treated after assessing in vitro apoptosis, adding up to a total of 73 treatment outcomes, which prospectively could be correlated with the in vitro results. The in vitro assay forecasted the clinical response correctly in 37/44 patients treated with Chl, in 14/17 patients treated with FdA and in 12/12 patients treated with splenic irradiation resulting in an overall predictive accuracy of 86%.

All these results indicate that the Annexin V-7-AAD apoptosis assay is a valuable tool to test drug sensitivity in CLL. Old and new prognostic factors as well as treatment history can modify spontaneous, irradiation or drug induced apoptosis. A good correlation between in vitro results and in vivo responses is translated in a high overall predictive accuracy .

Chapter VIII - The considerably variable clinical course (indolent to aggressive) of B-cell chronic lymphocytic leukemia (B-CLL) addressed clinical scientists to the search for biological prognostic indicators. The tumor burden-based staging systems developed almost three decades ago was followed by a quest involving a vast array of clinical and biological parameters. More recently, the analysis of IgVH mutation status provided the most powerful prognostic indicator in B-CLL. Beside Ig, the expression of CD38 was reported as having an independent prognostic value and to correlate – at least partially - with the absence of mutations in the IgVH genes. The consequence is that CD38 was proposed as an easy test, and an inexpensive surrogate of the Ig status. However, the validity of CD38 expression as a prognostic marker was not confirmed by other investigators and gave origin to debate still open. This chapter represents a contribute to the analysis of issues and problems linked to the use of CD38 as a prognostic marker in B-CLL.

Chapter IX - The identification of the non-random genetic alterations associated with cancer has great importance not only to basic research but also to disease management as diagnostic and prognostic markers. In contrast to other types of lymphomas, translocations are rare events in chronic lymphocytic leukemia/small lymphocytic lymphoma (CLL/SLL) and not any CLL/SLL-specific translocation has been found yet. Moreover, translocations have been reported to be a prognostic marker for an unfavorable clinical outcome. The purpose of our study was to contribute to the understanding of the chromosome alterations associated with this group of lymphoid malignancies. The combination of conventional cytogenetics with FISH (fluorescence *in situ* hybridization) analysis allowed us a more accurate definition of the genomic aberrations involved in CLL/SLL. Novel and uncommon chromosome rearrangements were found in our series. In the present work, we describe the cytogenetic and FISH analyses of 12 CLL/SLL patients out of 150 cases with successful cytogenetic study, that interestingly showed in their karyotypes new or infrequent clonal chromosome rearrangements, all of them associated to adverse clinical course with partial or null response to different chemotherapeutic agents. We identified 16 different chromosome rearrangements and translocations were the most frequent aberration (7/16; 43,8%). Some of these findings were recently reported by us as a new alterations in SLL/CLL patients: t(2;17)(p21;q23); der(17)t(12;17)(q13;q25); der(14)(t(11;14)(q13;q24), t(5;12;19) (q15;p11; q13); psu dic (17;2)(p11.2;p21), psu dic (12;21)(q24;q10), t(10;14)(q22;q32), del(3)(p11) and del(5)(q15). Translocation t(11;13)(p15;q22) and del(18)(p11) were not previously described, meanwhile, dic(17;18)(p11.2;p11.2), dup(12)(q13q24), del(7)(q34), t(2;14)(p11;q32) and del(7)(q11) were infrequently reported in the literature. Translocation t(2;17)(p21;q23) is recurrent in our series and t(2;14)(p11;q32) became recurrent from our data. Chromosomes 17, 2, 12 and 14 were the most frequently involved. Twenty one different breakpoints were found and 38.1% (8/21) of them, such as 2p21, 5q15, 17p11.2, 17q23, 12q13, 12q24, 14q32 and 18p11 were recurrent in our series. All but one patient of our series showed an unfavorable clinical outcome. A detailed analysis and a review of the cytogenetic and molecular cytogenetics findings in CLL/SLL were performed. Our findings may facilitate the understanding of genetic events associated with development and progression of CLL/SLL and may have implications in the clinical management of patients with this disease.

In: Chronic Lymphocytic Leukemia Research Focus
Editor: Chadi Nabhan, pp. 1-15

ISBN: 978-1-60021-526-1

Chapter I

Risk Stratification in the Management of Chronic Lymphocytic Leukemia (CLL)

Chadi Nabhan[*]
Oncology Specialists, S. C.; Lutheran General Cancer Center; Park Ridge, Illinois, USA

Abstract

For years, chronic lymphocytic leukemia (CLL) has been viewed as an indolent disease that is rarely lethal. Therapeutic options have included purine analogues, alkylating agents, and steroids. Since 1999, there have been a number of publications on the biologic aspects of CLL and various prognostic factors that influence survival. The challenge that hematologists face lies in trying to integrate this information in the management of patients with CLL. Genomic aberrations, mutational status of the immunoglobulin variable heavy chain gene (*IgVH*), CD 38 expression, and overexpression of a tyrosine kinase, ZAP-70, have all been reported to be potential prognostic markers that may identify a subset of patients whose disease course is likely to be aggressive and potentially fatal. CLL patients with poor prognosis usually have the unmutated *IgVH*, chromosomal abnormalities such as 17p or 11q, and overexpress ZAP-70 and CD 38. The important question is what is the best therapeutic strategy for such patients.

Data suggest that patients with 17p deletion and/or p53 mutations fail traditional therapy with purine analogues and rituximab. Several retrospective studies suggested that the humanized anti-CD 52 antibody alemtuzumab may be an attractive option in these patients. Furthermore, additional studies have suggested that patients who achieve molecular disease negativity could attain prolonged duration of remission. In this review the novel prognostic factors in CLL will be highlighted and suggestion on how they should be integrated in the therapeutic armamentarium of CLL will be explored. Finally, we will review some data on minimal residual disease and how this concept is being applied in future clinical trial design.

[*] Chadi Nabhan; Phone: 847-268-8200; Fax: 847-268-8030; E-Mail: cnabhan@oncmed.net

Introduction

Chronic Lymphocytic Leukemia (CLL) is the most common leukemia in the western hemisphere. Over 10,000 new cases are being diagnosed and almost 4,660 deaths are anticipated for 2006 in the United States.[1] The widespread use of complete blood count (CBC) testing in asymptomatic patients has increased the rate of detection and has certainly identified newly diagnosed patients at a younger age. The use of flow cytometry and immunophenotyping has also helped in separating CLL from other lymphoprolifertaive disorders and has eliminated previous cumbersome ambiguity.

Although the majority of CLL patients present asymptomatically, some have constitutional symptoms or complaints related to lymphadenopathy, splenomegaly, and cytopenias. Traditionally, RAI staging system has been used to separate patients into low, intermediate, and high risk groups.[2] Asymptomatic early stage patients are traditionally observed without early therapy since early intervention in these patients has not been shown to improve survival. Advanced stage patients are usually treated even if asymptomatic. Many agents have shown activity in CLL, but purine analogues-based therapy has traditionally provided patients with higher response rates and improved time to progression.[3] Although, never superior to alkylating agents in terms of overall survival, purine analogues have become the standard front-line therapy. Antibody-based therapies have become common in CLL with the identification of two target antigens (CD 52 and CD 20) for which commercially available antibodies exist. This has paved the way to incorporate antibodies in combination with traditional chemotherapies, with invariably higher response rates and acceptable toxicity.

The past several years have witnessed an explosion of information on prognostic factors in CLL. These factors have divided CLL into favorable and unfavorable entities. The challenge that faces clinicians and researchers lies in defining how to best approach these high-risk patients and in developing therapeutic strategies that are effective in this setting. In this review, we summarize the recent prognostic factors that have been in use over the past 5 years and discuss different potential therapeutic strategies that maybe effective in this setting. We also review recent evidence that suggest individualizing goals of therapy in CLL to incorporate eradicating minimal residual disease (MRD) in some patients, especially with some data suggesting improved survival in patients with MRD negativity.

Novel and Traditional Prognostic Factors in CLL

Over the years, assessable markers for disease activity have been used to separate patients into different risk categories. These include absolute lymphocyte count, short lymphocyte doubling time, serum LDH, diffuse pattern of bone marrow infiltration, poor performance status, and male gender. Recently, additional prognostic factors are being utilized. These factors have further enhanced our knowledge of the biology of CLL and will most likely aid in developing newer therapeutic options. Table 1 lists the common prognostic factors currently in use.

Table 1. Traditional and Novel prognostic factors associated with outcomes in CLL

Traditional	Novel
Age	IgV_H Status
Performance Status	CD 38 (>20-30% expression)
Diffuse bone marrow infiltrarion	ZAP-70 (>20% cells with expression)
Beta-2-microglobulin	Genomic Aberrations (CLL FISH Panel)
Percent of prolymphocytes	
Male gender	
Short lymphocyte doubling time	

IgV_H: Immunoglobulin variable heavy chain gene, ZAP-70: Zeta-associated protein, FISH: Fluroscence in-situ hybridization.

ß-2-Microglobulin

This glycoprotein has been shown to be an important prognostic factor in many hematologic malignancies. It is associated with the major histocompatibility complex (MHC) class I on nucleated cells, and usually reflects high tumor burden and turnover. The prognostic significance of this protein is well-established in non-Hodgkin's lymphoma (NHL) and multiple myeloma. Data suggest similar importance in CLL. Recently, Weirda and colleagues reported a retrospective analysis on 659 treatment-naïve and 1062 previously treated CLL patients. The authors found that baseline serum β-2-microglobulin and treatment choice were consistent and independent factors for predicting overall survival (OS) (P<0.05) in both treatment-naïve and previously treated patients.[4] Since other investigations questioned the prognostic value of this protein,[5] large scale prospective studies are required to assess its validity.[6]

IgV_H Gene Mutational Status

Recent evidence suggests that CLL can be divided into two separate groups; one with the unmutated V_H genes assumed to originate from pre-germinal center cells and another with mutated V_H genes originating from post-germinal center cells.[7,8] The mutational status is defined arbitrarily as less than 98% homology to the germ line gene.[9] Importantly, patients with the mutated CLL carry a favorable prognosis while those with the unmutated gene follow an unfavorable curse with rapid progression.[8,10,11] In addition, specific V_H genes such as V3-21 may be associated with an inferior outcome regardless of the mutational status.[12]

Although IgV_H mutation status is analyzed as part of clinical trial research, assessment of mutational status is only beginning to become available for daily clinical practice. This is due to the complexity in performing and interpreting the sequencing data for determination of IgV_H status.[13] Consequently, investigators have been evaluating various surrogate markers that can consistently and dependably predict the pathophysiological disease course in individual CLL patients. Such surrogate markers for IgV_H mutation status that have been most thoroughly tested include CD38 and zeta-associated protein-70 (ZAP-70) expression of CLL B cells.

CD 38

CD 38 is a transmembrane glycoprotein with variable expression on normal B cells depending on maturation, type of activation, and other factors. The interest in CD 38 overexpression in CLL stems from the fact that this glycoprotein could be a surrogate marker to the mutational status of the immunoglobulin variable heavy chain gene. Damle and colleagues confirmed the prognostic utility of unmutated IgV_H genes, and showed that CD38 expression can be used to categorize CLL patients.[7] According to their findings, both IgV_H gene mutation status and CD38 expression accurately predict clinical outcome. In a much larger study, Ibrahim and colleagues used flow cytometry to analyze CD38 expression on CLL cells from 218 patients.[14] They used a cut-off of 20% where patients were considered to have high CD38 expression if at least 20% of their CLL cells were positive for CD38. High CD38 expression was indicative of patients with aggressive disease and decreased survival (P=.00005). In addition, studies on clinical trials have shown that the cutoff of 20% or higher was also associated with poor response to therapy.[15]

Despite the initial enthusiasm to use CD 38 as a prognostic marker in CLL, many limitations precluded its wide acceptance. First, its expression may change over the course of the disease, although it has not been determined if this change represents progression to more aggressive disease.[13] Second, there is no consensus on how to best utilize CD38 as a prognostic factor or on the ideal cut-off point for expression; the cutoff of 30% expression to classify patients as having CD38 negative or positive status is arbitrary.[13] Finally, it is estimated that 10% to 30% of cases show discordant status for CD 38 and the V_H status making this glycoprotein much less than an ideal and a reliable surrogate marker.[16]

ZAP-70

Recently, gene expression profiling has been used to identify additional entities that may be of prognostic significance in CLL. In particular, the focus has been on genes whose expression correlates with the V_H mutational status. One such candidate is ZAP-70, a tyrosine kinase protein that is normally expressed in T cells and natural killer cells. ZAP-70 plays an important role in T-cell signaling, but its role in B-CLL is unknown. Several methods are used to study the expression of ZAP-70 in patients with CLL, including flow cytometry, Western blot, immunohistochemistry, and gene expression by analysis of messenger RNA, but none of these has been adequately standardized for general use. It is recognized that the assessment of ZAP-70 in B cells is problematic,[17] nevertheless, several pivotal studies have shown the prognostic value of ZAP-70 measurements in CLL.

Crespo and colleagues studied ZAP-70 expression in CLL cells from 56 patients using immunohistochemistry, Western blotting, and flow cytometry, and demonstrated that flow cytometric detection (with $\geq$20% ZAP-70-positive cells considered "positive") is relatively reliable and readily applicable for routine use in clinical laboratories.[18] These investigators also sequenced the IgV_H genes to correlate their mutational status with ZAP-70 expression. Among their 56 patients, 35 were ZAP-70 positive and had CLL B-cell clones that were unmutated. Of the remaining 24 patients who were ZAP-70 negative, a mutated status was identified in 21($P<.001$). ZAP-70 expression was shown to be consistent over time. Multiple regression analysis showed that ZAP-70 expression ($P<.001$), but not CD38 expression, correlated with IgV_H mutational status in CLL.

Rassenti and colleagues confirmed and extended these findings in a study that included 307 CLL patients from multiple institutions.[19] They found ZAP-70 expression in 117 of 164 individuals whose IgV_H status was unmutated (71%), but only in 24 of 143 CLL patients with mutated IgV_H status (17%; $P<.001$). The usefulness of ZAP-70 as a surrogate marker for IgV_H status is thus less clear, but the clinical significance of this observation was particularly impressive. ZAP-70 positive patients experienced a median time from diagnosis to initial therapy of 2.9 years, as compared with 9.2 years between diagnosis and initiation of therapy for ZAP-70 negative patients ($P<.001$).[19]

Although ZAP-70 holds significant promise as a prognostic marker, the 20% cutoff to categorize cells as having ZAP-70 positive status is arbitrary. Optimal methods of detection that permit rapid and inexpensive analysis also need to be addressed.[13]

Genomic Aberrations

The prognostic significance of chromosomal abnormalities has been recognized for over 15 years since Juliusson and colleagues reported on their observation that over 50% of patients with CLL carry genomic aberrations.[20,21] In their report, these investigators noted that 218 out of 381 patients had chromosomal translocations. The introduction of Fluroscence In-Situ Hybridization (FISH) has allowed detecting additional translocations in non-dividing cells, which in turn allowed the identification of additional genomic aberrations In CLL. While conventional cytogenetic analyses detected chromosomal aberrations in 40 to 50% of CLL patients, FISH has identified chromosomal aberrations in approximately 80% of late stage and 60% of early stage CLL patients.[22]

Döhner and colleagues reported on a FISH-detectable cytogenetic analysis they performed in a retrospective analysis of a cohort of 325 CLL patients.[23] They evaluated mononuclear blood cells with a panel of FISH probes, based on findings from the Juluisson investigation,[21] to look for a variety of chromosomal abnormalities, including deletions in 6q21, 11q22-23, 13q14, and 17p13; trisomy in 3q26, 8q24, and 12q13; and translocations involving 14q32. Genomic aberrations were found in 82% of cases. In addition, marked differences in clinical course according to the FISH-detectable cytogenetic findings were recognized. Median survival for patients with 17p deletions was only 32 months, in contrast to median survival of 111 months for patients with normal karyotypes. The 13q deletion seemed to confer a relative protective effect, with median survival of 133 months. Similar differences between these two risk groups were noted for the rate of disease progression.[23] These findings were later confirmed by another investigation conducted and reported by Dewald and colleagues.[24] Based on these two reports, a hierarchical model of FISH-detectable defects in CLL could be constructed, with the most benign course probably occurs in patients with 13 defects while more aggressive disease is seen in patients with deletion 17p13.1 or deletion 11q22-q23.

p53 and ATM Mutations in CLL

Recently, it has been recognized that deletions in 17p are related to the tumor suppressor gene p53.[23,25] This is important as it identifies a group of CLL patients characterized by an aggressive clinical course, with advanced disease stage and resistance to chemotherapy.[26,27,28]

Of significance, p53 dysfunction is present in more patients with CLL than cytogenetically detectable chromosomal 17 abnormalities, as identified by FISH analysis. Therefore, some researchers have suggested that functional assays of p53 in CLL B cells may provide better prognostic information.[23,13] For instance, in the recent interim report of the UK CLL02 trial, a phase II study of subcutaneous alemtuzumab plus fludarabine in patients with fludarabine-refractory CLL, 21 of 36 valuable patients demonstrated 11q and/or 17p deletions by FISH analysis. However, 4 of 8 patients with normal FISH evidenced p53 dysfunction as demonstrated by a p53 functional assay.[29] Although the exact role of the p53 gene in CLL remains to be defined, it is important to identify this subset of patients early on in the course of CLL as it is foreseeable to treat these patients differently.

Additionally, the ataxia telangiectasia mutated (ATM) gene is mutated in a subset of patients with 11q22-q23 deletions.[22] This gene is an upstream regulator of p53 and may play a role in the pathogenesis of CLL.[30] Thus, CLL patients with 11q deletions have marked lymphadenopathy and accelerated disease progression. Austen and colleagues analyzed 155 tumors from patients with CLL.[31] Twelve percent of the tumors evidenced ATM mutations, 4% evidenced p53 mutations, and 2 tumors showed mutations in both genes. Patients with ATM mutations had significantly decreased overall survival and treatment-free survival compared with patients with wild-type ATM/p53. ATM mutations were also found to be associated with other poor prognostic factors, including unmutated IgV_H genes, and demonstrated independent prognostic information in multivariate analysis. While this represents a very important finding, prospective validation of the importance of ATM mutations is required before this is widely applied to risk stratification of CLL.

The above findings argue in favor of risk stratifying patients into separate categories and suggest that developing different therapeutic interventions for each category is the rational future directive in CLL therapy.

Is Risk Stratification in CLL Justified?

Impact of p53 Mutation in CLL Therapy

It has been recognized that CLL patients with p53 mutation have an aggressive clinical course and resistance to chemotherapy. El Rouby and colleagues found p53 gene mutations in 15% of 53 patients with previously untreated CLL.[32] Among the 45 individuals with intact p53, treatment was required for 29 (64%). In contrast, therapy was necessary for 7 of 8 patients with p53 mutations, and the best response achieved among these patients was a partial remission in only one (*P*=.00009). Relative to other prognostic factors such as age, race, gender, and clinical stage, the risk of death was 13-fold greater for CLL patients with p53 mutations (*P*=.013). Although this investigation suggested the ominous prognostic significance for patients with p53 mutation, it did not correlate between these mutations and 17p deletion.

These findings were confirmed by Dohner and colleagues who identified 17 patients out of 100 examined ones with p53 mutation.[28] Time to initiate therapy was shorter for patients with p53 mutations compared with those with intact p53 (59 days vs. 396 days, *P*<.01), and overall survival was also significantly different (*P*<.001). Although half of the patients in this group were treated with purine analogues, none of the 12 patients with p53 mutations responded to treatment (*P*<.001). This confirms that mutations of p53 results in an estimated 7-fold greater risk of death for B-cell leukemia patients compared with patients who have intact p53. Others validated these findings opening the way to explore alternative therapeutic options for this subset of patients.[27]

Impact of IgV_H on Therapy

Recent studies have examined the effects of IgV_H mutational status, cytogenetic analysis, and p53 mutations collectively on treatment. Byrd and colleagues evaluated the impact of IgV_H mutation status, interphase cytogenetics (including 17p and 11q deletion) and p53 mutation status on a cohort of CLL patients treated with chemoimmunotherapy containing fludarabine and rituximab.[33] Eighty-eight patients who were treated with concurrent or sequential fludarabine and rituximab had undergone detailed prognostic factor assessment. Progression-free survival and overall survival were significantly shorter in patients with unmutated IgV_H status than in patients with mutated status. Progression-free survival significantly improved in patients with low-risk (e.g., deletion 13q) compared with patients with high-risk cytogenetics (e.g., deletion 17p and deletion 11q) (*P*=0.005). The frequency of p53 mutations that occurred independent of del(17p) were limited and prevented assessment of their independent prognostic significance.

Collective Impact of Prognostic Factors on Therapy

Grever and colleagues evaluated a comprehensive set of prognostic markers in a randomized phase III trial of 278 previously untreated, symptomatic CLL patients who received fludarabine and cyclophosphamide (FC) versus fludarabine alone (F).[34] Although FC allowed for higher complete response, overall response (OR), and median progression-free survival, the CR and OR did not differ based on interphase cytogenetics or V_H mutational status for the entire group of patients or when analyzed by arm of treatment. However, patients with poor cytogenetics achieved shorter PFS regardless of the treatment arm. A Cox proportional hazards analysis was performed to assess the impact of multiple molecular features and treatment arm on PFS. Treatment with F, presence of del (17p) or p53 mutation, and presence of del (11q) were highly significant in decreasing PFS. This suggests that-high risk genomic features including p53 mutations, del (17p), and del (11q) are highly predictive of shortened PFS, supporting a risk-stratified therapeutic approach in CLL.

A prospective cytogenetic analysis of the GCLLSG study that prospectively randomized untreated CLL patients to FC versus F demonstrated that median OS was significantly decreased in patients with 17p deletion ($P<0.001$.).[35,36]

Finally, investigators on behalf of the UKLRFCLL4 trial that evaluated the effects of chlorambucil, fludarabine, and FC in previously untreated patients confirmed that risk stratification is justified.[37] A significant correlation between IgV_H mutation status, CD38 expression, ZAP-70, and interphase FISH results was found. Regression analysis of age, stage, gender, IgV_H mutational status, and cytogenetic factors found that stage ($P=0.002$), IgV_H mutational status ($P=0.05$), p53 loss ($P=0.02$), and 11q deletion ($P=0.01$) all significantly affected response to therapy. In multivariate analysis, age, stage, and p53 loss had a significant impact on survival. When CLL patients with p53 loss were excluded, unmutated IgV_H status had a significant impact on survival ($P=0.007$), while CD38 or ZAP-70 expression did not.

The Concept of Minimal Residual Disease (MRD) in CLL

The current National Cancer Institute Working Group (NCI-WG) guidelines define responses based on morphological assessment of the peripheral blood and bone marrow and a physical examination of palpable disease.[38] The definition of complete response (CR) allows for up to 30% of malignant lymphocytes to be present in the bone marrow, which is clearly suboptimal, as this would overlook substantial residual disease. About 25% of patients with CR by NCI-WG criteria have been shown to have detectable MRD as assessed by sensitive four-color multiparametric flow cytometry assays that are capable of detecting one malignant cell in 10^4 to 10^5 leukocytes.[39] Moreover, among patients achieving NCI-WG CR following autologous SCT or immunotherapy, the presence of MRD was significantly predictive of worse event-free and overall survival.

The prognostic significance of MRD has been established in other hematologic malignancies such as acute promyelocytic leukemia (APL) and chronic myeloid leukemia (CML).[40,41] Similarly, emerging studies are beginning to report on the prognostic impact of MRD in patients with CLL.

In a study conducted by the Spanish GELCAB group, combination therapy with fludarabine, cyclophosphamide, and mitoxantrone (FCM) was evaluated in 60 patients with resistant or relapsed CLL. The study included MRD assessment as part of the response criteria for patients achieving CR and defined MRD negativity based on the absence of detectable residual CLL cells using both consensus primer PCR assay (sensitivity 1×10^{-3} leukocytes) and a more sensitive multiparametric flow cytometry (sensitivity 1×10^{-4} leukocytes) assay.[42] Following FCM therapy, 50% of patients achieved CR (overall response rate [ORR] 78%), with 17% achieving MRD-negative CR. Although the median duration of response was not significantly different between patients achieving MRD negativity and CR (21 months vs. 17 months), median OS was significantly different between patients with MRD-negative CR or clinical CR and patients not achieving CR. Despite the small size of this study, the results argue that MRD status has an impact on overall outcome.

More recently, combination chemoimmunotherapy with fludarabine, cyclophosphamide, and rituximab (FCR) was evaluated in a large cohort of patients with previously treated CLL.[43] MRD assessment was performed in a subgroup of responding patients using a novel patient-specific PCR-based assay (sensitivity 1×10^{-5} mononucleocytes). Twenty-one percent of assessed patients were MRD negative. The median time to progression for patients who became MRD negative was 44 months compared with 27 months in patients with MRD positivity. In another study, single-agent alemtuzumab was administered to 91 patients with relapsed or refractory CLL.[44] As part of the response criteria, MRD was evaluated in all patients using a sensitive four-color flow cytometry method (sensitivity 5×10^{-5} leukocytes). After a median of 9 weeks of therapy, ORR was 54%, with CR in 35% and MRD negativity in 20% of all patients. The median treatment-free survival was significantly different between these groups. While it was not reached in patients with MRD-negative CR, it was 20 months for patients with MRD-positive CR, 13 months for those with a partial response (PR), and 6 months for non-responders ($P<0.0001$). These findings suggest that MRD status has a significant impact on survival in CLL and that MRD response potentially may be more important and meaningful than the NCI-WG criteria for assessing the outcomes of therapy. Similarly, other studies evaluated and confirmed that MRD negativity is an attainable goal in previously untreated patients, although it is unclear what the best modality to eradicate MRD is. Tables 2 and 3 summarize few studies that aimed at MRD eradication in the relapsed and untreated settings respectively.

The realization that patients respond differently to different therapeutic choices depending on their underlying risk categories and the recognition that some patients could achieve MRD negativity that may translate into better long-term outcome, argue in favor of risk stratifying patients and begs for continued research into defining specific treatments for each risk category.

Table 2. Clinical studies in relapsed CLL with MRD end points

10	Patients	Used regimen	MRD negative remission (%)	Comments
Bosch [42]	60	FCM(Fludarabine, Cytoxan, Mitoxantrone)	17(among the entire cohort	58% Binet B 35% Binet C
Wierda [43]	177	FCR (Fluadarabine, Cytoxan, Rituxan)	21 among 75 responding assessed patients	Modified patient-specific PCR used to detect MRD
Moreton [44]	91	Alemtuzumab	20 among the entire cohort	4-color flow cytometry used to detect MRD.

MRD: Minimal residual disease, PCR: Polymerase Chain Reaction.

Table 3. Trials with MRD end points in untreated CLL

Reference	Patients	Regimen used	MRD negative remission (%)
Bosch [53]	64	FCM (Fludarabine, cytoxan, mitoxantrone)	24
Keating [54]	224	FCR (Fludarabine, cytoxan, rituximab)	41% among 192 responding assessed patients

Risk Stratification in CLL

Recent studies have suggested that, in addition to response to treatment, additional factors are predictors of rate of disease progression. These include genomic aberrations, mutational status of the immunoglobulin variable heavy chain gene, and MRD status.

Specifically, the deletions of 17p and/or mutations in the p53 tumor suppressor gene have been associated with treatment failure whether purine analogues or alkylating agents are used.[45,46] This in turn prompted investigators to explore effective therapies in these high-risk patients. Stilgenbauer and Dohner reported in 2002 on a single case of refractory CLL patient that attained over 18 months of CR after treatment with alemtuzumab.[26] Further molecular studies in this patient confirmed the presence of 17p deletion and p53 mutation. This report encouraged others to explore the activity of alemtuzumab in patients with p53 mutation.

Lozanski and colleagues retrospectively evaluated a series of fludarabine-refractory CLL cases treated with intravenous alemtuzumab.[47] This therapy resulted in 6 responses out of 15 cases with p53 mutations or deletions (OR 40%), which was higher than responses in other risk categories. Moreover, the German CLL Study Group 2H that evaluated the efficacy of alemtuzumab in purine analogues-refractory patients revealed high incidence of patients with 17p deletions (27%), suggesting an underlining association of this abnormality with fludarabine-resistant disease.[48] An interim analyses of this ongoing study has shown a response in 10 of 21 with VH unmutated, 5 of 10 with 11q-, and 6 of 10 with 17p- cases, providing evidence from a controlled trial that alemtuzumab may be effective in CLL with 17p-/p53 mutation.[48]

Other therapeutic choices that are being investigated to overcome these poor risk factors in CLL include high-dose chemotherapy with stem cell rescue, using autologous or allogeneic sources. In medically fit patients allogeneic stem cell transplant (SCT) adds the immune-mediated graft-versus-leukemia effect although the morbidity and mortality of graft-versus-host-disease requires significant attention. It continues to be controversial when to recommend SCT as opposed to conventional or monoclonal antibodies-based interventions. Transplant-related mortality (TRM) was 1.5% in the Medical Research Council (MRC) study reported by Milligan et al.[49] In the multicenter prospective autologous SCT study conducted in Germany, TRM was 5% with a 2-year overall survival rate of 88% among 105 patients[50] These results appear promising considering the high-risk patients included in this study, although the molecular presence of disease suggest inevitable relapse. Although some investigators suggested that SCT could overcome poor risk prognostic features,[51] it remains debatable whether SCT is applicable in patients who achieve MRD negativity or if autologous SCT with MRD-negative cells could provide better long-term outcome.[52]

Summary and Conclusion

Our understanding of the biology of CLL has increased remarkably in the past few years. The belief that all patients behave similarly and carry an indolent course is no longer valid. CLL is divided into a favorable entity that is usually characterized by the mutated immunoglobulin variable heavy chain gene, less than 20% of cells expressing ZAP-70, and the absence of poor risk genomic aberrations such as 17p- or 11q-. Unfavorable CLL patients usually have poor risk cytogenetics and/or have an unmutated gene. These patients usually have shorter time from diagnosis to requiring therapy, and often are poor responders to traditional therapeutic interventions. In addition, the ability to eradicate CLL on the molecular level with sensitive assays, argues in favor of treating certain subset of patients to MRD negativity, but it remains questionable whether this strategy should be adapted to every CLL patient. Moreover, the fact that certain therapies have activity in high-risk patients suggest that applying these treatments early in the course of this disease may be advisable.

It is suggested that CLL patients are investigated to ascertain whether they have a favorable or unfavorable disease. Unfavorable CLL patients are encouraged to participate in clinical trials that look at novel therapies with potential activities in this setting. These could include SCT or monoclonal antibodies such as alemtuzumab with the possibility of achieving MRD negativity. Favorable CLL patients could be observed if asymptomatic, but early therapy could be investigated in the context of a well-designed clinical trial, especially in younger individuals for whom a simple nihilistic palliative approach is no longer optimal. Only through risk stratification and individualized therapy, we can envision a cure for CLL in the near future.

References

[1] Jemal A, Siegel R, Ward E, et al: Cancer statistics, 2006. *CA Cancer J. Clin.* 56:106-30, 2006

[2] Rai KR, Sawitsky A, Cronkite EP, et al: Clinical staging of chronic lymphocytic leukemia. *Blood.* 46:219-34, 1975

[3] Rai KR, Peterson BL, Appelbaum FR, et al: Fludarabine compared with chlorambucil as primary therapy for chronic lymphocytic leukemia. *N. Engl. J. Med.* 343:1750-7, 2000

[4] Weirda W WX, O'Brien S, Faderl S, Ferrajoli A, Thomas D, Ravandi F, Cortes J, Giles F, Koller C, Kantarjian H, Lerner S, Keating MJ: Serum beta-2-microglobulin: the universal independent prognostic factor for patients with CLL. *Blood.* 106:336b, 2005

[5] Molica S, Levato D, Cascavilla N, et al: Clinico-prognostic implications of simultaneous increased serum levels of soluble CD23 and beta2-microglobulin in B-cell chronic lymphocytic leukemia. *Eur. J. Haematol.* 62:117-22, 1999

[6] Binet JL, Caligaris-Cappio F, Catovsky D, et al: Perspectives on the use of new diagnostic tools in the treatment of chronic lymphocytic leukemia. *Blood.* 107:859-61, 2006

[7] Damle RN, Wasil T, Fais F, et al: Ig V gene mutation status and CD38 expression as novel prognostic indicators in chronic lymphocytic leukemia. *Blood.* 94:1840-7, 1999

[8] Hamblin TJ, Davis Z, Gardiner A, et al: Unmutated Ig V(H) genes are associated with a more aggressive form of chronic lymphocytic leukemia. *Blood.* 94:1848-54, 1999

[9] Krober A, Seiler T, Benner A, et al: V(H) mutation status, CD38 expression level, genomic aberrations, and survival in chronic lymphocytic leukemia. *Blood.* 100:1410-6, 2002

[10] Tobin G: The immunoglobulin genes and chronic lymphocytic leukemia (CLL). *Ups J. Med. Sci.* 110:97-113, 2005

[11] Tobin G, Rosenquist R: Prognostic usage of V(H) gene mutation status and its surrogate markers and the role of antigen selection in chronic lymphocytic leukemia. *Med. Oncol.* 22:217-28, 2005

[12] Tobin G, Thunberg U, Johnson A, et al: Somatically mutated Ig V(H)3-21 genes characterize a new subset of chronic lymphocytic leukemia. *Blood.* 99:2262-4, 2002

[13] Shanafelt TD, Geyer SM, Kay NE: Prognosis at diagnosis: integrating molecular biologic insights into clinical practice for patients with CLL. *Blood.* 103:1202-10, 2004

[14] Ibrahim S, Keating M, Do KA, et al: CD38 expression as an important prognostic factor in B-cell chronic lymphocytic leukemia. *Blood.* 98:181-6, 2001

[15] Jelinek DF, Tschumper RC, Geyer SM, et al: Analysis of clonal B-cell CD38 and immunoglobulin variable region sequence status in relation to clinical outcome for B-chronic lymphocytic leukaemia. *Br. J. Haematol.* 115:854-61, 2001

[16] Ghia P, Guida G, Stella S, et al: The pattern of CD38 expression defines a distinct subset of chronic lymphocytic leukemia (CLL) patients at risk of disease progression. *Blood.* 101:1262-9, 2003

[17] Gibbs G, Bromidge T, Howe D, et al: Comparison of flow cytometric methods for the measurement of ZAP-70 expression in a routine diagnostic laboratory. *Clin. Lab. Haematol.* 27:258-66, 2005

[18] Crespo M, Bosch F, Villamor N, et al: ZAP-70 expression as a surrogate for immunoglobulin-variable-region mutations in chronic lymphocytic leukemia. *N. Engl. J. Med.* 348:1764-75, 2003

[19] Rassenti LZ, Huynh L, Toy TL, et al: ZAP-70 compared with immunoglobulin heavy-chain gene mutation status as a predictor of disease progression in chronic lymphocytic leukemia. *N. Engl. J. Med.* 351:893-901, 2004

[20] Juliusson G, Gahrton G: Chromosome aberrations in B-cell chronic lymphocytic leukemia. Pathogenetic and clinical implications. *Cancer Genet Cytogenet.* 45:143-60, 1990

[21] Juliusson G, Oscier DG, Fitchett M, et al: Prognostic subgroups in B-cell chronic lymphocytic leukemia defined by specific chromosomal abnormalities. *N. Engl. J. Med.* 323:720-4, 1990

[22] Stilgenbauer S, Bullinger L, Lichter P, et al: Genetics of chronic lymphocytic leukemia: genomic aberrations and V(H) gene mutation status in pathogenesis and clinical course. *Leukemia.* 16:993-1007, 2002

[23] Dohner H, Stilgenbauer S, Benner A, et al: Genomic aberrations and survival in chronic lymphocytic leukemia. *N. Engl. J. Med.* 343:1910-6, 2000

[24] Dewald GW, Brockman SR, Paternoster SF, et al: Chromosome anomalies detected by interphase fluorescence in situ hybridization: correlation with significant biological features of B-cell chronic lymphocytic leukaemia. *Br. J. Haematol.* 121:287-95, 2003

[25] Stilgenbauer S, Lichter P, Dohner H: Genetic features of B-cell chronic lymphocytic leukemia. *Rev. Clin. Exp. Hematol.* 4:48-72, 2000

[26] Stilgenbauer S, Dohner H: Campath-1H-induced complete remission of chronic lymphocytic leukemia despite p53 gene mutation and resistance to chemotherapy. *N. Engl. J. Med.* 347:452-3, 2002

[27] Sturm I, Bosanquet AG, Hermann S, et al: Mutation of p53 and consecutive selective drug resistance in B-CLL occurs as a consequence of prior DNA-damaging chemotherapy. *Cell Death Differ.* 10:477-84, 2003

[28] Dohner H, Fischer K, Bentz M, et al: p53 gene deletion predicts for poor survival and non-response to therapy with purine analogs in chronic B-cell leukemias. *Blood.* 85:1580-9, 1995

[29] Sayala H MP, Richard JA, Rawstron AC, O'Conner S, Evans P, Carter A, Dearden C, Matutes E, Pettitt AR, Kennedy BD, Hillmen P: Interim report of the UKCLL02 trial: a phase II study of subcutaneous alemtuzumab plus fludarabine in patients with fludarabine-refractory CLL (on behalf of the NCRI CLL Trials Subgroup). *Blood.* 106:599a-560a, 2005

[30] Bullrich F, Rasio D, Kitada S, et al: ATM mutations in B-cell chronic lymphocytic leukemia. *Cancer Res.* 59:24-7, 1999

[31] Austen B, Powell JE, Alvi A, et al: Mutations in the ATM gene lead to impaired overall and treatment-free survival that is independent of IGVH mutation status in patients with B-CLL. *Blood.* 106:3175-82, 2005

[32] el Rouby S, Thomas A, Costin D, et al: p53 gene mutation in B-cell chronic lymphocytic leukemia is associated with drug resistance and is independent of MDR1/MDR3 gene expression. *Blood.* 82:3452-9, 1993

[33] Byrd JC, Gribben JG, Peterson BL, et al: Select high-risk genetic features predict earlier progression following chemoimmunotherapy with fludarabine and rituximab in chronic lymphocytic leukemia: justification for risk-adapted therapy. *J. Clin. Oncol.* 24:437-43, 2006

[34] Grever MR LD, Dewald GW, Neuberg DS, Flinn IW, Appelbaum FR, Larson RA, Tallman MS, Gribben JG, Byrd JC: Outcome of treatment with fludarabine versus fludarabine and cyclophosphamide in chronic lymphocytic leukemia (CLL) is adversly impacted by high-risk genetic features; results from ECOG 2997. *Blood.* 104, 2004

[35] Eichhorst BF, Busch R, Hopfinger G, et al: Fludarabine plus cyclophosphamide versus fludarabine alone in first-line therapy of younger patients with chronic lymphocytic leukemia. *Blood.* 107:885-91, 2006
[36] Stilgenbauer S KA, Busch R, Eichhorst B, Kienle D, Winkler D, Hopfinger G, Lichter P, Emmerich B, Hallek M, Dohner H: 17p deletion predicts for inferior overall survival after fludarabine-based first line therapy in chronic lymphocytic leukemia: First analysis of genetics in the CLL4 trial of the GCLLSG. *Blood.* 106:212a, 2005
[37] Oscier DG RS, Orchard J, Davis Z, Best G, Morilla A, Brito-Bapapulle V, Matutes E, Pettitt A, Catovsky D: Prognostic factors in the UK LRF CLL4 trial. *Blood.* 106:549a, 2005
[38] Cheson BD, Bennett JM, Grever M, et al: National Cancer Institute-sponsored Working Group guidelines for chronic lymphocytic leukemia: revised guidelines for diagnosis and treatment. *Blood.* 87:4990-7, 1996
[39] Rawstron AC, Kennedy B, Evans PA, et al: Quantitation of minimal disease levels in chronic lymphocytic leukemia using a sensitive flow cytometric assay improves the prediction of outcome and can be used to optimize therapy. *Blood.* 98:29-35, 2001
[40] Lo Coco F, Diverio D, Avvisati G, et al: Therapy of molecular relapse in acute promyelocytic leukemia. *Blood.* 94:2225-9, 1999
[41] Hochhaus A, Reiter A, Saussele S, et al: Molecular heterogeneity in complete cytogenetic responders after interferon-alpha therapy for chronic myelogenous leukemia: low levels of minimal residual disease are associated with continuing remission. German CML Study Group and the UK MRC CML Study Group. *Blood.* 95:62-6, 2000
[42] Bosch F, Ferrer A, Lopez-Guillermo A, et al: Fludarabine, cyclophosphamide and mitoxantrone in the treatment of resistant or relapsed chronic lymphocytic leukaemia. *Br. J. Haematol.* 119:976-84, 2002
[43] Wierda W, O'Brien S, Wen S, et al: Chemoimmunotherapy with fludarabine, cyclophosphamide, and rituximab for relapsed and refractory chronic lymphocytic leukemia. *J. Clin. Oncol.* 23:4070-8, 2005
[44] Moreton P, Kennedy B, Lucas G, et al: Eradication of minimal residual disease in B-cell chronic lymphocytic leukemia after alemtuzumab therapy is associated with prolonged survival. *J. Clin. Oncol.* 23:2971-9, 2005
[45] Dohner H, Stilgenbauer S, James MR, et al: 11q deletions identify a new subset of B-cell chronic lymphocytic leukemia characterized by extensive nodal involvement and inferior prognosis. *Blood.* 89:2516-22, 1997
[46] Krober A, Bloehdorn J, Hafner S, et al: Additional genetic high-risk features such as 11q deletion, 17p deletion, and V3-21 usage characterize discordance of ZAP-70 and VH mutation status in chronic lymphocytic leukemia. *J. Clin. Oncol.* 24:969-75, 2006
[47] Lozanski G, Heerema NA, Flinn IW, et al: Alemtuzumab is an effective therapy for chronic lymphocytic leukemia with p53 mutations and deletions. *Blood.* 103:3278-81, 2004
[48] Seiler T, Dohner H, Stilgenbauer S: Risk stratification in chronic lymphocytic leukemia. *Semin. Oncol.* 33:186-94, 2006

[49] Milligan DW, Fernandes S, Dasgupta R, et al: Results of the MRC pilot study show autografting for younger patients with chronic lymphocytic leukemia is safe and achieves a high percentage of molecular responses. *Blood.* 105:397-404, 2005

[50] Dreger P, Montserrat E: Autologous and allogeneic stem cell transplantation for chronic lymphocytic leukemia. *Leukemia.* 16:985-92, 2002

[51] Moreno C, Villamor N, Colomer D, et al: Allogeneic stem-cell transplantation may overcome the adverse prognosis of unmutated VH gene in patients with chronic lymphocytic leukemia. *J. Clin. Oncol.* 23:3433-8, 2005

[52] Nabhan C, Bitran JD: Chronic lymphocytic leukemia: to transplant or not to transplant... that is the question? *J. Clin. Oncol.* 23:8126-7; author reply 8127, 2005

[53] Bosch F FA, Villamor N, Gine E, Abella MA, Gardella S, Besalduch J, onzalez M, Altes A, Escoda L, Ferrer S, Gonzalez-Barca E, Perez-Ceballos E, Asensi A, Montserrat E: Combination chemotherapy with fludarabine, cyclophosphamide and mitoxantrone (FCM) induces a high response rate in previously untreated CLL. *Blood.* 106:213a, 2005

[54] Keating MJ, O'Brien S, Albitar M, et al: Early results of a chemoimmunotherapy regimen of fludarabine, cyclophosphamide, and rituximab as initial therapy for chronic lymphocytic leukemia. *J. Clin. Oncol.* 23:4079-88, 2005

In: Chronic Lymphocytic Leukemia Research Focus
Editor: Chadi Nabhan, pp. 17-34
ISBN: 978-1-60021-526-1

Chapter II

New Research in Chronic Lymphocytic Leukemia

Zhang J. Zhong and Maher Albitar*
Quest Diagnostics Nichols Institute, San Juan Capistrano, Callifornia, USA

Abstract

Chronic lymphocytic leukemia (CLL) is relatively common leukemia manifests by the uncontrolled accumulation of mature B-cells that express CD5. The most intriguing aspect of CLL disease is the heterogeneity in its clinical course and outcome. Studying the mutation status of the immunoglobulin heavy-chain variable region (IgV_H) in patients with CLL reveiled that patients with CLL could have leukemic cells originating from either pre-somatic hypermutation stage and have aggressive disease or from post-somatic hypermutation stage and have more indolent course. These findings have stimulated active research to better understand the biology of CLL and helped clinicians to better manage their patients. Here we review the most recent data accumulated on the biology of CLL and discuss the clinical relevance of this data on the prediction of clinical behavior and outcome as well as therapy. We focus on the how basic science discoveries can be used in the clinical management and stratification of CLL patients for various therapeutic approaches.

Chronic lymphocytic leukemia (CLL) is the most common leukemia in the Western world. It usually occurs in older people and is more common in males than in females. One of the most intriguing aspects of CLL is its clinical course, which varies significantly among patients [1]. Some patients have minimal symptoms, require no treatment for a decade or two, and may die from other causes, while others rapidly develop aggressive disease and die within a short period of time. This difference is seen even among patients who have the same stage.

* Corresponding author: Maher Albitar; Quest Diagnostics Nichols Institute; 33608 Ortega Highway, Rm#108B; San Juan Capistrano, CA 92690-6130; Tel: (949) 728-4784; Fax: (949) 728-4990; E-mail: maher.x.Albitar@questdiagnostics.com

What has previously been referred to as T-cell CLL is no longer considered CLL and should now be referred to as prolymphocytic leukemia. Prolymphocytic leukemia is significantly more aggressive than B-cell CLL and requires different therapy. CLL now refers solely to B-cell CLL.

Many questions remain unanswered in terms of CLL biology, prognosis, and treatment. We summarize herein recent developments in B-cell CLL research, which we will call CLL for simplicity.

The Nature and Origin of B-Cell CLL

CLL cells have a well-defined morphology (small lymphoid cells with clumped chromatin and scant cytoplasm) and a specific immunophenotype (CD5-positive, CD23-positive, FMC-7-negative, CD79b-negative, dim surface light chains) [2]. This immunophenotype is rarely seen in normal B-cells.

Historically, CLL was believed to arise from CD5-positive, naïve immune incompetent B-cells that remain in the resting stage (G0/G1), mainly due to derailed apoptosis, and accumulate in circulation. Recent developments in molecular technology and subsequent research in CLL cell biology have largely challenged this theory.

At least 50% of CLL patients show somatic mutations in the immunoglobulin gene variable region (IgV_H), indicating the B-cells are not that naïve and CLL cells may have arisen from marginal zone B-cells [3, 4]. A mixed population of cells, including some with and some without an IgV_H mutation, are found in the marginal zone and their CD5 may be obtained upon stimulation [4,5]. Gene profiling studies suggest that the gene expression pattern of CLL is similar to that of memory B-cells but not to that of circulating naïve CD5-positive B-cells [6]. CLL cells most likely underwent one common transformational event, irrespective of whether or not the cells have an IgV_H mutation.

Furthermore, immunophenotyping suggests that CLL cells are activated, antigen experienced, B-lymphoid cells resembling those observed normally upon antigen stimulation [7]. CLL cells from all cases studied express classical activation markers, ie, CD23, CD25, CD69, and CD71. Increased density of expression of CD5 and CD27 and reduced density of CD22, CD32, CD79b, and IgD are consistent with cellular activation.

Since most CLL cells express CD5, one might think that CLL cells arise in cells with CD5 expression. Nearly all human fetal B-cells express CD5; such expression gradually disappears in postnatal life [8]. Only a small portion of CD5-positive cells is normally found in peripheral blood and the mantle zone of lymphoid follicles [9]. However, expression of CD5 in B-cells can be induced by a variety of stimuli [5]. It is difficult to know whether CLL cells arise in one of these two CD5-positive cell populations or whether CD5 is gained after leukemia clones are formed. Thus, CD5 expression may not be used as a critical factor in the judgment of CLL precursors.

The presence of a somatic mutation in some CLL cases indicates at least a portion of cases had encountered an antigen. Even cases without a mutation show a remarkable skewing in IgV_H gene usage, such as IgV_H 1-69, which suggests an encounter with an antigen selection [10].

Although most circulating CLL cells are mitotically inactive, CLL cell proliferation has been found to occur in the proliferation centers of lymphoid tissues and bone marrow [11]. Proliferation centers are aggregates of cells that are larger than the average CLL cell, have open chromatin, and frequently have nucleoli. These proliferation centers are usually found in the middle of otherwise typical, large aggregates of CLL cells. More importantly, studies on telomere length and activity in circulating CLL cells showed that CLL cells have gone through numerous cell divisions [12]. The telomere length, which is a measure of a cell's proliferative history, is significantly shorter in CLL cells than in normal B-cells, suggesting that a considerable number of cell divisions must have occurred in the CLL cells. Measurements of CLL B-cell kinetics in vivo, using a non-radioactive isotopic labeling method, showed that CLL cells have a brisk birth rate of 0.1% to 1% of the entire clone per day [13]. Therefore, CLL cells cannot be viewed as static, simply resulting from the accumulation of malignant cells that have not undergone appropriate apoptosis. Rather, they can be viewed as dynamic cells with substantial proliferation potential.

A familial clustering of CLL is well documented, and it is believed that genetic factors play a major role [14]. This is supported by the low incidence of CLL in the Asian population that has immigrated to the United States [15]. However, the similarity in clinical and cytogenetic characteristics and the frequency of IgV_H somatic mutation in sporadic and familial CLL suggest that the sporadic CLL is biologically the same disease as the familial one [16]. Recent studies have found a small population of CD5-positive monoclonal B-cells in many healthy adults [17]. The immunophenotype of this population closely resembles that of CLL. Such populations have been detected at increased frequency among relatives of CLL patients. Whether these clones represent precursor CLL cells that require further environmental stimulation or simply an aging related narrowing of the B-cell repertoire is not clear.

In summary, evidence for accurately defining the origin of CLL cells is still lacking. However, the available data suggest that 1) CLL cells are antigen challenged, leukemic cells with dynamic stages of growth, resting, and apoptosis; 2) CLL cells likely arise from a common precursor or undergo a common transformational event; and 3) heterogeneous subgroups exist and are defined by molecular genetics and various clinical behaviors.

CLL: One or Two Diseases?

Though CLL cells appear morphologically and immunophenotypically homogenous, differences in the clinical behavior of CLL cells with and without an IgV_H mutation suggest they are heterogeneous [18]. Whether CLL with an IgV_H mutation is the same as CLL without the mutation is debatable, because the clinical outcomes are very different. CLLs without the mutation are believed to be derived from the pregerminal center, while CLLs with the mutation are believed to be derived from the post-germinal center [19]. Some of the commonalities and differences are listed in table 1.

Table 1. Characteristics of Chronic Lymphocytic Leukemia in the Presence or Absence of the IgV_H Mutation

	IgV_H Mutation	
	Present	Absent
Prognosis	Good	Poor
Chromosomal abnormalities	13q	17p, 11q
Telomere length	Long	Short
Polyreactive antibodies	Rare	Frequent
Zap70 expression	No	Yes
CD38 expression	Rare	Frequent
Rai stage	Early	Advanced
Thrombopoietin	Low	High
IL-8	Low	High
β2-Microglobulin	Low	High
Hepatosplenomegaly	Rare	Frequent
TNF-α	Low	High
IL-6	Similar	Similar
bFGF	Similar	Similar
VEGF	Similar	Similar
HGF	Similar	Similar
Age	Similar	Similar
Morphology	Similar	Similar
Immunophenotype	Similar	Similar
Gene expression profile	Similar	Similar

Zap, zeta-associated protein; IL, interleukin; TNF, tumor necrosis factor; bFGF, basic fibroblast growth factor; VEGF, vascular endothelial growth factor; HGF, hepatocyte growth factor.

The reported preferential occurrence of certain IgV_H genes (eg, V_H1-69 in CLL without a mutated gene and V_H3-07 and V_H4-34 in CLL with a mutated gene) has suggested different antigenic challenges [20]. As indicated in table 1, the frequency of certain genetic abnormalities is different in CLLs with and without an IgV_H mutation. Unfavorable genomic changes such as 11q or 17p deletion and ataxia telangiectasia mutated (ATM) gene mutation are associated with CLL without an IgV_H mutation, while favorable changes such as a single deletion of 13q are associated with CLL with an IgV_H mutation [21]. Telomeres are composed of a repetitive DNA sequence and the specific proteins bound to the DNA. Telomeres shorten 100-200 bp with every cell division. CLLs with mutated IgV_H have longer telomeres than those without, indicating the number of cell divisions is different [12]. Some CLL cells produce polyreactive antibodies; such antibodies are expressed frequently in CLL lacking mutated IgV_H gene but only rarely in CLL with an IgV_H mutation [22]. Zap70, a member of the Src family of protein tyrosine kinase (PTK), has been detected at high frequency in patients without an IgV_H mutation and has even been suggested as a surrogate marker for lack of the mutation [23, 24]. Patients with IgV_H mutation usually present with high-risk features (high β2-microglobulin levels, CD38 expression, hepatosplenomegaly, advanced Rai stage, and elevated thrombopoietin, TNF-α, and IL-8 levels).

Despite all the differences stated above, there is a significant overlap in the characteristics shown in CLL with and without IgV_H mutation; individual features can be seen in patients of both groups. Recent progress in global gene expression profiling showed that CLL cells, irrespective of their IgV_H mutational status, have a common gene expression signature and only a limited number of differentially expressed genes [6,23]. These findings suggest CLL is a disease derived from common precursors. While the IgV_H mutation status is important for the clinical management of these patients, it should not be used as the sole diagnostic test for predicting clinical behavior or stratifying patients. A combination of clinical, IgV_H, ZAP-70, β2-microglobulin, and cytogenetic and FISH findings should be used to stratify patients for clinical management.

Role of B-Cell Receptor (BCR) in CLL

B-cell receptor (BCR) is a disulfide-bonded heterodimer composed of heavy and light immunoglobulin (Ig) chains complexed with the CD79a/CD79b heterodimer [25, 26]. It is believed that BCR plays a role in receptor expression and signal transduction by linking the antigen-binding Ig chains to intracellular tyrosine kinases of the Src family. Engagement of BCR leads to phosphorylation of immunoreceptor tyrosine-based activation motifs within the cytoplasm and this, in turn, activates cascades of intracellular signaling pathways including phosphatidyl 3-kinase and Akt kinase pathways. This is believed to be important in B-cell survival and induction of anti-apoptotic proteins (Mcl-1 and XIAP) and inactivation of apoptotic induction agents (BAD and caspase-9) [25, 27]. However, it is believed that, in general, activation of BCR eventually leads to cellular proliferation, differentiation, or apoptosis depending on co-signals, cell type, and stage of differentiation [25, 26].

CLL cells express low or undetectable levels of BCR [28, 29]. This diminished expression of BCR is not seen in any other B-cell lymphomas. It is believed that this correlates with reduced protein kinase activity and leads to impaired responses in CLL cells when BCR is engaged. In one study, engagement of IgM induced a signal transduction that resulted in inhibition of caspase activity, activation of NF-kB, and expression of *mcl*-1, *bcl*-2, and *bfl*-1; all are powerful factors involved in survival of CLL cells. These findings suggest BCR is important to CLL cell survival [30].

CD79a/CD79b are critical and integrated parts of BCR and defects in these molecules may lead to aberrant expression of BCR and signal transduction. In one study, severe impairment of glycosylation and folding of mu and CD79a chains were noted. This observation was believed to result in defective assembly and retention of these molecules in endoplasmic reticulum and to account for low levels of BCR in CLL cells [29]. Interestingly, this abnormality is more common in CLL with mutated IgV_H. CD79b is typically diminished or absent in CLL patients. Overexpression of an alternative CD79b transcript that lacks the extracellular domain, in addition to full length CD79b, is observed in most CLL cases [31]. This variant of CD79b appears to interfere with BCR signal transduction and inhibit apoptosis and may account for lower expression of BCR [32]. All this suggests that failure to signal is a membrane associated event rather than a cellular associated event. This was confirmed in a study using phosphorylation of Syk as a signal transduction indicator [33].

This study found that the majority of CLL cases without IgV_H mutation were able to signal via IgM, while the majority of CLL cases with the mutation failed to do so but were able to transmit the signal via IgD and CD79a (in some cases). This suggests an operational downstream pathway of signal transduction in CLL.

CLL Natural History and Progression

Patients presenting with CLL at an early stage (Rai stage 0 and some Rai stage I and II) remain stable for years and do not need therapy; however, patients presenting with more advanced disease (Rai III and IV) require therapy. Therapeutic response rate varies from 75% to 100% when purine analogue-based regimens are used. Fludarabine, alone or in combination with rituximab and other agents, has improved response and outcome significantly [34]. Some patients, on the other hand, will not respond to therapy or will respond for a brief period and then progress to more aggressive disease. Most of these patients die of bone marrow or organ failure due to tumor infiltration or transformation. The cause of resistance or progression is unknown; however, progression is frequently associated with the development of new karyotypic or molecular abnormalities including p53 mutation or deletion, ATM deletion, or trisomy 12 [35]. Some progression is associated with transformation to Richter syndrome in which the small CLL cells transform into large cells identical to those seen in large cell lymphoma. CLL can also transform into prolymphocytic leukemia (PLL), large cell lymphoma (5%), and, rarely, into acute lymphoblastic leukemia [36]. Since the median survival of patients with Richter syndrome is only 5 to 8 months, even with intensive combined chemotherapy and immunotherapy, it is warranted to monitor patients for these transformations. Historically, the diagnosis of Richter transformation requires the presence of sheets of large cells in lymphoid tissue or in bone marrow. One study conducted at MD Anderson showed that patients with increased numbers of large cells (>7%) in the bone marrow and elevated serum β2-microglobulin (>5mg/L) had overall survival identical to that seen in Richter syndrome [37]. Thus, the number of large cells and β2-microglobulin can be used as surrogate markers. The Epstein-barr virus (EBV) has associated with Richter transformation [38].

Clinically Relevant Research

Active basic and clinical research in the field of CLL has provided important insights into the biology. Such research has provided new tools for predicting clinical behavior and stratifying patients for therapeutic approaches. Until recently, CLL was considered a chronic disease that patients must live with and adapt to. Now clinical researchers are accepting the possibility of curing or clinically eliminating the disease owing to advances in immunotherapy and the application of nucleoside analogues. A description of the most clinically relevant advances follow.

1. MicroRNA

MicroRNAs are small RNAs transcribed from noncoding areas of the genome or the introns of coding genes [39]. These RNAs are believed to regulate gene expression at the post-transcriptional level by degrading or repressing target gene mRNA. Each microRNA can regulate a large amount of target genes. About half of the microRNAs are located at fragile sites and genomic regions involved in cancer; this suggests they play an important role in the development or progression of cancer [40]. The first clue that microRNAs might be involved in CLL pathogenesis was the loss or decreased expression of two microRNAs (15a and 16-1) that are located in the 13q14 gene region [41]. A microRNA microarray capable of examining all known microRNA expression was used to compare the expression profiles among CLL patients, healthy people, tonsillar CD5-positive B-lymphocytes, and peripheral mononuclear cells [42]. This technique identified 13 genes that are clearly associated with CLL progression. Interestingly, germ-line and somatic mutations in miR-16-1-miR-15a primary precursors were found in some patients with CLL but not in normal individuals; this finding suggests possible relevance to genetic susceptibility for CLL [43]. The potential of using micro-RNAs for diagnosis or prediction of clinical behavior, or even using micro-RNA in therapeutic approaches, is very exciting and may result in new tools for the fight against cancer [44].

2. Lipoprotein Lipase

Lipoprotein lipase (LPL) is a heparin sensitive enzyme that mediates hydrolysis of chylomicrons and very low-density lipoprotein (VLDL) along with apolipoprotein C [45]. As such, it plays an important role in lipid metabolism. In a study of gene expression profiling in patients with CLL, the LPL gene was identified as one of the preferentially expressed genes associated with lack of an IgV_H mutation [46]. Similarly, the expression of the ADAM29 gene was associated with the expression of a mutated IgV_H gene in CLL [47]. A subsequent study using quantitative reverse transcriptase polymerase chain reaction proved that LPL is an independent prognostic factor in patients with both early and late stage CLL [48]. LPL expression levels, especially as a ratio of the ADAM29 gene expression, were associated with progression free and overall survival [49]. Moreover, LPL expression has been reported to be a better predictor of survival than ZAP-70 and equally as powerful as IgV_H [47]. Genes associated with high LPL expression are involved in fatty acid degradation and MTA3 signaling [50]. Further studies are needed to explore the possible roles of those genes in the pathogenesis of CLL.

3. IgV_H

Somatic mutation of IgV_H often occurs in germinal centers of lymphoid follicles after activation of antigen, which eventually leads to production of B-cells with high affinity antibody. It is believed that when leukemogenesis is initiated before cells are exposed to antigen in the pregerminal stage, the IgV_H gene is not mutated, and these cells have more aggressive behavior. In contrast, when leukemogenesis arises in cells that were already exposed to antigen and that already had undergone somatic mutations to accommodate the antigen, the cells are at a later stage of differentiation and have less aggressive tendencies. The role somatic mutations of IgV_H play in the biology of CLL is unknown. As shown in table 1, CLL characteristics differ based on the IgV_H mutation status; however, it is not clear whether the differences are caused by the mutation status or simply associated with the mutation status. Antigen selection may also play a role in CLL. Certain V_H genes are preferentially expressed over the others, suggest antigen selection. For example, V_H1-69 is statistically overrepresented, is predominantly expressed in CLL cells without IgV_H mutation, and is seen in advanced stages, whereas V_H3-07 and V_H4-34 are usually seen in CLL cells with mutated IgV_H [4, 20] Patients with V_H3-21 usually have poor prognosis irrespective of mutational status. The cells in these patients are usually clonal for lambda light chain, rearrange a V lambda 2-14 gene, and have homologous complementarity determining region 3s, suggesting the recognition of a common antigen [51, 52].

Mutations of IgV_H are detected by comparing DNA sequences of IgV_H in target cells with the corresponding region in the germ line. A sequence that differs from the germ line gene by 2% or more is defined as mutated.

4. Zap-70

Zap-70, a member of the Src family of protein tyrosine kinase (PTK), is a 70-kd protein associated with the T-cell receptor (TCR) zeta chain [53]. It can undergo tyrosine kinase phosphorylation after stimulation of TCR, causing activation of the Tec family of PTKs and downstream pathways. Zap-70 is normally expressed in T-cells and natural killer cells and sometimes in neoplastic B-cells, but it is not expressed in normal B-cells.

Global gene expression profiling of CLL cells with and without mutated IgV_H revealed some differentially expressed genes in these two groups. Zap-70 was the most differentially expressed gene at the mRNA level and was associated with absence of IgV_H mutation [24, 54]. These results were confirmed by later studies using immunohistochemistry, Western blot, and flow cytometry at the protein level. In aggressive disease, CLL cells usually lack an IgV_H mutation and express Zap-70, while patients with indolent disease usually do not express Zap-70 and have mutated IgV_H [55].

Zap-70 has been proposed as a surrogate marker for IgV_H mutations and correlates with IgV_H mutational status, disease progression, and survival. Some studies suggested that ZAP-70 adds additional prognostic value beyond that of IgV_H mutation status; however, testing for the expression of ZAP-70 remains controversial and reproducibility of results between various laboratories has been disappointing. Numerous attempts to standardize testing have proved to be very difficult. The most difficult aspect of ZAP-70 testing, however, is clinical validation since CLL remains a chronic disease and prospective studies require a long follow up. At this point, the best approach for validating a clinically relevant ZAP-70 test is to correlate results with IgV_H mutation status. This approach is not optimal because the correlation between ZAP-70 and the IgV_H mutation status is not perfect. Many factors play a role in ZAP-70 testing including the viability of tested cells, the method for making the cells permeable for intracellular staining, the type of antibody used, the type of ZAP-70 isoform expressed, and the type of controls used for distinguishing positive from negative. In contrast, testing for IgV_H mutation status has been standarized and results are reliable and reproducible; thus, it should be used as the standard test.

The mechanism underlying the differential expression of Zap-70 in the presence and absence of IgV_H mutation is still not understood. Few studies have attempted to explore the impact of Zap-70 overexpression in CLL cells lacking the mutation. CLL cells expressing high levels of Zap-70 have increased tyrosine phosphorylation after BCR activation, suggesting enhanced BCR signaling capacity and lowered threshold for that signaling [56]. Ligation of IgM in Zap-70-positive CLL cells induced higher levels of downstream phosphorylated Syk, B-cell linker protein, and phospholipase–C gamma which may suggest more aggressive behavior [57]. Expression of ZAP-70 is also reported to enhance the ability of CLL cells to respond to migratory and survival signals [58].

5. Cytokines and Growth Factors

Thrombopoietin (TPO) is an important growth factor for megakaryopoeisis and platelet production. A study of TPO levels in 127 patients with CLL showed that circulating TPO levels were significantly higher in CLL patients than in normal controls. TPO levels correlated with advanced clinical stage, high levels of β2-microglobulin, lack of IgV_H mutation, and shorter survival. Thus, the TPO level is an independent, powerful, prognostic factor [59].

Interleukins are pleiotropic cytokines produced by a variety of cells, and they play important roles in B-cell growth and differentiation. Higher serum or plasma levels of IL-6 are associated with advanced clinical stage, elevated β2-microglobulin, and overall survival in patients with CLL [60, 61]. CLL cells do not appear to be the source of circulating IL-6. One study showed that the dimeric form of IL-6 could prolong survival of CLL cells in vitro [62]. CLL cells also express IL-8 and IL-8 receptor and appear to be an autocrine target of IL-8 [63]. IL-8 prolongs the survival and prevents apoptosis of CLL cells apparently through up regulation of *bcl*-2 [64]. Higher IL-8 levels are associated with advanced stages, correlate with serum β2-microglobulin levels, and are associated with risk of death in CLL patients [65]. The importance of cytokines in the biology of the disease is most likely related to the disease itself as well as the host and ability of the host to fight the disease or keep it under control. More studies are needed to define this role and to develop new therapeutic approaches that specifically address the abnormal levels of cytokines.

6. Angiogenic Factors

Bone marrow microenvironment provides the growth factors and cytokines necessary for regulating growth and differentiation of hematopoietic cells and vascular formation. In malignancy, nonvascular formation is necessary for continued tumor growth and metastasis and is usually stimulated by angiogenic growth factors [66]. The two main growth factors involved are vascular endothelial growth factor (VEGF) and basic fibroblast growth factor (bFGF). VEGF primarily binds to high affinity receptors on endothelial cells and regulates growth of blood vessels, while bFGF is pleiotropic and acts on a variety of cells [67]. Some people think angiogenic factors may have direct effects on the growth and differentiation of hematopoietic cells. Early studies showed that, though detected in a minority of CLL cases, higher serum VEGF levels correlated with advanced clinical stage and predicted risk of disease progression in early stages of CLL [68]. Later, it was shown that serum levels of VEGF are significantly higher in patients with CLL than in normal controls [69]. The association with advanced clinical stage and increased risk of disease progression was confirmed; furthermore, high levels were associated with poor outcome [69]. When combined with β2-microglobulin, VEGF and β 2-microglobulin showed the highest prognostic power as compared with multiple known prognostic factors in CLL[70].

Though serum bFGF is not increased in patients with CLL, elevated intracellular levels have been found in intermediate or high-risk CLL and correlate with clinical stage [71]. Higher intracellular bFGF is associated with resistance to fludarabine-induced apoptosis in vitro.

7. FISH and Cytogenetics

Conventional metaphase chromosome banding analysis and fluorescence in situ hybridization (FISH) are two major tools used to identify chromosomal abnormalities [72, 73]. Since adequate metaphases frequently cannot be obtained from CLL cells, stimulation with mitogen or CD40 ligands has been employed to induce progression through the cell cycle and improve the chances of obtaining a useful karyotype. FISH analysis can sometimes provide information that cannot be obtained from routine karyotyping. For example, the 13q14 deletion is detected in 70% to 80% of CLL cases when using FISH but in <30% of cases when using routine karyotyping. Other deletions frequently detected include the 11q22, 17p13, and 6q27 [73]. Trisomy 12 is also frequently observed in CLL. Translocations are less common and sporadic. A recent study used CD40 ligand stimulation which led to identification of translocations in about 34% of patients [74]. In addition, the development of array-based comparative genomic hybridization (CGH) has discovered more recurrent genomic alterations [75].

Deletion of 13q14 is usually associated with good prognosis if it is the only abnormality detected [73]. Two microRNAs (15a and 16-1) appear to be involved, as discussed above. Since microRNAs have shown importance in regulatory control over target genes, the loss of a microRNA may contribute to the initiation or progression of CLL. Further studies for microRNA-15a and 16-1 target genes may yield fruitful results.

Deletion of 11q22 is associated with poor prognosis [73, 76]. One of the candidate genes located in this area is ATM. ATM is associated with an autosomal recessive disorder, ataxia telangiectasia, that is characterized by neurologic symptoms and immunodeficiency. Additionally, ATM is associated with malignancies, including lymphoid disorders [77]. In CLL, both somatic and germ line mutations have been detected, as well as the deletion [78,

79, 80]. Loss of or decreased ATM expression has been detected at both RNA and protein levels. ATM inactivation can be either homozygous or heterozygous (deletion in one allele and point mutation in the second). Most mutations happen in the PI-3-K domain, which is involved in activation and phosphorylation of p53 during DNA repair [79]. Loss of function of this domain may cause defects in p53 response to DNA damage, leading to tumor initiation, transformation, or progression.

Trisomy 12 is associated with poor prognosis in patients with CLL, atypical morphology, and increased incidence of prolymphocytes [81]. The biological importance and potential role of trisomy 12 have not been elucidated nor have the candidate oncogenes been identified. Although the MDM2 gene has been suggested, protein expression analysis of chromosome 12 candidate genes has failed to identify genes that are can be implicated as the cause for the selection of trisomy 12 in CLL [82, 83]. Such analyses have compared protein expression levels in CLL cases with and without trisomy 12 and CD19-positive B-cells as well as in various cell lines.

Deletion of 17p is associated with poor prognosis, and patients who have the deletion are frequently resistant to chemotherapy [84, 85]. These associations might be due to involvement of the p53 tumor suppressor gene. Although 17p hosts a few important functional genes, the clinical impact is by far most likely caused by p53 deletion. Mutations may also occur in patients with CLL. p53 is important in DNA repair and apoptotic elimination of cells with damaged DNA [86]. Inactivation leads to impaired DNA repair and dysfunctional control of cell death, which may contribute to accumulation of genetic defects and CLL progression. Since p53 plays an important role in maintaining genomic integrity and mediating the action of DNA-damaging drugs, it is not surprising that deletion of 17p results in resistance to important drugs such as the purine analogs [85].

Chromosomal translocations are rarely detected by conventional cytogenetic analysis. After one group stimulated CLL cells with CD40 ligands or a combination of CpG-oligodeoxynucleotides and IL-2, better metaphase spreads were obtained and they identified translocations in 34% of 96 cases [74]. Half of these were balanced translocations. The presence of translocations is associated with older age, advanced stage, and higher levels of β2-microglobulin, thymidine kinase, and CD38. However, there are no CLL-specific translocations.

Development of array-based CGH has contributed significantly to the discovery of genomic aberrations or imbalances. The technology can be used for clinical diagnosis and prediction of disease progression. This approach allows us to compare thousands of genes at the RNA or DNA level. In one study, a DNA microarray chip specifically dedicated to the analysis of recurrent genomic imbalances in CLL was constructed. Analysis of samples from 106 patients with CLL demonstrated the chip's sensitivity and specificity and led to the identification of two previously unknown recurrent genomic aberrations (trisomy 19 and MYCN gene amplification) [75]. In another study of 100 patients with CLL-associated abnormalities (eg, deletion of 17p, 11q, and 6q and trisomy 12), microarray analysis discovered a variety of differentially expressed genes associated with these genetic defects [87]. Detailed study of the function of these genes will certainly shed light on their contribution to CLL clinical behaviors.

Clinical Research and Therapy

Treatment with alkylating agents or nucleoside analogues may achieve remission but not necessarily improve overall survival [88]. Fludarabine-based therapy is becoming the standard for front-line therapy [89]. In previously untreated patients with active CLL, fludarabine demonstrated higher overall response compared with alkylator-based chemotherapy regimens [90]. To improve response rates, fludarabine has been combined with chemotherapeutic agents (cyclophosphamide) and/or monoclonal antibodies (anti-CD20, rituximab) [91, 92]. Such regimens have demonstrated response rates of 76% to 100%.

Alemtuzumab, a human unconjugated anti-CD52 monoclonal antibody, has an 87% response rate when used as front-line therapy in patients with CLL [93]. CLL cells were easily cleared from the blood in these patients. Alemtuzumab also has potential to eradicate minimal residual disease [94, 95]. Infection, however, is of concern with alemtuzumab therapy, since normal T-cells, almost all of which express CD52, are killed [95, 96].

Allogeneic myeloablative stem cell transplantation remains a valid option for treating younger patients and those with experiencing drug resistance [97, 98]. Vaccine therapy administered in the early stages of the disease is being explored [99, 100]. If successful in eradicating the disease in early stage, patient management will change radically. Thus, the chronic nature of the disease is being challenged and the lessons learned from basic and clinical research are likely to lead to improved overall survival and quality of life.

References

[1] Zwiebel JA and Cheson BD. Chronic lymphocytic leukemia: staging and prognostic factors. *Semin. Oncol.* 1998;25:42-59.

[2] Bennett JM, Catovsky D, Daniel MT, et al. Proposals for the classification of chronic (mature) B and T lymphoid leukaemias. French-American-British (FAB) Cooperative Group. *J. Clin. Pathol.* 1989;42:567-584.

[3] Hamblin TJ, Davis Z, Gardiner A, Oscier DG, Stevenson FK. Unmutated Ig V(H) genes are associated with a more aggressive form of chronic lymphocytic leukemia. *Blood.* 1999;94:1848-1854.

[4] Chiorazzi N and Ferrarini M. B cell chronic lymphocytic leukemia: lessons learned from studies of the B cell antigen receptor. *Annu. Rev. Immunol.* 2003;21:841-894.

[5] Morikawa K, Oseko F, Morikawa S. Induction of CD5 antigen on human CD5- B cells by stimulation with Staphylococcus aureus Cowan strain I. *Int. Immunol.* 1993;5:809-816.

[6] Klein U, Tu Y, Stolovitzky GA, et al. Gene expression profiling of B cell chronic lymphocytic leukemia reveals a homogeneous phenotype related to memory B cells. *J. Exp. Med.* 2001;194:1625-1638.

[7] Damle RN, Ghiotto F, Valetto A, et al. B-cell chronic lymphocytic leukemia cells express a surface membrane phenotype of activated, antigen-experienced B lymphocytes. *Blood.* 2002;99:4087-4093.

[8] Hardy RR, Hayakawa K. CD5 B cells, a fetal B cell lineage. *Adv. Immunol.* 1994; 55:297-339.

[9] Caligaris-Cappio F, Gobbi M, Bofill M, Janossy G. Infrequent normal B lymphocytes express features of B-chronic lymphocytic leukemia. *J. Exp. Med.* 1982;155:623-628.

[10] Johnson TA, Rassenti LZ, Kipps TJ. Ig VH1 genes expressed in B cell chronic lymphocytic leukemia exhibit distinctive molecular features. *J. Immunol.* 1997;158:235-246.

[11] Schmid C and Isaacson PG. Proliferation centres in B-cell malignant lymphoma, lymphocytic (B-CLL): an immunophenotypic study. *Histopathology.* 1994;24:445-451.

[12] Damle RN, Batliwalla FM, Ghiotto F, et al. Telomere length and telomerase activity delineate distinctive replicative features of the B-CLL subgroups defined by immunoglobulin V gene mutations. *Blood.* 2004;103:375-382.

[13] Messmer BT, Messmer D, Allen SL, et al. In vivo measurements document the dynamic cellular kinetics of chronic lymphocytic leukemia B cells. *J. Clin. Invest.* 2005;115:755-764.

[14] Sellick GS, Catovsky D, Houlston RS. Familial chronic lymphocytic leukemia. *Semin. Oncol.* 2006;33:195-201.

[15] Weiss NS. Geographical variation in the incidence of the leukemias and lymphomas. *Natl. Cancer Inst. Monogr.* 1979;(53):139-142.

[16] Ishibe N, Sgambati MT, Fontaine L, et al. Clinical characteristics of familial B-CLL in the National Cancer Institute Familial Registry. *Leuk. Lymphoma.* 2001;42:99-108.

[17] Rawstron AC, Green MJ, Kuzmicki A, et al. Monoclonal B lymphocytes with the characteristics of "indolent" chronic lymphocytic leukemia are present in 3.5% of adults with normal blood counts. *Blood.* 2002;100:635-639.

[18] Hamblin TJ, Davis Z, Gardiner A, Oscier DG, Stevenson FK. Unmutated Ig V(H) genes are associated with a more aggressive form of chronic lymphocytic leukemia. *Blood.* 1999;94:1848-1854.

[19] Damle RN, Wasil T, Fais F, et al. Ig V gene mutation status and CD38 expression as novel prognostic indicators in chronic lymphocytic leukemia. *Blood.* 1999;94:1840-1847.

[20] Fais F, Ghiotto F, Hashimoto S, et al. Chronic lymphocytic leukemia B cells express restricted sets of mutated and unmutated antigen receptors. *J. Clin. Invest.* 1998;102:1515-1525.

[21] Krober A, Seiler T, Benner A, et al. V(H) mutation status, CD38 expression level, genomic aberrations, and survival in chronic lymphocytic leukemia. *Blood.* 2002;100:1410-1416.

[22] Herve M, Xu K, Ng YS, et al. Unmutated and mutated chronic lymphocytic leukemias derive from self-reactive B cell precursors despite expressing different antibody reactivity. *J. Clin. Invest.* 2005;115:1636-1643.

[23] Rosenwald A, Alizadeh AA, Widhopf G, et al. Relation of gene expression phenotype to immunoglobulin mutation genotype in B cell chronic lymphocytic leukemia. *J. Exp. Med.* 2001;194:1639-1647.

[24] Crespo M, Bosch F, Villamor N, et al. ZAP-70 expression as a surrogate for immunoglobulin-variable-region mutations in chronic lymphocytic leukemia. *N. Engl. J. Med.* 2003;348:1764-1775.

[25] Niiro H and Clark EA. Regulation of B-cell fate by antigen-receptor signals. *Nat. Rev. Immunol.* 2002;2:945-956.

[26] Fuentes-Panana EM, Bannish G, Monroe JG. Basal B-cell receptor signaling in B lymphocytes: mechanisms of regulation and role in positive selection, differentiation, and peripheral survival. *Immunol. Rev.* 2004;197:26-40.

[27] Ringshausen I, Schneller F, Bogner C, et al. Constitutively activated phosphatidylinositol-3 kinase (PI-3K) is involved in the defect of apoptosis in B-CLL: association with protein kinase Cdelta. *Blood.* 2002;100:3741-3748.

[28] Payelle-Brogard B, Magnac C, Alcover A, Roux P, Dighiero G. Defective assembly of the B-cell receptor chains accounts for its low expression in B-chronic lymphocytic leukaemia. *Br. J. Haematol.* 2002;118:976-985.

[29] Vuillier F, Dumas G, Magnac C, et al. Lower levels of surface B-cell-receptor expression in chronic lymphocytic leukemia are associated with glycosylation and folding defects of the mu and CD79a chains. *Blood.* 2005;105:2933-2940.

[30] Bernal A, Pastore RD, Asgary Z, et al. Survival of leukemic B cells promoted by engagement of the antigen receptor. *Blood.* 2001;98:3050-3057.

[31] Alfarano A, Indraccolo S, Circosta P, et al. An alternatively spliced form of CD79b gene may account for altered B-cell receptor expression in B-chronic lymphocytic leukemia. *Blood.* 1999;93:2327-2335.

[32] Cragg MS, Chan HT, Fox MD, et al. The alternative transcript of CD79b is overexpressed in B-CLL and inhibits signaling for apoptosis. *Blood.* 2002;100:3068-3076.

[33] Lanham S, Hamblin T, Oscier D, Ibbotson R, Stevenson F, Packham G. Differential signaling via surface IgM is associated with VH gene mutational status and CD38 expression in chronic lymphocytic leukemia. *Blood.* 2003;101:1087-1093.

[34] Yee KW, O'Brien SM. Emerging drugs for chronic lymphocytic leukaemia. *Expert Opin. Emerg. Drugs.* 2006;11:167-189.

[35] Tsimberidou AM, Keating MJ. Richter syndrome: biology, incidence, and therapeutic strategies. *Cancer.* 2005;103:216-228.

[36] Tsimberisou AM, Keating MJ. Richter's transformation in chronic lymphocytic leukemia. *Semin. Oncol.* 2006;33:250-256.

[37] Ma Y, Mansour A, Bekele BN, et al. The clinical significance of large cells in bone marrow in patients with chronic lymphocytic leukemia. *Cancer.* 2004;100:2167-2175.

[38] Ansell SM, Li CY, Lloyd RV, et al. Epstein-Barr virus infection in Richter's transformation. *Am. J. hematol.* 1999;60:99-104.

[39] Bartel DP. MicroRNAs: genomics, biogenesis, mechanism, and function. *Cell.* 2004;116:281-297.

[40] Calin GA, Sevignani C, Dumitru CD, et al. Human microRNA genes are frequently located at fragile sites and genomic regions involved in cancers. *Proc. Natl. Acad. Sci. U S A.* 2004;101:2999-3004.

[41] Calin GA, Dumitru CD, Shimizu M, et al. Frequent deletions and down-regulation of micro- RNA genes miR15 and miR16 at 13q14 in chronic lymphocytic leukemia. *Proc. Natl. Acad. Sci. U S A.* 2002;99:15524-15529.

[42] Calin GA, Liu CG, Sevignani C, et al. MicroRNA profiling reveals distinct signatures in B cell chronic lymphocytic leukemias. *Proc. Natl. Acad. Sci. U S A.* 2004;101:11755-11760.

[43] Calin GA, Ferracin M, Cimmino A, et al. A MicroRNA signature associated with prognosis and progression in chronic lymphocytic leukemia. *N. Engl. J. Med.* 2005;353:1793-1801.

[44] Chen CZ. MicroRNAs as oncogenes and tumor suppressors. *N. Engl. J. Med.* 2005;353:1768-1771.

[45] Mead JR, Irvine SA, Ramji DP. Lipoprotein lipase: structure, function, regulation, and role in disease. *J. Mol. Med.* 2002;80:753-769.

[46] Vasconcelos Y, De Vos J, Vallat L, et al. Gene expression profiling of chronic lymphocytic leukemia can discriminate cases with stable disease and mutated Ig genes from those with progressive disease and unmutated Ig genes. *Leukemia.* 2005;19:2002-2005.

[47] vann't Veer MB, brooijmans AM, Langerak AW. The prediction value of lipoprotein lipase for survival in chronic lymphocytic leukemia, *Haematologica.* 2006;91:56-63.

[48] Nuckel H, Huttmann A, Klein-Hitpass L, et al. Lipoprotein lipase expression is a novel prognostic factor in B-cell chronic lymphocytic leukemia. *Leuk. Lymphoma.* 2006;47:1053-1061.

[49] Oppezzo P, Vasconcelos Y, Settegrana C, et al. The LPL/ADAM29 expression ratio is a novel prognosis indicator in chronic lymphocytic leukemia. *Blood.* 2005;106:650-657.

[50] BilbanM, Heintel D, Scharl T, et al. Deregulated expression of fat and muscle genes in B-cell chronic lymphocytic leukemia with high lipoprotein lipase expression. *Leukemia.* 2006;20:1080-8.

[51] Tobin G, Thunberg U, Johnson A, et al. Chronic lymphocytic leukemias utilizing the VH3-21 gene display highly restricted Vlambda2-14 gene use and homologous CDR3s: implicating recognition of a common antigen epitope. *Blood.* 2003;101:4952-4957.

[52] Falt S, Merup M, Tobin G, et al. Distinctive gene expression pattern in VH3-21 utilizing B-cell chronic lymphocytic leukemia. *Blood.* 2005;106:681-689.

[53] Chan AC, Iwashima M, Turck CW, Weiss A. ZAP-70: a 70 kd protein-tyrosine kinase that associates with the TCR zeta chain. *Cell.* 1992;71:649-662.

[54] Wiestner A, Rosenwald A, Barry TS, et al. ZAP-70 expression identifies a chronic lymphocytic leukemia subtype with unmutated immunoglobulin genes, inferior clinical outcome, and distinct gene expression profile. *Blood.* 2003;101:4944-4951.

[55] Rassenti LZ, Huynh L, Toy TL, et al. ZAP-70 compared with immunoglobulin heavy-chain gene mutation status as a predictor of disease progression in chronic lymphocytic leukemia. *N. Engl. J. Med.* 2004;351:893-901.

[56] Chen L, Widhopf G, Huynh L, et al. Expression of ZAP-70 is associated with increased B-cell receptor signaling in chronic lymphocytic leukemia. *Blood.* 2002;100:4609-4614.

[57] Chen L, Apgar J, Huynh L, et al. ZAP-70 directly enhances IgM signaling in chronic lymphocytic leukemia. *Blood.* 2005;105:2036-2041.

[58] Richardson SJ, Matthews C, Catherwood MA, et al. ZAP-70 expression is associated with enhanced ability to respond to migratory and survival signals in B-cell chronic lymphocytic leukemia (B-CLL). *Blood.* 2006;107:3584-3592.

[59] Koller C, Bekele BN, Zhou X, et al. Plasma thrombopoietin compared with immunoglobulin heavy-chain mutation status as a predictor of survival in chronic lymphocytic leukemia. *Blood.* 2006;108:1001-1006.

[60] Lai R, O'Brien S, Maushouri T, et al. Prognostic value of plasma interleukin-6 levels in patients with chronic lymphocytic leukemia. *Cancer.* 2002;95:1071-1075.

[61] Fayad L, Keating MJ, Reuben JM, et al. Interleukin-6 and interleukin-10 levels in chronic lymphocytic leukemia: correlation with phenotypic characteristics and outcome. *Blood.* 2001;97:256-263.

[62] Moreno A, Villar ML, Camara C, et al. Interleukin-6 dimers produced by endothelial cells inhibit apoptosis of B-chronic lymphocytic leukemia cells. *Blood.* 2001;97:242-249.

[63] di Celle PF, Carbone A, Marchis D, et al. Cytokine gene expression in B-cell chronic lymphocytic leukemia: evidence of constitutive interleukin-8 (IL-8) mRNA expression and secretion of biologically active IL-8 protein. *Blood.* 1994;84:220-228.

[64] di Celle PF, Mariani S, Rieva L. Interleukin-8 induces the accumulation of B-cell chronic lymphocytic leukemia cells by prolong survical in an autocrine fasion. *Blood.* 1996;874382-9.

[65] Wierda WG, Johnson MM, Do KA, et al. Plasma interleukin 8 level predicts for survival in chronic lymphocytic leukaemia. *Br. J. Haematol.* 2003;120:452-456.

[66] Folkman J. Angiogenesis in cancer, vascular, rheumatoid and other disease. *Nat. Med.* 1995;1:27-31.

[67] Klagsbrun M and Moses MA. Molecular angiogenesis. *Chem. Biol.* 1999;6:R217-24.

[68] Molica S, Vitelli G, Levato D, Gandolfo GM, Liso V. Increased serum levels of vascular endothelial growth factor predict risk of progression in early B-cell chronic lymphocytic leukaemia. *Br. J. Haematol.* 1999;107:605-610.

[69] Gora-Tybor J, Blonski JZ, Robak T. Circulating vascular endothelial growth factor (VEGF) and its soluble receptors in patients with chronic lymphocytic leukemia. *Eur. Cytokine Netw.* 2005;16:41-46.

[70] Molica S, Vitelli G, Levato D, Ricciotti A, Digiesi G. Clinicoprognostic implications of increased serum levels of vascular endothelial growth factor and basic fibroblastic growth factor in early B-cell chronic lymphocytic leukaemia. *Br. J. Cancer.* 2002;86:31-35.

[71] Menzel T, Rahman Z, Calleja E, et al. Elevated intracellular level of basic fibroblast growth factor correlates with stage of chronic lymphocytic leukemia and is associated with resistance to fludarabine. *Blood.* 1996;87:1056-1063.

[72] Juliusson G and Gahrton G. Cytogenetics in CLL and related disorders. *Baillieres Clin. Haematol.* 1993;6:821-848.

[73] Dohner H, Stilgenbauer S, Benner A, et al. Genomic aberrations and survival in chronic lymphocytic leukemia. *N. Engl. J. Med.* 2000;343:1910-1916.

[74] Mayr C, Speicher MR, Kofler DM, et al. Chromosomeal translocations are associated with poor prognosis in chronic lymphocytic leukemia. *Blood.* 2005;107:742-51.

[75] Schwaenen C, Nessling M, Wessendorf S, et al. Automated array-based genomic profiling in chronic lymphocytic leukemia: development of a clinical tool and discovery of recurrent genomic alterations. *Proc. Natl. Acad. Sci. U S A.* 2004;101:1039-1044.

[76] Dohner H, Stilgenbauer S, James MR, et al. 11q deletions identify a new subset of B-cell chronic lymphocytic leukemia characterized by extensive nodal involvement and inferior prognosis. *Blood.* 1997;89:2516-2522.

[77] Taylor AM, Metcalfe JA, Thick J, Mak YF. Leukemia and lymphoma in ataxia telangiectasia. *Blood.* 1996;87:423-438.

[78] Starostik P, Manshouri T, O'Brien S, et al. Deficiency of the ATM protein expression defines an aggressive subtype of B-cell chronic lymphocytic leukemia. *Cancer Res.* 1998;58:4552-7.

[79] Schaffner C, Stilgenbauer S, Rappold GA, Dohner H, Lichter P. Somatic ATM mutations indicate a pathogenic role of ATM in B-cell chronic lymphocytic leukemia. *Blood.* 1999;94:748-753.

[80] Stankovic T, Stewart GS, Fegan C, et al. Ataxia telangiectasia mutated-deficient B-cell chronic lymphocytic leukemia occurs in pregerminal center cells and results in defective damage response and unrepaired chromosome damage. *Blood.* 2002;99:300-309.

[81] Matutes E, Oscier D, Garcia-Marco J, et al. Trisomy 12 defines a group of CLL with atypical morphology: correlation between cytogenetic, clinical and laboratory features in 544 patients. *Br. J. Haematol.* 1996;92:382-388.

[82] Winkler D, Schneider C, Krober A, et al. Protein expression analysis of chromosome 12 candidate genes in chronic lymphocytic leukemia (CLL). *Leukemia.* 2005;19:1211-1215.

[83] Merup M, Juliusson G, Wu X, et al. Amplification of multiple regions of chromosome 12, including 12q13-15, in chronic lymphocytic leukaemia. *Eur. J. Haematol.* 1997;58:174-180.

[84] Lin K, Sherrington PD, Dennis M, Matrai Z, Cawley JC, Pettitt AR. Relationship between p53 dysfunction, CD38 expression, and IgV(H) mutation in chronic lymphocytic leukemia. *Blood.* 2002;100:1404-1409.

[85] Dohner H, Fischer K, Bentz M, et al. p53 gene deletion predicts for poor survival and non-response to therapy with purine analogs in chronic B-cell leukemias. *Blood.* 1995;85:1580-1589.

[86] Vogelstein B, Lane D, Levine AJ. Surfing the p53 network. *Nature.* 2000;408:307-310.

[87] Haslinger C, Schweifer N, Stilgenbauer S. Microarray gene expression profiling of B-cell chronic lymphocytic leukemia subgroups defined by genomic aberrations and VH mutation status. *J. Clin. Oncol.* 2004;22:3937-49.

[88] Chemotherapeutic options in chronic lymphocytic leukemia: a meta-analysis of the randomized trials. CLL trialists' collaborative group. *J. Natl. Cancer Inst.* 1999;91:861-8.

[89] Rai Kr, Peterson BL, Appelbaum FR, et al. Fludarabin compared with chlorambucil as primary therapy for chronic lymphocytic leukemia. *N. Engl. J. Med.* 2000;343:1750-7.

[90] Leporrier M. Chevret S, Cazin B, et al. Randomized comparison of fludarabin, CAP, and ChOP in 938 previously untreated stage B and C chronic lymphocytic leukemia patients. *Blood.* 2001;98:2319-25.

[91] O'Brien SM, Kantarjian HM, Cortes J, et al. Results of the fludarabine and cyclophosphamide combination regimen in chronic lymphocytic leukemia. *J. Clin. Oncol.* 2001;19:1414-20.

[92] Keating MJ, O'Brien S, Albitar M, et al. Early results of a chemoimmunotherapy regimen of fludarabine, cyclophosphamide, and rituximab as initial therapy for chronic lymphocytic leukemia. *J. Clin. Oncol.* 2005;23:4079-88.

[93] Lundin J, Kimby E, Bjorkholm M, et al. Phase II trial of subcutaneous anti-CD52 monoclonal antibody alemtuzumab (Campath 1H) as first-line treatment for patients with B-cell chronic lymphocytic leukemia (B-CLL). *Blood.* 2002;100:768-73.

[94] O'Brien SM, Kantarjian HM, Thomas DA, et al. Alemtuzumab as treatment for residual disease after chemotherapy in patients with chronic lymphocytic leukemia. *Cancer.* 2003;98:2657-63.

[95] Wendtner CM, Ritgen M, Schweighofer CD, et al. Consolidation with alemtuzumab in patients with chronic lymphocytic leukemia (CLL) in first remission-experience on safety and efficacy within a randomized multicenter phase III triak if the German CLL Study Group (GCLLSG). *Leukemia.* 2004;18:1093-101.

[96] Thursky KA, Worth LJ, Seymour JF, et al. Spectrum of infection, risk and recommendations for prophylaxis and screening among patients with lymphoproliferative disorders treated with alemtuzumab. *Br. J. Haematol.* 2006;132:3-12.

[97] van Besien K, Keralavarma B, Devine S, et al. Allogeneic and autologous transplantation for chronic lymphocytic leukemia. *Leukemia.* 2001;15:1317-25.

[98] Thomson KJ, Mackinnon S. Role of allogeneic transplantation in low-grade lymphoma and chronic lymphocytic leukemia. *Curr. Opin. hematol.* 2006;13:273-9.

[99] Hus I, Rolinski J, Tabarkiewicz J, et al. Allogeneic dendritic cells pulsed with tumor lysates or apoptotic bodies as immunotherapy for patients with early-stage B-cell chronic lymphocytic leukemia. *Leukemia.* 2005;19:1621-7.

[100] Wahl U, Nossner E, Kronenberger K, et al. Vaccination against B-cell chronic lymphocytic leukemia with trioma cells: preclincal evaluation. *Clin. Cancer Res.* 2003;9:4240-6.

In: Chronic Lymphocytic Leukemia Research Focus
Editor: Chadi Nabhan, pp. 35-75
ISBN: 978-1-60021-526-1

Chapter III

Transplantation Strategies for the Treatment of Chronic Lymphocytic Leukemia

Edgardo S. Santos[*1] *and Mohamed A. Kharfan-Dabaja*[2*]
[1] Division of Hematology-Oncology,
Tulane University School of Medicine, New Orleans, Louisania, USA
[2] Division of Blood and Marrow Transplantation, H. Lee Moffitt
Cancer Center and Research Institute, Tampa, Florida, USA

Abstract

Novel biologic and molecular markers have been recently described in chronic lymphocytic leukemia that are likely to challenge the traditional approach of "watchful waiting" in asymptomatic patients and poor-risk prognostic factors. Molecular techniques such as microarray DNA, gene expression profile, polymerase chain reaction, and others, are capable of predicting differences in survival even when patients are stratified to the same clinical stage at diagnosis. Several ongoing clinical trials are evaluating whether early therapeutic intervention for chronic lymphocytic leukemia (CLL), compared to later treatment at onset of disease progression has an effect on survival. This paradigm becomes more complex when patients are diagnosed at a younger age. Also, integrating these markers into early therapeutic decision-making could be uniquely challenging, particularly when selecting the optimal therapeutic approach. Substantial progress has been made in developing new targeted therapies that are not only capable of inducing responses by non-specific cytotoxicity, but by activating different apoptotic mechanisms

[*] Edgardo S. Santos, M.D., FACP. Division of Hematology-Medical Oncology; Tulane University School of Medicine; 1430 Tulane Avenue, SL-78; New Orleans, LA 70112; Phone: 504-988-6352; Facsimile: 504-988-5483; Email: esantos1@tulane.edu; Job Title: Assistant Scientific Director, Office of Clinical Research; Associate Director, Hematology-Oncology Fellowship Program; Assistant Professor of Medicine

[*] Mohamed A. Kharfan-Dabaja, M.D., FACP. Division of Blood and Marrow Transplantation; H. Lee Moffitt Cancer Center and Research Institute; 12902 Magnolia Drive. SRB-4; Tampa, FL 33612; Phone: 813-745-8248; Facsimile: 813-979-3071; KharfaMA@moffitt.usf.edu; Job Title: Assistant Professor of Medicine

and other immunomodulatory effects as well. High dose chemotherapy (HDCT) followed by autologous hematopoietic cell transplantation (HCT) induces clinical and molecular responses in patients with CLL; however, it has not yet translated into cure. Conversely, allogeneic HCT appears to be a reasonable alternative option, especially when targeted strategies are incorporated into modern reduced-intensity preparative regimens. In this chapter, we discuss a variety of transplantation strategies that have been explored in the treatment of chronic lymphocytic leukemia, and evaluate the impact of novel prognostic factors in that perspective.

Introduction

Chronic lymphocytic leukemia (CLL) is the second most common leukemia diagnosed in the United States of America. It is estimated that 10,020 new cases and 4,660 deaths of CLL will occur in 2006 [1]. Our review will focus on B-cell CLL which is characterized by a classic phenotype of CD5, CD19, and CD23 receptors positive on the lymphocyte cell surface membrane. The clinical course of B-CLL is very variable, and its survival ranges from months to decades [2, 3]. Thus, B-CLL may exhibit either an indolent course in which the disease remains "stable" without any therapeutic intervention, or a rapidly progressive disease which require prompt therapeutic intervention. Why does B-CLL have these remarkable differences in clinical behavior? When the presence of mutations in the immunoglobulin heavy chain variable region gene (IgV_H) were first elucidated, it became clear to understand why these somatic changes make these clonal cells less susceptible to the effect of "self-antigens" on the B-cell receptor (BCR) [4, 5]. When the BCR is intact, B-CLL clonal cell is constantly stimulated by the binding of auto-antigens to BCR. The balance between negative and positive signals delivered by BCR and the survival signals transduced by IgD and delivered by other cells and cytokines determine whether the CLL clone proliferates or dies by apoptosis (figure 1). This and other discoveries were the beginning of an "avalanche" of basic research focused on CLL tumor biology. The results of those studies have been fruitful, providing more insights into the tumor biology of B-CLL, identifying novel prognostic tools, and being the anchor for development of future clinical trials which may be tailored to a specific group of B-CLL. We refer to a "specific group" of B-CLL because have already learned that even if a group of patients are clinically classified into a similar clinical stages, huge differences exist among them at the molecular level [6].

To date, we have been able to identify various poor-prognostic factors which include unfavorable cytogenetics (17p- and 11q-) [7], adverse biologic factors (unmutated IgV_H gene, a subset of mutated IgV_H named V3-21) [8, 9], overexpression of certain cell receptors (CD38, zeta-associated protein 70 [ZAP-70]) [10, 11], among others (table 1). Furthermore, the advent of these markers has allowed us to retrospectively analyze several clinical trials, correlate them with response to therapy, time to treatment, and survival. We now know that 17p- confers resistant to fludarabine-based regimens, and perhaps this may be an indication to seek out therapeutic alternatives which may overcome this acquired resistant mechanism [12, 13].

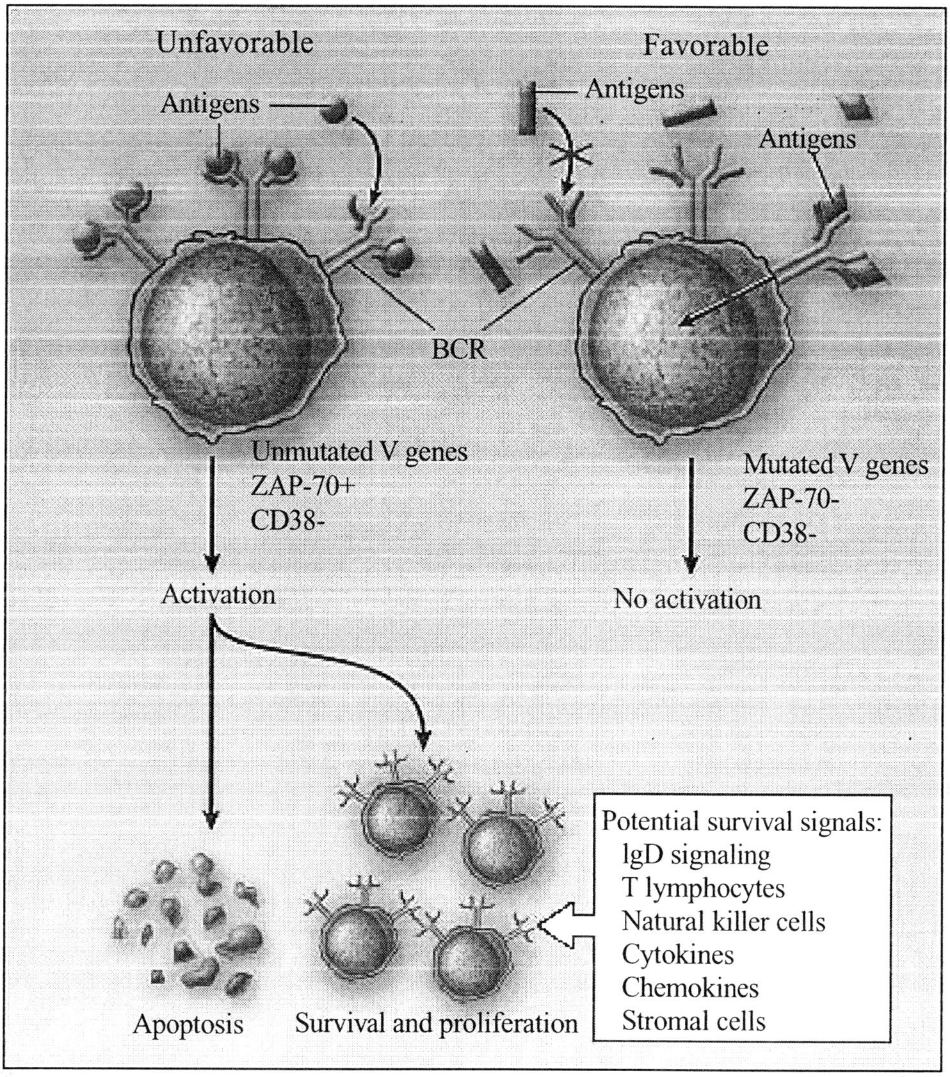

Figure 1. The Promoting Roles of Antigen Stimulation and Accessory Signals from the Microenvironment in CLL. B cells in patients with CLL who have unfavorable prognostic markers (left side of figure) are stimulated by the binding of self-antigens to the B-cell receptor. The dynamic balance of negative and positive signals delivered by the B-cell receptor and the survival signals transduced by IgD and delivered by other cells, cytokines, and chemokines determine whether the leukemic cell proliferates or dies by apoptosis. B cells from patients with CLL who have favorable prognostic markers (right side of figure) are less capable of triggering apoptosis, survival, or proliferation owing to an inability to bind antigen because of changes in the shape of B-cell receptors, because of V-gene mutations (blue rectangular antigen does not fit B-cell receptors, a condition known as "clonal ignorance"), or because of defective B-cell receptor signaling (despite an adequate fit of green diamond-shaped antigen in B-cell receptors). This lack of receptor stimulation may be a factor associated with a less aggressive form of the disease. From Chiorazzi N et al. Chronic lymphocytic leukemia. N Engl J Med. 2005; 352(8):804-815.

Table 1. Markers associated with Adverse Prognosis in Chronic Lymphocytic Leukemia

Clinical Markers	Gender: male Advanced clinical stage: ➢ Rai III or IV (anemia and/or thrombocytopenia) ➢ Binet B or C
Hematologic Markers	Doubling time (absolute lymphocyte count double < 12 months) Diffuse bone marrow infiltration
Serum markers	Increased serum lactate dehydrogenase (LDH) level Increased serum β-2 microglobulin level Low albumin level Increased plasma level of soluble CD52 Increased serum soluble CD23 Increased serum thymidine kinase level Increased plasma level of tumor necrosis factor-α. Increased serum level of vascular endothelial growth factor receptor-2 (VEGF-2) Increased plasma level of interleukin-8
Cell membrane receptor markers	CD38 expression > 30% malignant B-cell lymphocyte ZAP-70 expression > 20% malignant B-cell lymphocyte
Genetic or chromosomal markers	Unmutated IgVH gene Deletion of 11q Deletion of 17p Trisomy 12 Absence of 13q changes Presence of p53 mutation

We have previously quoted the traditional "watchful wait" approach in asymptomatic, early stage B-CLL patients. What has been the background for this consensus approach? For many decades since its introduction 1950s, chlorambucil was the "standard of care" for the treatment of B-CLL. Two main reasons led to abandon such an aggressive and proactive management in B-CLL: most of the cases have an "indolent" course and the lack of effective therapies which may impact its survival curve. It was not until late 1990s when single agent fludarabine established itself as an agent which induces better responses and higher complete remission (CR) rate than chlorambucil albeit not offering a survival advantage to patients [14-20]. Nonetheless, fludarabine was accepted as the new standard and the backbone for the development of several combined cytotoxic regimens; more recently in combination with targeted therapies such as monoclonal anti-CD20 and –CD52 antibodies [21-28]. These combinations have improved progression-free survival (PFS) in those patients who achieved CR (figure 2). As an example of this progress, Byrd et al. published a landmark study in which they retrospectively compared the results of 2 studies conducted by the Cancer and Leukemia Group B (CALGB), 9011 and 9712 [29]. Although it was a retrospective comparison, noteworthy all end points analyzed (overall response, CR, PFS, and OS) were in favor of fludarabine plus rituximab rather than fludarabine alone [29]. However, despite all advances in understanding B-CLL biology and therapy, it remains an incurable disease. Fortunately, the growing therapeutic armamentarium has parallel novel laboratory techniques capable of measuring the presence of minimal residual disease (MRD), which may guide the decision-making in the near future.

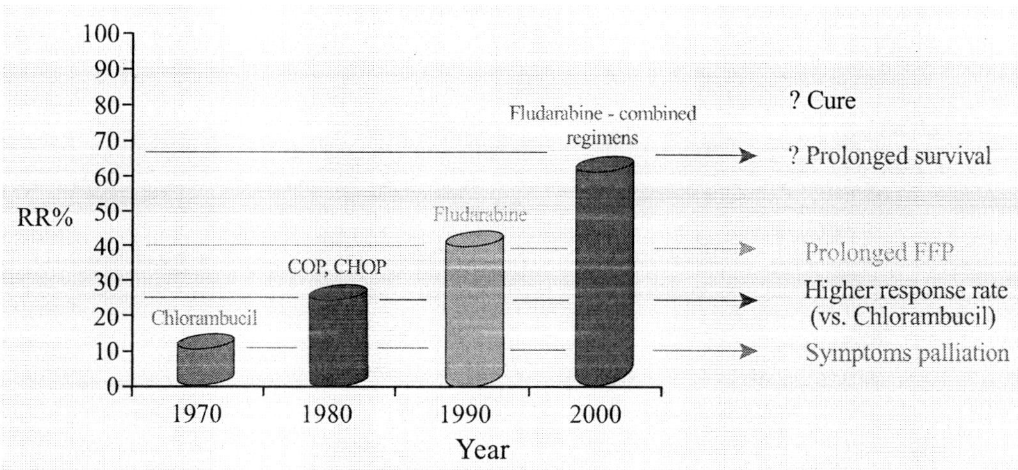

Figure 2. Chronic Lymphocytic Leukemia: Response rate and treatment goals over the years. From Monserrat E. CLL Therapy: progress at last! Blood. 2005; 105(1):2-3. © 2005 by The American Society of Hematology.

With this panoramic view on the evolution of B-CLL treatment, an important question remains: can we improve the response rates seen with chemoimmunotherapy approaches?, is there a chance for cure?, and how can we manage or treat MRD?. A therapeutic alternative which may address those issues is the use of hematopoietic cell transplantation (HCT). Several studies have been conducted using either autologous (with or without purging procedures), "non-myeloablative" allogeneic transplantation, or reduced-intensity conditioning (RIC) regimens.

HDCT followed by autologous HCT have failed to demonstrate thus far a "plateau" in overall survival (OS) curves in CLL. However, this modality of transplantation has been able to induce molecular remissions (MR) in patients who harbor poor-prognostic features when an anti-leukemic "purging" procedure has been incorporated into the transplant regimens [30, 31]. Conversely, allogeneic HCT offers another anti-leukemic effect beyond the classic cytotoxicity produced by HDCT: a graft-versus-leukemia (GVL) effect. Nevertheless, this latter therapeutic approach has been associated with a high treatment-related mortality (TRM) around 40% in some series. As a result, other strategies such as non-myeloablative ("mini-allo") and RIC preparative regimens have been developed. The outlook of allogeneic HCT looks more promising than anytime in the past.

Incorporating Novel Prognostic Markers in CLL Management

To understand how novel prognostic markers may impact the selection of transplantation approach, assessment of treatment efficacy, and decision-making in the post-transplant setting, we must bear in mind several facts:

a) Novel prognostic markers such as IgV_H mutational status, ZAP-70, genomic aberrations have been recently recognized.
b) HCT, either autologous or allogeneic, has been used not only as the last resource to treat a refractory/relapsed CLL patients, but also without a prospective randomized analysis whatsoever of these procedures themselves (conditioning regimens used, *"ex-vivo"* or *"in-vivo"* purging methods, and others) and these novel markers.
c) The introduction of biologic therapies has contributed to making CLL therapy more effective, and these same agents have been incorporated into transplantation regimens; thus, these novel therapies exert an antitumor effect still not well understood on CLL biology and may also impact appropriate assessment of these novel prognostic factors.
d) Well-designed, prospective randomized clinical trials taking into consideration genomic abnormalities, gene mutations, overexpression of certain receptors, and others are necessary to define the role of HCT in CLL.

Thus, making an assessment of what role ZAP-70 play in a patient who undergoes HCT does, it is still very prematurely. However, in retrospective analyses and in some open-label clinical trials recently published, we find "fingerprints" which display the significance and importance of these novel markers in CLL.

IgVH Mutational Status

In a multivariate analysis which included 205 CLL patients at or shortly after being diagnosed with disease, only unmutated IgV_H gene, loss or mutation of p53 gene, and clinical stage retained prognostic significance [32]. A clinical relevance finding was the difference in median survival found in those patients who harbored unmutated IgV_H or loss/mutation of p53 gene (119 and 47 months, respectively) than in those individuals who had mutated IgV_H (median survival of 310 months) (figure 3). Time to first treatment has also been correlated with IgV_H mutational status [33]. A study in which 115 CLL patients were included showed that unmutated IgV_H gene and/or positive expression of CD38 antigen urged for an earlier treatment for CLL than in those individuals who were negative for both factors (1.75 versus 18 years, respectively; $P < 0.0001$). This study is crucial to support the scientific rationale of ongoing clinical trials where early versus late treatment is being compared based on the presence of these poor-risk factors, as the survival of these individuals is significantly shorter.

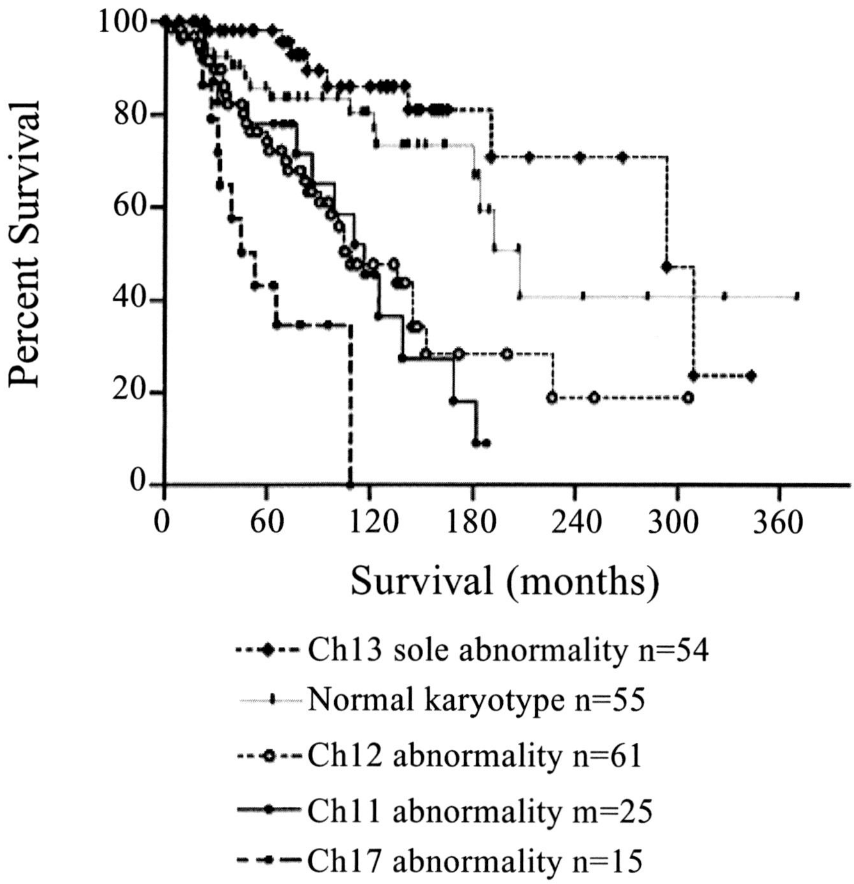

Figure 3. Survival of patients. A comparison of the survival of patients with del/t 13q14 as the sole abnormality detected by karyotype or Southern blotting; +12 or del 11q23 detected by karyotype analysis or interphase FISH; loss or mutation of p53 detected by interphase FISH or mutation analysis; and patients with no abnormality using any of the above techniques. The median survival of patients with no abnormality, del/t13q14 as the sole abnormality, +12, del 11q23, or p53 loss/mutation was 292, 209, 122, 117, and 47 months, respectively. From Oscier DG et al. Multivariate analysis of prognostic factors in CLL: clinical stage, IGVH gene mutational status, and loss or mutation of the p53 gene are independent prognostic factors. Blood 2002; 100(4):1177-1184. © 2002 by The American Society of Hematology.

ZAP-70

By comparing gene expression profiles of CLL cells with IgV_H mutational status, other important proteins have been identified such as the signaling molecule ZAP-70 [34]. Further studies using flow cytometry analyzes to evaluate ZAP-70 status from peripheral blood were able to reveal subgroups of patients with different clinical outcome and survival [35]. In a landarmark study, Crespo et al. found that a majority of patients who had > 20% leukemic cells positive for ZAP-70 also had IgV_H unmutated. Conversely, 21 patients out of 24 (88%) who had < 20% of their leukemic cells ZAP-70 positive were found to have IgV_H mutated.

Most interestingly, the level of ZAP-70 was invariably over the time in sequential samples from 30 CLL patients, and hence becoming a good surrogate marker not only for prognosis, but also for treatment response assessment as we discuss later on. Even patients initially staged as Binet A showed a rapid progression and poorer survival outcome when > 20% of their leukemic cells expressed ZAP-70.

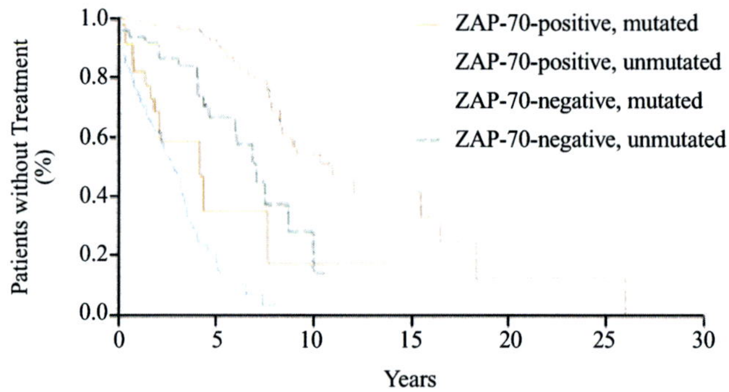

No./No. at Risk						
ZAP-70-positive, mutated	11/24	1/2	0/1	0/0	0/0	0/0
ZAP-70-positive, unmutated	77/117	6/11	0/0	0/0	0/0	0/0
ZAP-70-negative, mutated	8/119	23/82	3/18	3/5	0/1	1/1
ZAP-70-negative, unmutated	12/47	7/17	0/1	0/0	0/0	0/0

Figure 4. Relationship between ZAP-70 expression and IgV_H mutational status and the time from diagnosis to initial therapy. Kaplan–Meier curves depict the proportion of untreated patients with CLL according to the time since diagnosis. The patients are grouped with respect to their IgV_H mutational status and whether they did or did not express ZAP-70. From Rassenti LZ et al. ZAP-70 compared with immunoglobulin heavy-chain gene mutation status as a predictor of disease progression in chronic lymphocytic leukemia. N Engl J Med. 2004; 351(9):893-901.

In a large study involving 307 CLL cases, Rassenti et al. analyzed the relationship between ZAP-70 and IgV_H mutational status [11]. They were able to corroborate the presence of unmutated IgV_H with high-level expression of ZAP-70, and to add ZAP-70 as a prognostic factor for treatment need. They divided their study population into 4 different groups depending on the percentage of leukemic cells expressing ZAP-70 (0 to $\leq$ 10%; 10% to $\leq$ 20%; 20 to $\leq$ 30%; > 30%) and evaluate time from initial diagnosis to initial treatment. For the first two groups, these periods were 9.2 years and 9.0 years, respectively ($P = 0.23$). For the other two groups (20-30% and >30%), this period was 3.2 and 2.6 years, respectively ($P = 0.70$) (figure 4). The threshold level of 20% leukemic cells positive for ZAP-70 was considered to be optimal as it was shown in previous studies [11]. When this cut-off level was used to define ZAP-70 as positive or negative, the median time from diagnosis to initial therapy for those who were ZAP-70 negative was 9.2 years in comparison with 2.9 years for those who were ZAP-70 positive ($P < 0.001$). Importantly, in those patients who had a mixed phenotype (either ZAP-70 positive + mutated IgV_H or ZAP-70 negative + unmututed IgV_H) and which also represented 23% of the study population, ZAP-70 was found to be a stronger predictor than IgV_H status. To date, this discordance is still under investigation and it poses some challenge in daily clinical practice.

CD38 Overexpression

The CD38 antigen, a member of the family of the leukemic multi-lineage markers, mediates several functions including signal transduction and cell adhesion as well as a pivotal role in lymphocyte homing [36-38]. The ligation of CD38 has been reported to inhibit normal human B-cell lymphopoiesis while CD38 signaling mediated by agonistic antibodies prevents apoptosis of germinal center B-cells [39, 40]. It was suggested that CD38 level of expression may adversely impact survival of B-cell leukemic clones. Its expression has been linked to disease progression and outcome of leukemias including B-CLL [6, 41, 42]. In this regard, Damle et al. published their pivotal study in which a strong correlation was found between IgV_H mutational status and CD38 expression [6]. The authors concluded that those patients who harbored unmutated IgV_H also had leukemic clone expressing a level of $\geq$ 30% CD38. CD38 expression was also associated with unmutated IgVH in the cohort of patients studied by Oscier et al. ($P < 0.001$) (table 2) [32]. Furthermore, CD38 overexpression identified a subgroup with inferior outcome and was able to sort out patients with more aggressive clinical behavior in the Rai intermediate risk group. Damle el al. proposed the use of IgV_H gene mutation analysis and CD38 expression not only as prognostic markers for disease progression, but also as indicators for outcome in CLL patients. Indeed, they also considered CD38 expression a surrogate marker for IgV_H gene status. Hamblin reported 30% discordance between IgV_H mutation status and CD38 expression [43]. Other studies have also shown discrepancy between CD38 expression and IgV_H status. Consequently, these observations coined CD38 expression as not a good predictor of IgV_H mutational status [44-49]. Recent studies have correlated CD38 expression with OS, PFS, event-free survival (EFS), time to first treatment, and response to therapy [10, 50-57]. Recently, CD38 positivity in association with increased expression of its ligand (CD31) has been reported to identify a subset of cases with inferior outcome than those with CD38 expression alone [58].

In general, the percentage of CD38 expression may be considered independent prognostic marker, and measurement of CD38 may be used in the routine work-up of B-CLL to identify the poor-risk cases as it has become available to most of clinical laboratories. However, there are several caveats for using this marker by itself including its fluctuation during the disease course. The percentage of CD38 expression may increase with disease progression or during infections [43, 59]. Hence, it is relevant to measure CD38 expression close to or at the time of clinical presentation. To date, the significance of CD38 in CLL is still debatable, and ongoing clinical trials continue studying this marker.

Table 2. Association between IgV$_H$ gene status and other prognostic factors

Factor	Mutated N = 123	Unmutated N = 82	Odds ratio (95% CI)	*P*
Sex-male	49% (60)	71% (58)	2.5 (1.4, 4.6)	.003
Abnormal karyotype	60% (74)	83% (67)	3.2 (1.6, 6.3)	.001
Karyotype evolution present	18% (19)	29% (18)	1.9 (0.9, 3.9)	.15
Atypical morphology	18% (21)	59% (47)	6.6 (3.5, 12.7)	< .001
CD38≥30%	25% (27)	72% (44)	8.0 (3.9, 16.2)	< .001
Disease stage				< .001
A	93% (114)	67% (55)	1.0	
B	2% (3)	22% (18)	12.4 (3.5, 44.0)	
C	5% (6)	11% (9)	3.1 (1.1, 9.2)	
Chromosome 13 abnormal	53% (64)	29% (24)	0.4 (0.2, 0.7)	.002
Chromosome 12 abnormal	22% (26)	45% (35)	2.9 (1.6, 5.5)	.001
Chromosome 11 abnormal	8% (9)	20% (16)	3.1 (1.3, 7.4)	.02
Chromosome 17 abnormal	5% (6)	14% (9)	2.8 (1.0, 8.4)	.09

From Oscier DG et al. Multivariate analysis of prognostic factors in CLL: clinical stage, IGVH gene mutational status, and loss or mutation of the p53 gene are independent prognostic factors. Blood 2002; 100(4):1177-1184. © 2002 by The American Society of Hematology.

Cytogenetic Abnormalities

During the early stages of the disease, cytogenetic lesions are uncommon in the CLL clone. However, as the disease progresses, cytogenetic lesions start to appear. Among them, the most common is 13q14.3 deletion which occurs in more than 50% of CLL cases [7]. The most ominous alterations are deletions at 11q22-23, 17p13, and 6q21 [7]. It is believed that p53 and ataxia-telangiectasia mutated (ATM) genes are involved in deletions at 17p13 and 11q22-23, respectively, but which genes are the main players in these chromosomal alterations is still an unanswered question [7, 60]. The clinical relevance of the former deletions is the fact that both genes confer chemoresistance and regulate apoptosis, and both have been described in unmutated IgV$_H$ CLL cells which *per se* carry a poor prognosis [5, 60].

As mentioned before, there are a percentage of CLL cases in which ZAP-70 positive is not correlated with unmutated IgV$_H$ status. To elucidate this discordance, Krober et al. studied other possible variables which may help to understand these differences [61]. By a multivariate regression analysis, Krober et al. developed two models which included both high-risk genomic aberrations and, alternatively, IgV$_H$ mutational status and V3-21 usage or ZAP-70 as independent predictors for clinical outcome [61]. The study found different modes of discordance between ZAP-70 and IgV$_H$ depending on the presence or absence of high-risk genomic alterations such as 11q and 17p deletions or rearrangement of the specific variable region gene, V3-21 gene. In this study, the expression of ZAP-70 predicted an unfavorable clinical course in those cases where V3-21 was present regardless of the IgV$_H$ mutational status. However, ZAP-70 was unable to predict the clinical outcome in most of the cases which had unmutated IgV$_H$ associated with 11q or 17p deletion. Then, it seems that in the absence of high-risk chromosomal abnormalities, IgV_H status and expression of ZAP-70 may have similar prognostic impact, and hence, be used alternatively as predictor factors [61]. Thus, the authors proposed that high-risk genomic aberrations being considered as part of risk stratification for CLL together with IgV_H mutational status and ZAP-70 expression. Similar observations regarding the importance of genomic aberrations in CLL were presented by Byrd et al. in 2004 ASH meeting. The status of IgVH gene was assessed in 88 out of 104 patients enrolled into CALGB 9712 study. Herein, untreated CLL patients received fludarabine with or without rituximab. When genomic abnormalities were associated with clinical outcome, they conferred a shorter PFS in patients who harbor 17p or 11q deletions (table 3) [62].

In addition to the association between OS and genomic aberrations described by Dohner et al. [7], the presence of these chromosomal abnormalities may impact response to therapy. In 2004 ASH meeting, the preliminary genetic data from CLL4 trial conducted in the United Kingdom was presented [63]. A total of 730 chemonaive CLL patients were treated with chlorambucil, single-agent fludarabine or combined fludarabine and cyclophosphamide. The analyzed data included genomics from 400 patients. It showed that those patients who had 17p deletion also had the worst outcome. Almost half of them did not respond to treatment, and the 2-year survival rate was 67% in comparison with 87% for those patients with no chromosomal abnormalities (table 4) [63].

Table 3. Cytogenetic Abnormalities and Clinical Outcomes of Untreated CLL patients after Fludarabine and Rituximab based-therapy

Cytogenetic Abnormalities	17p-	11q-	6q-	Trisomy 12	13q-	Normal
No. (%) pts	3 (3%)	15 (17%)	1 (1%)	23 (26%)	24 (27%)	22 (25%)
CR (%)	0	53	100	25	38	64
PFS (months)	18	25	42	42	45	49

Pts: patients; "-": deletion. "Reprinted with permission from Byrd JC". Blood. 2004; 104(11):139a.

Table 4. FISH analysis by response (non-hierarchical) and Survival

Abnormality	CR/nPR	PR	NR/PD	p values*	2 yr survival (95% CI)
17p del	33%	18%	48%	0.004	67% (50%-85%)
11q del	29%	44%	27%	NS	73% (59%-86%)
Trisomy 12	41%	31%	29%	NS	72% (57%-87%)
6q del**	39%	29%	32%	NS	75% (56%-95%)
13q del	42%	37%	21%	NS	91% (86%-95%)

*Analysis of individual abnormalities versus the rest; **: 326 patients. "Reprinted with permission from Catovsky D". Blood. 2004; 104(11):8a.

The Principles of HCT Modalities

HCT is a complex therapeutic approach which involves different types of transplantation each having their specific mechanism of action. For educational purposes, we may divide the HCT approaches into: 1) autologous HCT, 2) myeloablative allogeneic HCT, 3) non-myeloablative allogeneic HCT ["mini" allogeneic HCT and reduced intensity conditioning (RIC) regimen], and 4) myeloablative syngeneic HCT. Autologous HCT involves harvesting patient's own hematopoietic cells that are then re-infused in the patient at transplant. It invariably involves the use of myeloablative conditioning regimen prior to transplant in order to achieve bone marrow ablation and thus, maximum tumor cell killing. Numerous studies have shown that a peripheral stem cell source offers an advantage with earlier engraftment of platelets and neutrophils [64, 65].

In the CLL, most patients undergoing autologous HCT go through different steps leading to transplant. Usually, the median time to relapse post-autologous HCT for CLL patients is around 5 years. In a study reported by Trisolini et al., for instances, 60% of the patients relapsed after a median of 31.5 months, and 11 out of 18 relapsed patients were able to continue further therapy, attained a response, and being alive after a median of 37 months post-relapse [66]. The projected OS at 5-year was 40%. Nonetheless, it was obvious that autologous HCT does not lead to cure in CLL. Thus, how may we improve the results seen with autologous HCT? Although not standardized, researchers have introduced *"in vivo"* purging using the novel MoAb anti-CD52 (alemtuzumab) prior to stem cell collection. If nopurging is applied at this step, then patient moves into stem cell mobilization/collection which is done after salvage cytoreductive chemotherapy is administered, in order to reduce the burden of leukemia cells and to minimize the presence of tumor cells in the harvest. Another alternative to accomplish this, it is the use of "*ex-vivo*" purging in which patient's hematopoietic cells are exposed to immunomagnetic beads, among others, allowing the collection of "non-contaminated" $CD34^+$ cells. Then, patients receive a preparative myeloablative regimen with HDCT (with or without radiation) called conditioning. Collected stem cells are then re-infused. Typically, it takes 2-3 weeks for those re-infused cells to start reproducing and maintaining hematopoeisis (engraftment). Although autologous HCT has resulted in high CR rates, most patients eventually relapse [67]. In order to decrease these events, purging approach either before or after stem cell collection has been introduced as part of transplant programs. In this sense, alemtuzumab seems to be a promising agent. Nevertheless, purging procedure has not yet been standardized and more work is needed to improve its effectiveness in decreasing relapse.

Conversely, allogeneic HCT utilized other principles: the infusion of "tumor-free" source of cells for transplant and/or donor-lymphocyte infusion (DLI) in case of relapses. The stem cells are collected from an HLA-matched donor (a person who may be related or unrelated to the patient). Furthermore, the donor may be HLA matched, or partially mismatched. This "external" source was thought to provide better outcome because of its capability of reducing relapse rate by re-infusing a harvest free of leukemic cells. However, allogeneic HCT comes at an equally great price related to the high rate of TRM and morbidity, including graft rejection, graft versus host disease (GvHD), and infections. TRM of allogeneic HCT in patients with CLL has been reported as high as 36% even in well-experienced centers [68]. A report obtained from the European Blood and Marrow Transplantation (EBMT) registry which included 188 allografted CLL patients disclosed a TRM of 49% at 36-month post-transplant [69]. These discouraging numbers were more than those seen after standard allografting indications such as acute leukemias or chronic myeloid leukemia. However, relatively few relapses occur after allogeneic HCT (15% within the first 3 years in the EBMT series), and the survival curve appears to reach a plateau long-term [70-72]. Thus, the benefit obtained with allogeneic HCT called graft-versus-leukemia (GVL) effect has become a central part for the success of this procedure. The importance of GVL effect was evident after clinical trials involving syngeneic transplantation (between identical twins) showing evidence of significant increase in rate of relapse when compared to its allogeneic counterpart [73].

GvHD, a serious complication of allogeneic HCT, is mainly due to donor T-cell effect against the host cells and organs leading to severe immunologic and inflammatory reactions [74, 75]. To decrease the risk of GvHD, T-cell depletion of donor cells has shown to be effective [76]. Unfortunately, once T-cell depletion has been performed, a significantly higher rate of relapse occurred in these T-cell depleted allogeneic transplants [77]. These observations led to the belief that T-cells are also the main effector cells implicated in GVL. In 1994, Korngold and Slavin showed that T-cells infused with hematopoietic cells had a direct anti-tumour effect on murine leukemic cells [78, 79]. Subsequently, numerous clinical trials have confirmed GVL in allogeneic HCT. In one of those first trials by Weiden et al., the association between GvHD and rate of relapse was clearly evident [80]. The relative relapse rate was 2.5 less in allogeneic marrow recipients with GvHD than in recipients without it ($P < 0.01$). This anti-leukemic effect was attributed to GVL effect, an immune mechanism regulated by the T-cell system.

An interesting question is whether T-cells in GVL recognize the leukemic cells because of specific leukemia-antigens expressed on them or is it part of the non-specific allogeneic reaction against all host hematopoietic cells, including the leukemic cells that share many antigens on their cell surface. Studies done by Sprent et al. showed that donor T-cells have an anti-leukemic effect as well as a direct cytotoxic effect in allogeneic transplant [81]. It is clearly evident that GvHD is associated with GVL and decrease rate of relapse of leukemia, and both phenomenon are triggered and regulated by T-cell system. Unfortunately, it is unclear what makes this immune reaction to favor one effect over the other. Numerous trials are underway to try and find immunomodulatory intervention that would tilt this graft effect against leukemic cells. This may include manipulation of immunosuppressive therapy in order to allow for a certain degree of mild GvHD that in turn may reflect ongoing GVL effect [82]. Another mechanism being used is changing the amount of transfused stem cells that have been shown also to alter the risks of getting GvHD. A novel arising concept in such manipulations involves the supportive cells being transfused namely bone marrow mesenchymal stem cells which only recently have been shown to play a cardinal role both in engraftment and GvHD [83]. Also, it has been postulated that GvHD is an immune response accentuated, and perhaps stimulated, by injury resulting from the preparative regimen used before transplantation [84]. It seems that this injury is confined to gastrointestinal tract where Peyer's patches play an important chemotaxis role after the injury, attracting donor T-cells, and hence, significantly contributing to GvHD development. Thus, using less intensive conditioning regimens prior to allogeneic HCT may help to ameliorate this undesirable effect.

Noteworthy, GVL phenomenon is also disease dependent, and for instances is more prominent in CLL, CML, and multiple myeloma than in acute lymphoblastic leukemia.

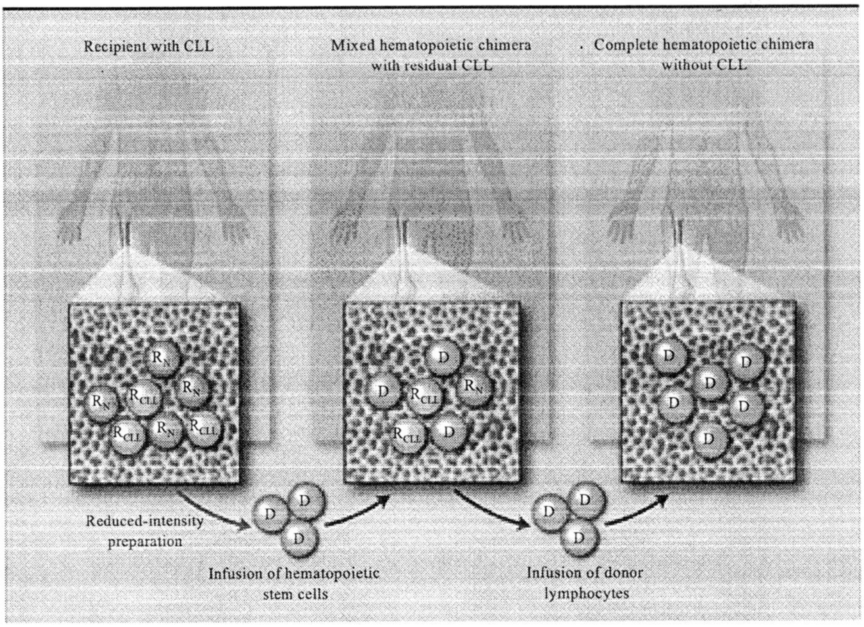

Figure 5. Transplantation Involving Reduced-Intensity Preparative Regimens. A recipient with CLL has normal (R_N) and malignant (R_{CLL}) cells in the marrow. A reduced-intensity regimen provides sufficient immunosuppression to permit the engraftment of hematopoietic stem cells from an allogeneic donor (D). Mixed hematopoietic chimerism results: normal donor cells and normal and malignant recipient cells coexist. The extent of donor chimerism can be ascertained by molecular analysis. Infusions of donor lymphocytes are given to eradicate residual recipient normal and malignant cells for complete donor chimerism. From Copelan EA. Hematopoietic Stem-Cell Transplantation. N Engl J Med. 2006; 354(17):1813-1826. Copyright © 2006 Massachusetts Medical Society. All rights reserved.

In allogeneic HCT, there are two different conditioning regimens used: myeloablative and non-myeloablative. The myeloablative allogeneic HCT uses dose intense chemotherapy with or without radiation in preparation for stem cell re-infusion. The steps followed are similar to myeloablative autologous HCT except for the source of stem cells. At transplant, recipient undergoes myeloablative conditioning followed by donor stem cell infusion, and is also started on immunosuppressive therapy to minimize the chance of GvHD. As mentioned before, this therapeutic option is associated with a high TRM close to 40% in patients with advanced cancer such as CLL patients. Hence, hematologists are shying away from this procedure for CLL therapy, and more inclined to offer RIC. A better understanding of GvHD led to the development of RIC regimens in the late 1990s. RIC differs from the previous myeloablative allogenic HCT because it uses reduced doses of chemotherapy to facilitate engraftment, and it depends on the graft to eradicate the malignant cells (GVL effect). With RIC regimens, we initially try to establish a temporary mixed hematopoietic chimerism. Then, using DLI we attempt to eradicate residual recipient normal and malignant cells which finally will induce a full donor chimerism state (figure 5). This approach will maximize on the benefit of GVL effect. Some studies have shown evidence of GVL effect on malignant clone of CLL [85, 86]. For instances, Schetelig et al. studied the outcome of 30 patients with advanced CLL who underwent RIC regimen. After RIC, patients received HCT from related (n = 15) or unrelated donor (n = 15). Peripheral blood samples of all patients with CR were assessed for MRD using PCR. After a median follow-up of 2 years, 23 patients were alive. Acute GvHD grade II to IV was observed in 17 patients (56%), and chronic GvHD was observed in 21 patients (75%). Twelve patients (40%) achieved CR, and 16 patients (53%) achieved PR (table 5). Interestingly, late CR occurred up to 2 years after HCT [85]. Four out of 12 patients with CR were not able to go through MRD assessment due to unsuccessful construction of a clone-specific primer. Noteworthy, MR was confirmed in the remnant 8 patients who achieved CR. MR was achieved after a median of 9 months (range, 1 to 18 months) post-transplantation (figure 6) [85]. OS, PFS, and non-relapse mortality at 2 years was 72%, 67%, and 15%, respectively (figure 7). This study re-emphasizes the significant improvement in TRM (4 patients) and demonstrates MR after long follow-up likely secondary to GVL.

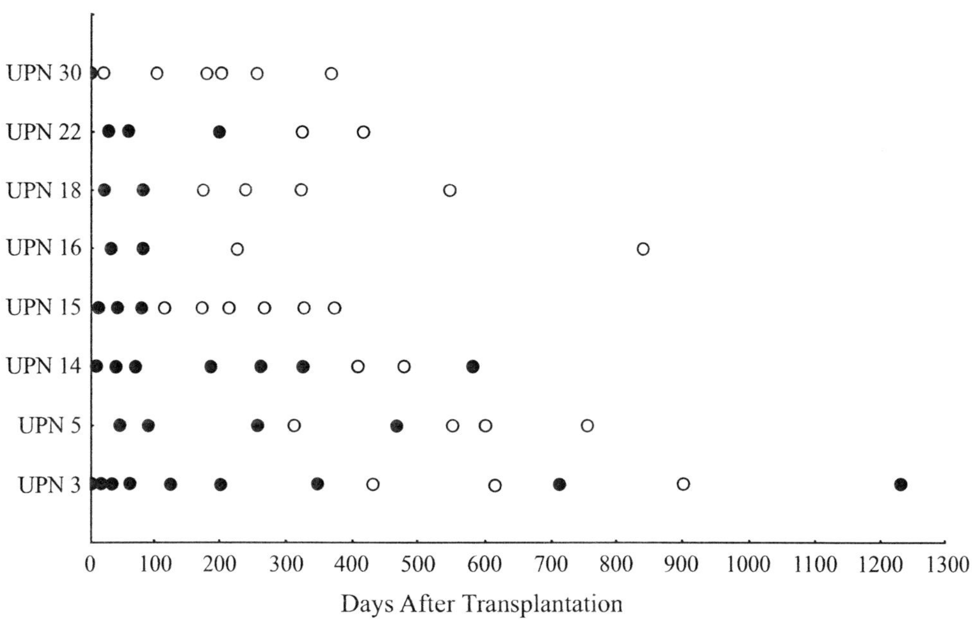

Figure 6. Molecular monitoring. (•) Detection of clone-specific DNA; (○) a negative PCR result. "Reprinted with permission from the American Society of Clinical Oncology" Schetelig J et al. Evidence of a Graft-Versus-Leukemia Effect in Chronic Lymphocytic Leukemia after Reduced-Intensity Conditioning and Allogeneic Stem-Cell Transplantation: The Cooperative German Transplant Study Group. J Clin Oncol. 2003; 21(14):2747-2753. 2003 by American Society of Clinical Oncology.

Table 5. Selected clinical trials using RIC/Non-Myeloablative Regimen to treat CLL

Reference	N	Conditioning	GVHD Prophylaxis	GvHD	OS/EFS	TRM/NRM
Schetelig [85] (Prospective; multicenter)	30	Flu/BU/ATG	CyA alone: 7 CyA + MTX: 11 CyA + MMF: 12	aGVHD: 56% cGVHD: 75%	2-yr probability of OS: 72% 2-yr probability of PFS: 67%	At 2 yrs: 15% NRM
Sorror [86] (Prospective; multicenter; Phase I)	64	TBI (2Gy) + Flu: 53 TBI (2Gy): 11	CyA + MMF	aGVHD: 61% cGVHD: 50%	2 yr OS: 60% 2 yr DFS: 52%	At 100 days: 11% At 2 years: 22%
Caballero [104] (Prospective; multicenter)	30	Flu/Mel: 19 Flu/Mel/Campath: 1 Flu/BU/ATG: 5 Flu/TBI/ATG: 5	First 20 pts: CyA and MTX (short course) Last 10 pts: CyA alone	aGvHD: 19 pts at a median of 33 days. cGvHD: 17 of 26 evaluable pts	At 6-year: OS: 70% EFS: 72%	At 68 months: probability of NRM: 22%.
Dreger [105] (Retrospective-data from 29 BMT centers	77	*Intensive:* 34/77 Flu/BU Flu/Mel +/- Campath Flu/BCNU/Mel +/- ATG BEAM *Moderate:* 43/77 TBI 2GyTBI 2Gy + Flu Flu/CY/Thiotepa Flu/CY +/- ATG Others	CyA +/- MTX +/- MMF	aGVHD: 34% cGVHD: 58%	At 2-year: EFS: 56% OS: 72%	At 1 yr: 18% TRM
Delgado [106] (Prospective)	41	Campath Fludarabine Melphalan	CyA	aGVHD: 41% cGVHD: 33%	2 yr OS: 51%	26%

ATG: anti-thymocyte globulin; MMF: mycophenolate mofetil; NRM: non relapse mortality.

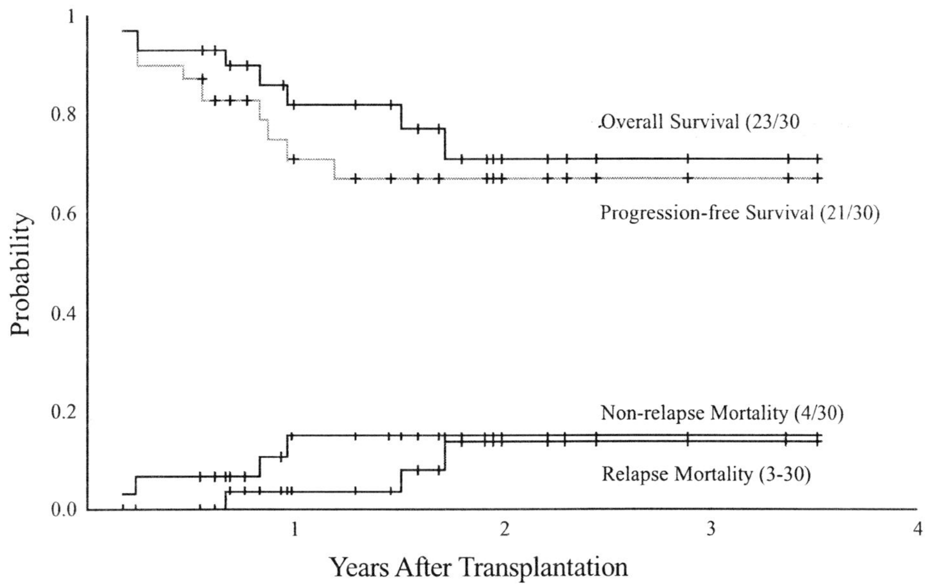

Figure 7. The 2-year probability of OS and PFS was 72% (95% confidence interval [CI], 54% to 90%) and 67% (95% CI, 49% to 85%), respectively. Estimates of non-relapse and relapse mortality were 15% (95% CI, 1% to 29%) and 16% (95% CI, 0% to 32%), respectively. "Reprinted with permission from the American Society of Clinical Oncology" Schetelig J et al. Evidence of a Graft-Versus-Leukemia Effect in Chronic Lymphocytic Leukemia after Reduced-Intensity Conditioning and Allogeneic Stem-Cell Transplantation: The Cooperative German Transplant Study Group. J Clin Oncol. 2003; 21(14):2747-2753. © 2003 by American Society of Clinical Oncology.

This milder form of allogeneic HCT (RIC) is also accompanied by a relatively lower morbidity and mortality incidence, and is rapidly gaining momentum in the treatment of a wide variety of hematologic malignancies, especially CLL.

When allogeneic HCT is elected as a therapeutic approach for CLL, DLI becomes a valuable intervention which may be used in case of relapsed CLL. Indeed, DLI has also has been used when T-cell depleted allogeneic transplant is performed. T-cell depletion has been used in some allogeneic HCT in order to minimize the risk GvHD and marrow hypoplasia. Nevertheless, this approach has been associated with clear increase in the risk of relapse as previously discussed, and that is the rationale for DLI subsequent to engraftment [74]. Donor lymphocytes (CD4/CD8) are collected from the same donor of stem cells, and then infused in the patient. In the case of relapsed CLL, DLI is used to exert GVL effect in the relapse setting by the same mechanism discussed earlier.

The last type of allogeneic HCT is syngeneic transplantation where the donor is the patient's identical twin. Theoretically, there is no allogeneic effect, no leukemic cells in the graft, and no graft exposure to therapy.

Transplantation in CLL Treatment: The Clinical Experience

Autologous HCT

Thus far, HDCT followed by hematopoietic cell rescue represents a complex therapeutic decision considering the impact that novel prognostic factors have produced on the clinicians in the last years. Aberrant genomics, mutation analysis and certain receptor overexpression levels are challenging the traditional "watchful observation" approach that has characterized the treatment of CLL for the last decades. Furthermore, the lack of prospective randomized studies comparing HCT versus best induction chemotherapy or chemo immunotherapy for patients with high-risk CLL leave any kind of therapeutic decision on the clinical experience of treating physician or the availability of a clinical research in which the patient may be enrolled.

As clinicians, we are facing difficult decisions such as when is appropriate to introduce a transplantation procedure as therapy within the algorithm of CLL management, what are the best pre-transplant induction and conditioning regimens, whether "purging" needs to be part of the therapeutic schema, what type of purging ("*in vivo*" or "*ex vivo*"), and how these patients will be followed-up after transplantation. Also, the capability of measure MRD poses another challenge: should MRD be the end-point to assess either conventional chemotherapy, or chemo immunotherapy or transplantation therapeutic efficacy? The role of autologous HCT in CLL treatment has been studied in retrospective and nonrandomized prospective studies, some of them with or without purging approaches.

One of the first studies involving autologous HCT in CLL was conducted by Rabinowe et al. [87]. Herein, 20 patients with poor-risk CLL underwent autologous HCT. Patients received uniform HDCT consisting of cyclophosphamide and total-body irradiation (CY/TBI) followed by rescue with multiple MoAb-purged autologous bone marrow (12 patients) or T-cell depleted allogeneic bone marrow from HLA-identical siblings (8 patients). All patients achieved hematologic engraftment. Seventeen out of 19 (89%) evaluable patients achieved CR and 2 patients (1 allogeneic and 1 autologous) had persistent bone marrow disease. Toxicity was similar amongst both treatment modalities. The TRM was 10%. These results demonstrate that autologous HCT is feasible and relatively safe, supporting a possible role for autologous HCT as a therapeutic option for patients with high-risk CLL. In another study, a mixed cohort (in terms of transplantation modality used) revealed important observations and differences between both transplantation modalities [88]. Also, in this cohort, most of the patients had unmutated IgV_H mutation. The outcome of 162 CLL patients who underwent autologous or allogeneic HCT was analyzed. The patients had poor prognosis defined as either advanced stage or intermediate stage with poor prognostic IgV_H mutational status. All patients received CY/TBI regimen followed by infusion of allogeneic or autologous bone marrow. The results from the autologous HCT arm revealed the following: 137 patients received B-cell purged autografts, TRM was 9%, 4 patients failed to engraft and 1 had a very delayed engrafment [88]. They had significantly delayed neutrophil engrafment ($P = 0.03$) and a trend toward delayed platelet engraftment when compared with patients who received other regimens. At the moment of report, 44 patients had died post- autologous

HCT, and 59 have relapsed. The study concluded that HCT may result in prolonged survival, but because of its high morbidity and mortality, it should be reserved to high poor-risk CLL patients. Noteworthy, it is the fact that allogeneic HCT arm did not show statistical difference in terms of OS, EFS, or probability of relapse in comparison with autologous HCT arm [88].

A retrospective study analyzed the outcome between autologous HCT and standard therapy based on IgV_H gene mutational status [30]. In this risk-matched comparison, 66 patients who have undergone autologous HCT [conditioning regimen: fractionated TBI: 12-14.4 Gy and cyclophosphamide (2 x 60 mg/kg)] were matched against 291 patients treated conventionally. Variables such as age, Binet stage, IgV_H mutational status and lymphocyte count were found to be fully-matched in 44 pairs of patients. The groups were also balanced for other factors such as CD38 expression and adverse genomic aberrations. Survival was significantly superior in favor of the autologous HCT group compared to those treated with conventional therapy, when calculated from diagnosis (hazard ratio [HR] 0.39; $P = 0.03$) or from study entry (HR 0.32; $P = 0.006$) [30]. The benefit of autograft remains statistically significant when the analysis was restricted to those who had unmutated IgV_H. Cox regression analysis showed that autologous HCT is an independent favorable prognostic factor for survival. In this study, autologous HCT proofed to offer a survival benefit for patients with poor-risk CLL. Nonetheless, the question that remains unanswered is when to transplant these high-risk patients in the course of their disease.

In a retrospective study, Trisolini et al. evaluated whether or not patients who relapsed after autologous HCT may be retreated with intensive approaches. In their cohort, autologous HCT was used as “consolidation” after induction therapy with fludarabine. Thirty patients who achieved CR prior to autologous HCT were included in the analysis [66]. All patients had advanced disease (Binet B or C), and most of them (n = 24) received BEAM regimen as conditioning. At a median follow-up of 31.5 months, 18 (60%) patients have relapsed post-transplant. Interestingly, 61% of those who relapsed were able to receive further therapy, and remain alive after a median follow-up of 37 months from relapse [66]. These findings raise questions regarding the effect of HDCT on the CLL malignant clone and its microenvironment.

Herein, we discuss some of the most relevant prospective studies which had analyzed autologous HCT in CLL. Dreger et al. studied 18 patients diagnosed with CLL who underwent autologous HCT [89]. The median age of the cohort was 49 years, and patients with Binet stages A, B, and C were included in the analysis; however, most of them (89%) had adverse prognostic factors such as high lymphocyte count and/or diffuse bone marrow infiltration. All patients had demonstrable molecular disease by PCR. Stem cell mobilization and harvesting was successful in 14 patients using Dexa-BEAM regimen (table 6). All grafts were purged using immunomagnetic methods. Following conditioning with chemoradiation, 13 of 14 patients were re-infused with purged autologous hematopoietic cells. Engraftment occurred faster in patients receiving mobilized peripheral blood stem cells (PBSC) (n=10) than in those receiving bone marrow cells (n = 3). TRM was reported as 0% [89]. At a median follow up of 48 months, only one patient relapsed and 2 additional had evidence of molecular disease by PCR amplification of the complementary determining region III (CDRIII) gene, but remained in clinical CR. This study supported autologous HCT as a safe and an effective treatment for patients with CLL [89]. Two caveats in this study were the relatively younger age population and the inclusion of earlier clinical stages. Sutton et al. evaluated the feasibility and efficacy of autologous HCT in a heavily pretreated CLL population in a prospective study [90]. Twenty patients underwent autologous HCT. Stem cell collection was performed following treatment with the ESHAP regimen and, surprisingly, only 8 (40%) of 20 had sufficient harvested stem cells to proceed with autografting. Six (75%) patients remained in CR at a median follow-up of 30 months after stem cell infusion (table 6). Thus, autologous HCT showed to be a feasible procedure, however, it was emphasized that transplantation must be considered at earlier clinical stage to avoid cumulative hematologic toxicity resulting from prior therapies [90].

In one of the most important clinical trials conducted recently in CLL using HCT as therapeutic approach, the German Chronic Lymphocytic Leukemia Study Group (GCLLSG) reported a prospective multicenter trial evaluating the feasibility and efficacy of: 1) early HCT procedure in patients with poor-risk CLL and 2) in a CLL population where most of them (86%) were chemonaive at study entry [91]. Patients underwent PBSC mobilization using Dexa-BEAM regimen, followed by myeloablative therapy with CY/TBI and then re-infusion of purged stem cells. Cytoreduction regimens (CHOP, fludarabine alone, or fludarabine and cyclophosphamide) were optional. There were 128 of 179 evaluable patients. Pre-transplant documented responses were as follows: CR: 28%, PR: 64% [91]. Ninety eight of 128 patients (77%) proceeded to autologous HCT. At 36 months of follow-up, median progression-free survival (PFS) of all 128 patients, on intent-to-treat, was 59 months (54 months for those with unmutated IgV_H). OS at 4 years for all 128 patients was 84% [16]. Grade III/IV non-hematological toxicities were mainly due to infections (18% pre-transplant; 7% post-transplant). Two of 6 fatal complications were seen in post-transplant setting, corresponding to a TRM of 5% (table 5). HDCT followed by autologous HCT is feasible and promising for poor-risk CLL patients. Likewise the previous study, Milligan et al. conducted a prospective study in which previously untreated patients were treated with autologous HCT. This landmark study was the first ever to only enroll chemonaive CLL patients for HCT treatment as up front line after initial treatment with fludarabine [92]. A total of 115 chemonaive patients received fludarabine therapy followed by stem cell mobilization and collection. Sixty-five (56%) patients had adequate stem cell collection and were able to proceed to autologous HCT. Only one patient died as a result of therapy. Forty-eight patients (74%) achieved CR, and 26 of 41 (63%) patients who were not in CR at the time of their transplantation achieved CR after transplantation [92]. The 5-year OS and DFS from transplantation were 77.5% and 51.5%, respectively. Sixteen of 20 (80%) evaluable patients achieved MR by PCR in the first 6 months following transplantation (table 5). Subsequently, 5 of 65 (8%) developed acute myeloid leukemia/ myelodysplastic syndrome (AML/MDS), at a surprisingly high incidence [92]. This landmark study raises several questions and confirm some facts: 1) is it fludarabine the best induction therapy in CLL, considering that only 56% of patients were able to go through HCT rescue due to poor stem cell collection?, 2) HDCT followed by stem cell rescue improves clinical responses seen in the induction setting, 3) may purging procedure improve the results seen in this study?, 4) "conventional" (conditioning regimen followed by stem cell re-infusion only) autologous HCT is not curative in CLL, and 5) why a high incidence of AML/MDS in this chemonaive cohort of patients?

Table 6. Selected prospective clinical trials using Autologous HCT in CLL patients

Investigator	N pts	Conditioning Regimen	Purging (Y= yes; N= no)	Response/PFS/OS	TRM
Dreger [89]	18	TBI/CY	Y Immunomagnetic methods	At 48 mo: 1 relapse; 2 pts. with evidence of molecular disease	0%
Sutton [90]	20	TBI/CY	N	At 30 mo: 6/8 in CR.	NR
Dreger [91]	179 (98 proceed to HCT)	TBI/CY	Y	At 36 months: median PFS (all 128 patients intended to treat): 59 months 4-year OS: 84%.	5%
Milligan [92]	115 (65 proceed to HCT)	TBI/CY	N	74% CR 5-year OS: 77.5% DFS: 51.5%. 16/20 evaluable pts achieved MR.	1 pt. died. 8% developed AML/MDS
Tedeschi [94] Montillo [95]	20	TBI/CY or Melphalan	Y *In vivo*: Campath	At 10 months: 100% CR.	0%
Montillo [96]	34 (18 proceed to HCT)	Cytosine arabinoside	Y *In vivo*: Campath	At 14.5 months post-transplant: 94% CR (17/18).	NR

N pts: number of patients; TBI/CY: total body irradiation plus cyclophosphamide.

In the next three studies, we discuss autologous HCT where purging procedures have been added to the complex treatment schema. Dreger et al. analyzed the efficacy of "*ex-vivo*" purging of PBSC [93]. Twenty patients with poor-risk CLL underwent uniform stem cell mobilization with double B-cell depletion of the harvested PBSC using immunomagnetic $CD34^+$ cell selection (Isolex300i Nexell, Irvine, CA) followed by a negative step with anti-CD19/20/23/37-labeled immunomagnetic beads. The purified PBSCs were reinfused after myeloablative treatment with chemoradiation (CY/TBI). CLL cells were undetectable in 15 of 25 using flow cytometric testing. Rapid and durable engraftment developed in all cases. At a median follow-up of 20 months, 17 patients were in CR, and 2 had recurrent disease [93]. Purging did not appear to negatively impact engraftment, and may have a positive role in decreasing relapse; however, longer follow-up is needed to draw more solid conclusions.

Using an "*in vivo*" purging procedure, the Milano group evaluated through a phase I trial 14 CLL patients who were pretreated with fludarabine-containing regimens and subsequent alemtuzumab subcutaneously (10 mg x 3 weekly for 6 weeks) [94]. A fludarabine-containing regimen was given as first-line therapy to 12 of these patients, and a median of 6 cycles were administered. At the time of transplant the median age was 57.7 years; all patients were in CR, and 9 patients showed polyclonal rearrangement of the IgV_H by PCR. Post-transplant evaluation showed no grade III/IV non-hematological toxicity, and no TRM. No CMV reactivation was seen following HCT, even in those patients (n=6) who developed CMV reactivation during alemtuzumab administration [94]. At a median follow up of 10 months (range, 3-30), all patients were in CR. This study concluded that sequential therapy with "*in vivo*" purging prior to AHCT is feasible (table 6). The majority of patients (13 of 14) achieved molecular remission (MR) proved by the presence of polyclonal IgV_H rearrangement post-transplant [94]. An update results were presented in the 2005 ASH meeting where the outcome of 20 patients treated with fludarabine-containing regimens, and then subcutaneous alemtuzumab prior to autologous HCT [95]. Prior to transplantation, 100% of the patients were in clinical CR by NCI-WG criteria, and only 65% of them were on MR by PCR. At 9- and 12-months post-transplant follow-up, 16 and 12 patients were evaluated, respectively. All but one patient was on MR by PCR. This combination did not result in increased infection-related adverse events, and produced a high and sustained MR [95]. Most recently, Montillo et al. published the results of a phase II trial where 34 patients received alemtuzumab as consolidation therapy (*'in vivo"* purging), to improve prior response obtained with a fludaribine-based induction, and before PBSC harvest [96]. Consolidation therapy consisted of alemtuzumab at 10 mg subcutaneously thrice weekly for 6 weeks. Mobilization was performed using cytosine arabinoside (Ara-C) and granulocyte-colony stimulating factor. CR rates improved from 35% (attained with induction therapy) to 79% using consolidation, including 56% MRD negativity by PCR. CMV reactivation occurred in 18 patients, all of whom were successfully treated with ganciclovir. Hematopietic cell collection was successful in 24 of 26 patients, but only 18 were able to undergo autologous HCT. The median follow-up for all patients was 25.5 months since alemtuzumab administration [96]. Seventeen (94%) out of 18 patients who underwent autografts were in CR at 28 months (table 6). In the group which did not proceed to autologous HCT, 7 showed progression of disease after a median of 7 months.

Purging in CLL treatment has not been analyzed in a randomized fashion, and_hence, no final conclusion can be drawn. However, single-arm studies have shown that *"in vivo"* purging prior to stem cell collection improves the quality of response in these patients. How does it affect the transplant outcome is still unknown. The study presented by Montillo et al. [95] make us to think that even with the use of purging pre-harvest, HDCT plays an important role and has a real impact on the CLL clone and its microenvirnonment.

Myeloablative Allogeneic HCT

The major benefits obtain from allogeneic HCT are the infusion of tumor-free hematopoietic cells and induction of GVL effect. However, these positive features come with a high price to pay: a substantial morbidity and a high TRM. Some reasons for these high TRM observed may have been the age of CLL patients who have been enrolled into these clinical trials (in comparison with the age of patients who have undergone HCT for other hematologic indications) and the fact that most of the patients have either advanced disease or heavily pre-treated CLL, and/or relapsed/refractory CLL clones. A database registry study of 54 patients (IBMTR and EBMT) with a median age of 41 years showed that allogeneic HCT is feasible with a non-relapse mortality of 46%, hematologic remission of 70% and a 3-year probability of survival of 46% [97]. Most patients received CY/TBI followed by infusion of bone marrow from an HLA-identical sibling (table 7). Another analysis of 187 patients from EBMT registry data showed a TRM of 50% and a 3-year OS probability of 45% demonstrating a "plateau" effect [98]. Doney et al. reported the outcome of 25 patients (median age: 46.6 years) with CLL who received related donor HCT (HLA-identical siblings: 21; syngeneic: 3; DRB1 locus mismatched sibling: 1) [99]. The two major preparative regimens included BUCY (busulfan total dose: 14 mg/kg orally, cyclophosphamide total dose: 150 mg/kg IV) or CY/TBI (cyclophophamide total dose: 120 mg/kg, TBI: 10-16 Gy). Non-relapse mortality (NRM) at 100 days was 57% for the 7 patients conditioned with BUCY and 17% for the patients conditioned with TBI-containing regimens (table 7). Reasons for such increased mortality observed with BUCY were attributed due to lack of busulfan pharmacokinetics to target dose levels and a high dose of cyclophophamide (150 mg/kg) used. Veno-occlusive disease (VOD) of the liver was seen in 11 patients and renal failure in 6 [99]. An actuarial survival at 5 years for the entire group was 32%.

In 2000, Pavletic et al. reported 23 previously treated CLL patients who were transplanted between 1988 and 1997 [71]. The median age of the patients was 46 years. Most patients (22/23) received chemoradiation for conditioning (TBI plus cyclophosphamide or etoposide and cyclophosphamide). GvHD prophylaxis was standard for all patients (methotrexate and cyclosporine A). Twenty patients (87%) achieved CR. Fifty-four percent of the patients developed grade II-IV acute GvHD. At 26 months, 14 patients (61%) are alive and disease-free [71]. Nine patients have died, one of whom had progressive B-CLL. The projected 5-year failure-free survival, OS, and relapse rates after allogeneic HCT were 65%, 62%, and 5%, respectively (table 7). The low relapse rate may reflect GVL effect [71]. Toze et al. evaluated outcome of 30 patients with CLL who received myeloablative allogeneic HCT using related (n =20, 67%) or unrelated (n =10) donors at two transplant centers in Canada [100]. The median interval from diagnosis to transplantation was 4.8 (0.3-13) years, median number of prior therapies was three and median age 48 years. The preparative regimen included TBI in 15 (50%) patients. Fourteen of 30 patients (47%) were alive at a median follow-up of 4.3 (2.4-10.5) years. Actuarial OS at 5 years is 39% (OS 48% for related donor and 20% for unrelated donor); cumulative incidence of NRM and relapse is 47% and 19%, respectively (table 7). Unfortunately, the survival after allogeneic HCT has been low even with a low relapse rate [100]. One reason may be the high toxicity associated with allografting in CLL likely related to factors mentioned before. Hence, clinicians have focused into the development of other allogeneic approaches in which allografting advantages can be obtained with less morbidity and mortality. One of the alternatives to reach this goal has been the development of RIC regimens.

Table 7. Selected clinical trials using Myeloablative Allogeneic HCT in CLL

Reference	Clinical Trial Design	N pts	Conditioning Regimen	GvHD Prophylaxis	Response	TRM
Michallet [97]	Restrospective, multicenter, international trial.	54	Varied Most common used: TBI/CY	CyA or MTX or both	70% achieved HR. *At 24 mo: 44% alive. *At 10 year: 19 were still alive.	46%
Michallet [98]	EBMTR Retrospective	380 pts total Allogeneic: 187	NR	72%: CyA + MTX 19%: T-cell depletion	3-yr OS probability 45%	TRM
Doney [99]	Single center	25 21 HLA-identical 1 mismatch 3 identical twin	BU/CY or TBI-containing regimen (TBI/CY or CY/E/TBI)	Most: CyA + MTX	13 CR by marrow morphology. 6/7 evaluable pts. achieved CR by flow cytometry. All 9 pts. with chimerism studies achieved 100% donor engraftment.	TRM 57% for HDCT (BU/CY). TRM 18% for chemo/TBI.
Pavletic [71]	Retrospective study in 2 centers	23	All received TBI CY/TBI (6 pts) E/CY/TBI (16 pts)	CyA + MTX	20 pts achieved CR. At 26 mo: 14 pts. alive and disease-free.	9 pts (39%) have died; 1 of whom had progressive B-CLL
Toze [100]	Two Canadian Institutions	30	TBI (50% pts)	CyA + MTX (28 pts) T-cell depletion (2 pts)	Median follow-up of 4.3 yrs: 47% alive. 5-yr acturial OS: 39% 5-yr EFS: 39% No deaths/events/relapses have occurred more than 5 years post BMT.	NRM: 47% (cumulative incidente, CI 95%)

CyA: cyclosporine; BU: busulfan; E: etoposide.

Reduced Intensity/Non-Myeloablative Allogeneic HCT

Allogeneic HCT remains the only treatment modality with potential for inducing long-term DFS in CLL at expense of a high TRM. In a very interesting retrospective analysis, Dreger et al. compared RIC versus standard myeloablative HCT in patients diagnosed with CLL [101]. During the same period of time, 73 cases of CLL were treated with the following RIC regimens: fludarabine/cyclophosphamide: 21/73, fludarabine/melphalan: 17/73, fludarabine/busulfan: 13/73, fludarabine/TBI 2Gy: 10/73, TBI 2Gy: 7/73, and other: 5/73. Eighty-two CLL cases underwent standard myeloablative conditioning using the following regimens: cyclophosphamide/TBI: 69/82, melphalan/TBI: 4/82, busulfan/cyclophosphamide: 4/82, and other: 5/82. An unadjusted comparison of RIC and standard myeloablative conditioning, with median follow-up of 22 and 29 months respectively, did not reveal significant differences in OS, EFS and TRM amongst the groups [101]. There were fewer relapses in the standard myeloablative group (28% versus 11%, $P = 0.008$). After adjustment for age, gender, donor/cell source and disease-remission status, a significant reduction in TRM was observed in the RIC population (HR: 0.4; 95% confidence interval 0.18-0.9; $P = 0.003$) [101]. The favorable influence of RIC regimens on TRM incidence, and the possible unfavorable effects on risk of relapse need to be confirmed in prospectively randomized studies.

Several studies have shown that allogeneic HCT modality of treatment may overcome the presence of unmutated IgV_H gene in the CLL clone. In this regard, Ritgen et al. demonstrated that GVL-mediated immunotherapy is effective in this high-risk CLL population [102]. Nine patients with unmutated IgV_H who received allogeneic HCT following RIC (fludarabine at 30 mg/m^2 and cyclophosphamide at 500 mg/m^2 over 5 days), and 26 patients who underwent autologous HCT (control) were included in the study. The kinetics of MRD was measured by quantitative allele-IgV_H PCR. MRD became undetectable in 7 of 9 (78%) post-allogeneic (non myeloablative) HCT patients from day +100 onwards subsequent to chronic GvHD) or DLI, and those 7 patients remained in CR and MR at a median follow-up of 25 months. On the other hand, PCR negativity was achieved in only 6 of 26 patients (23%) after autologous HCT for unmutated IgV_H, and was not durable. Importantly, this study suggests that elimination of MRD in CLL post-allogeneic HCT is dependent to a great extent on a *bona fide* GVL effect rather than the sole myeloablative effect of the conditioning regimen [102]. Similarly, Moreno et al. evaluated 50 patients with poor-risk CLL (defined as advanced clinical stage, diffuse bone marrow infiltration, short doubling time, massive or progressive lymphadenopathy, or poor-risk cytogenetics) who underwent autologous (n =27) and allogeneic HCT (n =23) and had available data on IgV_H sequences [103]. Thirty-four (68%) patients (allogeneic: 14; autologous: 20) had unmutated IgV_H genes, whereas 16 patients (32%) had somatically hypermutated IgV_H genes (allogeneic: 9; autologous: 7). There were no significant differences in the clinical characteristics of patients such as age, gender, and time from diagnosis to transplantation. Risk of relapse at 5 years was significantly higher following autologous HCT (61%; 95% confidence interval: 44% to 84%) than after allogeneic HCT (12%; 95% confidence interval: 3% to 44%; $P < 0.05$). Indeed, EFS at 5 years was superior for allogeneic HCT patients compared with autologous HCT patients (post-autologous HCT: 17% versus post-allogeneic HCT: 71%, $P = 0.0002$) [103]. In CLL patients with unmutated IgV_H, analysis of 5-year risk of progression showed that 13 of 20 autologous HCT patients (66%; 95% confidence interval: 47% to 93%) experienced disease progression with a median time to progression (TTP) of 28 months, compared with 2 of 14 patients undergoing allogeneic HCT (17%; 95% confidence interval: 5% to 60%, $P = 0.01$) (figure 8A). Furthermore, there was a strong trend for longer EFS with a "plateau" following allogeneic HCT (median: 22 months versus not reached; $P = .0.06$) (figure 8B) [103]. However, no difference in OS was observed among unmutated CLL patients between the two modality of HCT, with median survivals of 6 and 4 years after allogeneic and autologous HCT, respectively ($P = 0.33$) (figure 8C). In patients with mutated IgV_H, all end points (risk of progression, EFS, and OS) were significantly better in the allogeneic HCT group than in those who underwent autologous HCT. In 2004 ASH meeting, Sorror et al. presented the results of a multicenter phase I clinical trial in which 64 refractory CLL patients who were ineligible for myeloablative allogeneic HCT due to age or co-morbidities received RIC regimen prior to related (n= 44) or URD (n= 20) allografts between 1997-2003 [86]. Median age was 56 years (range, 44-69); prior therapies were 4; and patients were refractory to at least 1 regimen (56 to fludarabine, 19 to alkylators, 14 to rituximab, 4 to alemtuzumab, 2 had failed autologous HCT). Conditioning regimen consisted of TBI 2 Gy alone (n =11) or combined with fludarabine (n =53). Postgrafting immunosuppression consisted of mycophenolate mofetil and CyA (table 5). Post-transplant: 3 patients had graft rejection; 1 died of aplasia; 2 are alive with disease relapse. The incidence of acute and chronic GvHD at 2-year were 27% and 50%, respectively [86]. At 24 months follow-up, the ORR was 67% (50% in CR), whereas 5% had stable disease. URD-patients had significantly higher CR rate than matched related donors-patients, suggesting more effective GVL activity. All 11 responding patients had MR. Estimated 2-year incidence of relapse/progression, DFS, and OS were 26%, 52%, and 60%, respectively [86]. The interesting and intriguing observation seen in URD allografts has provoked an intentional HLA-mismatch investigation which currently ongoing; perhaps, it may be a possible tool to increase GVL, and thus, decrease relapse rate. These findings suggest again that GVL effect may overcome the negative impact of unmutated IgV_H mutational status in patients with poor-risk CLL [103]. In another study which included 30 patients (median age 53 years old; median of 3 previous line of therapy; 37% of the patients being fludarabine-refractory), RIC allogeneic HCT was capable of overcoming not only unmutated IgV_H gene but also the presence of adverse chromosomal abnormalities such as 17p- and 11q- [104]. The patients underwent HLA matched-sibling RIC allogeneic HCT across 8 centers in Spain (table 5). RIC regimens consisted of: fludarabine/melphalan: 63.3%; fludarabine/melphalan/alemtuzumab: 3.3%; fludarabine/ busulfan/ATG (rabbit): 16.7%; fludarabine/TBI (200 cGy)/ATG: 16.7%. Fourteen (60%) of the 23 patients had an unmutated IgV_H status; 8 (32%) of 25 patients had 11q– with four of them also displaying unmutated IgV_H; six (24 %) had 17p– aberration (five were also unmutated for IgV_H) [104]. All patients received mobilized PBSC from matched-related donors. Twenty-five (83.3%) of 30 patients in the study had active disease at the time of transplant, including six with refractory/progressive disease. Twenty-one (78%) of 27 evaluable patients achieved CR by day 100. At a median follow-up of 47.3 months (range, 12-74.6), 22 (81%) patients were alive and in remission. The 5- year actuarial DFS, EFS, and

OS were 93%, 70%, and 72%, respectively [104]. Thus, it seems that RIC allogeneic HCT is an effective modality of therapy in high-risk CLL patients including those with aberrant chromosomal abnormalities.

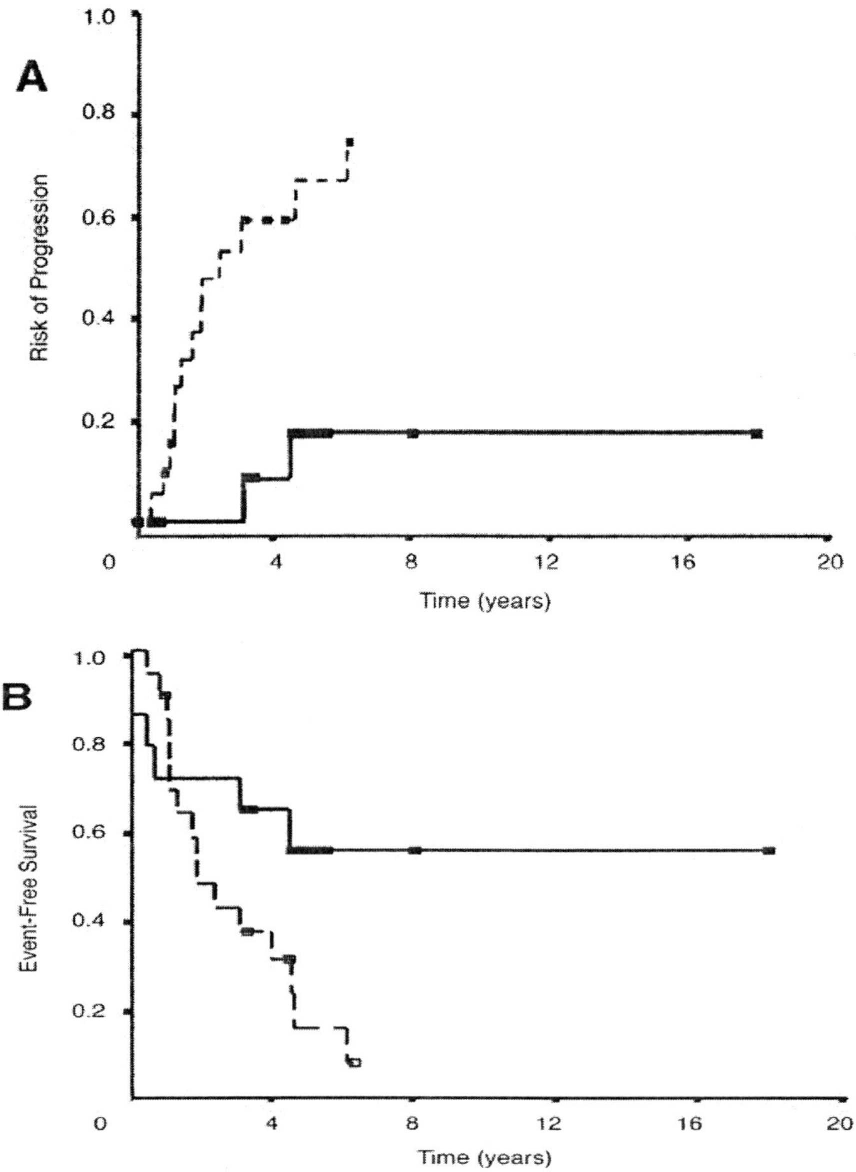

Figure 8 continued

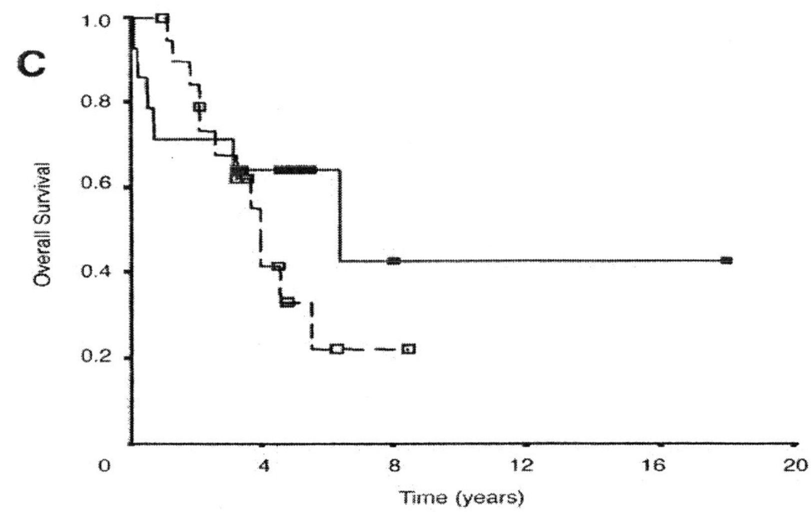

Figure 8. (A) Risk of Progression, (B) EFS, (C) and OS in CLL patients with unmutated V_H genes according to the type of transplantation. Allogeneic stem cell transplantation (SCT) patients (———) showed a lower risk of progression (P = .01) and better EFS (P = .06) compared with autologous SCT patients (- - -), with no differences in OS (P = .33). "Reprinted with permission from the American Society of Clinical Oncology" Moreno et al. Allogeneic Stem-Cell Transplantation may overcome the Adverse Prognosis of Unmutated VH Gene in patients with Chronic Lymphocytic Leukemia. J Clin Oncol. 23(15):3433–3438. © 2005 by American Society of Clinical Oncology.

RIC allogeneic HCT has taken the attention of clinicians and researchers to treat CLL, and most recent studies include this modality as treatment of choice. Dreger et al. evaluated 77 patients with CLL who received RIC allogeneic HCT at 29 centers in Europe [105]. The median age was 54 (range: 30-66) and the median number of prior chemotherapy regimes was 3 (range: 0-8). Donors consisted of HLA-identical siblings in 81% of the cases. The conditioning regimens were designated as "intensive" (n =34) or "moderate" (n =43): intensive: fludarabine/busulfan, fludarabine/melphalan +/- alemtuzumab, fludarabine/BCNU/ melphalan +/- ATG, and BEAM; moderate: TBI 2Gy, TBI 2Gy/fludarabine, fludarabine/cyclophosphamide/thiotepa, fludarabine/cyclophosphamide +/- ATG, and others (table 5). Regimens were defined as intensive if busulfan doses of more than 8 mg/kg, or melphalan doses of more than 110 mg/m^2 were used. At 2-year, EFS and OS are 56% and 72%, respectively [105]. Recently, the outcome in 41 consecutive CLL patients treated with fludarabine, melphalan, and alemtuzumab conditioning regimen followed by allogeneic HCT was reported [106]. Donors were as follows: HLA-matched siblings (n =24) and unrelated (n =17). Four out of 17 unrelated donors were mismatched (single-allele) with the recipients. The regimen was administered as follows: alemtuzumab 20 mg/day on days –8, -7, -5, -5 and -4 (cumulative dose, 100 mg), fludarabine 30 mg/m^2 daily on days –7, -6, -5, -4 and -3 (cumulative dose, 150 mg/m^2), and melphalan 140 mg/m^2 on day -2. GvHD prophylaxis consisted of intravenous infusion of cyclosporine over 1 hour at 3 mg/kg daily starting on day -1, with a target level of 150 to 200 ng/mL [106]. The 2-year OS and TRM were 51% and 26%, respectively. Acute and chronic GVHD was observed in 41% and 33% of the patients, respectively [106]. These findings demonstrate the feasibility of alemtuzumab-containing regimens with a relatively low incidence of GvHD (an effect that has been observed by other investigators), *albeit* at the expense of increased TRM from viral and fungal infections.

The use of alemtuzumab as *"in vivo"* purging has shown to improve the quality of response obtained prior to harvest process, and thus, it may impact on the HCT outcome. Nevertheless, there has not been a prospective randomized study conducted in which we compare purging versus not purging followed by HCT, and the impact of this novel therapy on poor-prognostic factors and MRD.

Reduced Intensity and T-Cell Depletion Strategy In CLL

We have discussed how T-cell depletion represents an effective mean of reducing the incidence and severity of acute or chronic GvHD, albeit at the expense of increased rates of stem cell graft rejection or disease relapse. Several methods for depleting T-cells have been developed and implemented over the years. An "*in vitro*" study evaluated alemtuzumab for T-lymphocyte depletion from human HLA-matched bone marrow sibling-allografts prior to transplantation in 11 high-risk patients [107]. Herein, there were no signs of GvHD in any of the patients who were observed for a maximum period of 12 months; furthermore, no anti-GvHD prophylaxis was administered. However, late graft failure was observed in 2 patients (18%). Due to its capability of alter the immune system, alemtuzumab has been explored not only in CLL but also in other hematologic malignancies (non-Hodgkin's lymphoma, multiple myeloma, chronic myeloid leukemia, AML, etc) which have been subject to HCT approaches [108]. The rationale for alemtuzumab use was to suppress the recipient immune system allowing donor stem cell engraftment without using an excessive toxic conditioning regimen and to decrease GvHD incidence. In several studies in which nonmyeloablative allogeneic HCT or RIC have been used to treat different lymphoproliferative disorders (including CLL), the use of alemtuzumab as part of conditioning regimen has been associated with low incidence of GvHD in general, less or no grade III/IV GvHD reported at all, facilitation and sustained donor allogeneic engraftment, and low TRM [109, 110]. One of the major problems encountered with the use of alemtuzumab in the treatment of CLL either as induction therapy (single-agent or in combination), purging, or conditioning regimen is the rate of infections, especially CMV reactivation. Chakrabarti et al. evaluated the incidence of CMV infections in 101 patients treated with alemtuzumab-based regimen in the setting of HCT [111]. At a median of 27 days, 51 patients (50.5%) developed CMV infection after transplantation with an increased probability of 84.8% in patients at risk of CMV infection. Also, the probability of recurrence of CMV infection, before and after 100 days, was 53.6% and 46.6%, respectively. CMV disease was fatal in 3 (3%) cases. CD4 recovery was slow across all the subgroups, with none achieving a normal $CD4^+$ T-cell count at 1 year after transplantation [111]. This high incidence of CMV infections may be secondary to alemtuzumab use in the conditioning regimen, or perhaps a synergistic T-cell-induced dysfunction and depletion when combined with fludarabine. A noteworthy fact is a low incidence of CMV disease despite a high incidence of CMV reactivation, presumably due to an effective CMV prophylaxis regimen (either oral ganciclovir or valacyclovir or valganciclovir).

The Role of Purging in CLL Management

Although autologous HCT results in a high CR rate, early studies showed that most patients eventually relapse [112]. Purging approaches either "*in vitro*" or "*in vivo*" have been developed in CLL treatment and other hematologic malignancies with the goal and need of reducing stem cell contamination with leukemic cells. For instances, "*ex vivo*" B-cell depletion from stem cell grafts using modern CD34+ selection devices (e.g., Isolex300i or Clinimacs) may eliminate 3-4 log of clonal CLL cells [113]. This result can be improved to 6 log reduction if a step of negative B-cell depletion is introduced into the purging procedure [114]. To date, other purging methods have been tried such as the use of MoAbs. Although, purging procedure has not yet been standardized and further refining is needed to improve its effectiveness that may translate in decreasing incidence of relapse, most recent studies have shown preliminary data suggesting a role for purging stem cell harvest.

Alemtuzumab has become the cornerstone of several eradication-based treatments due to its ability to achieve high CR and to successfully purge MRD from both blood and bone marrow in CLL patients [115]. Alemtuzumab also provides the option of "*in vivo*" purging of blood and bone marrow prior to stem cell harvest collection. We know that CR in CLL is not enough to secure durability of response. Consequently, the newer molecular techniques which have higher level of sensitivity to detect MRD are critically necessary in the management of CLL. Obtaining undetectable MRD with the use of agents such alemtuzumab and rituximab has become a huge step forward in CLL treatment. This approach may provide further beneficial long-term outcome.

Is There a Role for Post-Transplant Maintenance Therapy in CLL

MRD is a new concept in CLL management. When the data obtained through sophisticated methods such as PCR, four-colour flow cytometry, and others have been analyzed, we have started to understand the critical role of MRD in CLL. Moreno et al. analyzed MRD using three different methods in samples taken from 40 patients who underwent HCT (autologous or allogeneic) [116]. Several conclusions were derived from this study: a. quantitative methods such as quantitative PCR and flow cytometry have a very good correlation among them and are accurate for MRD detection, b. detection of MRD on patients who underwent autologous HCT was able to predict clinical relapse, and c. patients having detectable MRD before 6 months post-transplant were identified as "high-risk" for relapse [116]. Interestingly, MRD detection did not impact allograft outcome in CLL. The authors support the use of quantitative methods for MRD detection in the assessment of response in CLL. In general, the presence of MRD after any kind of therapy in CLL translates into an imminent risk for relapse at any time. Studies have suggested that MRD level may have an impact on duration of response.

The outcomes seen thus far with autologous HCT strongly correlate with disease status prior to transplantation as well as persistence or losing MRD negativity post-transplantation. In this regard, the GCLLSG-CLL3 protocol addressed the impact of myeloablative therapy followed by autograft rescue on MRD in relation to the predictive value of post-transplant MRD for clinical outcome, and the influence of IgV_H mutational status on MRD [117]. The report included 476 blood samples from 46 CLL patients. Autograft resulted in a significant reduction of disease load (pretransplant versus postransplant levels) at 90-180 days post-transplant ($P < 0.001$), with no significant difference between patients with mutated and unmutated IgV_H, respectively. MRD levels did not have a prognostic impact during the first 6 months post-transplant, but stable or decreasing MRD kinetics 6-12 months post-transplant were strongly predictive for favorable outcome (4-year PFS 100%) [117]. PFS for those with more detectable residual disease was significantly inferior (37%; $P < 0.02$). Interestingly, increasing MRD levels early post-transplant was observed almost exclusively in patients with unmutated IgV_H gene. One may conclude from this study that a worse outcome in patients with unmutated IgV_H may be due to a faster regrowth of the leukemic clone rather than less effective reduction of MRD levels [117]. The GCLLSG study emphasizes the importance of measuring MRD pre- and post-transplant. Perhaps, one may argue that patients who do not achieve a substantial decrease on MRD level at 6 months post-transplant may benefit from additional therapy, and hence, there may be place for maintenance therapy in this setting. With novel target therapies such as anti-CD20 and –CD52 MoAbs, MR in this setting could be our future objective goals. Studies using these agents pre- and post-transplant have shown their efficacy in reducing leukemic cell levels, and hence inducing better MRD. Whether these observations will eventually translate into survival advantage still remains to be seen.

HDCT followed by autologous HCT appear to add benefit to responses following induction therapy. Following autologous HCT, the numbers of CRs and MRs post-transplant are higher [92, 96, 117, 118]. Our ability to evaluate MRD level using methods such as PCR will provide the ground to analyze different treatment strategies in the context of clinical trials. *In vivo* purging using MoAb prior to HDCT and stem cell rescue seems to be an innovative and promising approach. As longer as clinicians continue learning on common adverse events and how to deal with them (such as increased incidence of infections) as a result of using this novel therapeutic approach, purging may become an important part of the future conditioning regimens.

More studies are needed to evaluate purging as integral part of complex pre-transplant approaches prior to stem cell harvest as well as the future role of HDCT rescue by autologous HCT on MRD and how MRD evaluation pre- and post-transplant may direct our future management in CLL patients. Indeed, an important consideration will be the effect and the application of the novel prognostic factors in CLL using also the novel combination therapies such as FCR and CFAR which have resulted in high CR rate with MRD or no detectable MRD in untreated or previously treated patients with CLL. Similar responses have been seen with the use of alemtuzumab in the setting of relapse/refractory disease [110, 115]. To date, MRD is becoming an attractive objective goal to assess response to therapy in future clinical trials, especially when transplantation procedures either autologous or allogeneic are used to treat CLL.

Conclusion

The cumulative data in HCT for CLL suggests that this modality of therapy is a feasible procedure to be performed and offers a chance of "possible" cure in CLL. Although we are making a lot of progress in CLL tumor biology and treatment, the lack of prospective randomized clinical trials does not allow us to make definitive conclusions. However, some of the prospective studies performed indicate that HCT may have an important role in the future treatment of CLL especially in those who have very poor adverse factors.

The clinical data indicates that HDCT followed by autograft have demonstrated a low TRM, but a lack of a "plateau" effect in the EFS or relapse-free survival curves. Although the component of HDCT used in this approach has an impact on poor-prognostic factors such as IgV_H mutational status, the procedure is still subject of contamination, and it remains a major concern. As a result, purging procedure has become as an alternative to decrease this "contamination". Thus far, the ultimate role, dosing, sequence, and setting for purging have been determined yet. Autologous HCT can deliver MR in a significant number of patients; however, the latter are not permanent and relapses are to be anticipated, especially when undetectable MRD is not reached or patient persists with "pre-transplant" MDR levels.

On the other hand, allogeneic HCT has a chance for CLL cure. Nonetheless, allografting carries a high TRM and morbidity. Besides the cytotoxic effect delivered by the conditioning regimen, GVL effect plays a crucial anti-leukemic role in CLL. In this sense, non-myeloablative and RIC regimens are gaining attention and becoming more popular in the research arena, giving the clinicians a chance to eliminate the leukemic clone through "free-tumor" stem cells, GVL effect, and avoiding the high TRM and morbidity incidence seen with standard ("full") allogeneic transplantation. A major interest of ongoing research is to define the pre- and post-transplant immunosuppression needed to establish a temporary mixed chimerism, and later a full donor chimerism following RIC, to maximize on the benefit of GVL effect. Although the results found so far with the use of RIC in CLL are encouraging, it is still an investigational approach. Data regarding chronic GvHD after long-term follow-up post-RIC is currently missing and is crucial because of the strong association of GVL with chronic GvHD. However, as it was discussed, the introduction of agents such as alemtuzumab has shown to decrease the incidence of GvHD in the setting of RIC regimen in CLL.

To date, it is is difficult to recommend any HCT modality as first-line therapy even in patients with poor-risk features outside the context of clinical trials, and also having more effective combined chemoimmunotherapy apply now in the upfront arena. The ongoing clinical trials in CLL will provide insights of the true role of novel adverse markers either individually or collectively. HCT in CLL remains an "experimental" treatment alternative; if HCT is considered, the patients need to be educated about the risks and benefits of the procedure prior to decision-making.

References

[1] American Cancer Society: Cancer Facts and Figures 2006. Retrieved August 2006 from www.cancer.org.

[2] Jemal A, Siegel R, Ward E, et al. Cancer statistics, 2006. *CA Cancer J. Clin.* 2006; 56(2):106-130.

[3] Rai KR, Sawitsky A, Cronkite EP, Chanana AD, Levy RN, Pasternack BS. Clinical staging of chronic lymphocytic leukemia. *Blood.* 1975; 46(2):219-234.

[4] Hashimoto S, Dono M, Wakai M, et al. Somatic diversification and selection of immunoglobulin heavy and light chain variable region genes in IgG+ CD5+ chronic lymphocytic leukemia B cells. *J. Exp. Med.* 1995; 181(4):1507-1517.

[5] Oscier DG, Thompsett A, Zhu D, Stevenson FK. Differential rates of somatic hypermutation in V(H) genes among subsets of chronic lymphocytic leukemia defined by chromosomal abnormalities. *Blood.* 1997; 89(11):4153-4160.

[6] Damle RN, Wasil T, Fais F, et al. IgV gene mutation status and CD38 expression as novel prognostic indicators in chronic lymphocytic leukemia. *Blood.* 1999; 94(6):1840-1847.

[7] Dohner H, Stilgenbauer S, Benner A, et al. Genomic aberrations and survival in chronic lymphocytic leukemia. *N. Engl. J. Med.* 2000; 343(26):1910-1916.

[8] Fais F, Ghiotto F, Hashimoto S, et al. Chronic lymphocytic leukemia B cells express restricted sets of mutated and unmutated antigen receptors. *J. Clin. Invest.* 1998; 102(8):1515-1525.

[9] Chiorazzi N, Rai KR, Ferrarini M. Chronic lymphocytic leukemia. *N. Engl. J. Med.* 2005; 352(8):804-815.

[10] Ibrahim S, Keating M, Do KA, et al. CD38 expression as an important prognostic factor in B-cell chronic lymphocytic leukemia. *Blood.* 2001; 98(1): 181-186.

[11] Rassenti LZ, Huynh L, Toy TL, et al. ZAP-70 compared with immunoglobulin heavy-chain gene mutation status as a predictor of disease progression in chronic lymphocytic leukemia. *N. Engl. J. Med.* 2004; 351(9):893-901.

[12] Lozanski G, Heerema NA, Flinn IW, et al. Alemtuzumab is an effective therapy for chronic lymphocytic leukemia with p53 mutations and deletions. *Blood.* 2004; 103(9):3278-3281.

[13] Stilgenbauer S and Dohner H. Campath-1H-induced complete remission of chronic lymphocytic leukemia despite p53 gene mutation and resistance to chemotherapy. *N. Engl. J. Med.* 2002; 347(6):452-453.

[14] Keating MJ, Kantarjian H, Talpaz M, et al. Fludarabine: a new agent with major activity against chronic lymphocytic leukemia. *Blood.* 1989; 74(1):19-25.

[15] Cheson BD. Current approaches to the chemotherapy of B-cell chronic lymphocytic leukemia: a review. *Am. J. Hematol.* 1989; 32(1):72-77.

[16] De Rossi G, Mauro FR, Caruso, R, Monarca B, Mandelli F. Fludarabine and prednisone in pretreated and refractory B-chronic lymphocytic leukemia (B-CLL) in advanced stages. *Haematologica.* 1993; 78(3):167-171.

[17] Spriano M, Clavio M, Carrara P, et al. Fludarabine in untreated and previously treated B-CLL patients: a report on efficacy and toxicity. *Haematologica.* 1994; 79(3): 218-224.

[18] Johnson S, Smith AG, Loffler H, et al. Multicentre prospective randomised trial of fludarabine versus cyclophosphamide, doxorubicin, and prednisone (CAP) for treatment of advanced-stage chronic lymphocytic leukaemia. The French Cooperative Group on CLL. *Lancet.* 1996; 347(9013):1432-1438.

[19] Keating MJ, O'Brien S, Lerner S, et al. Long-term follow-up of patients with chronic lymphocytic leukemia (CLL) receiving fludarabine regimens as initial therapy. *Blood.* 1998; 92(4):1165-1171.

[20] Leporrier M, Chevret S, Cazin B, et al. Randomized comparison of fludarabine, CAP, and ChOP in 938 previously untreated stage B and C chronic lymphocytic leukemia patients. French Cooperative Group in Chronic Lymphocytic Leukemia. *Blood.* 2001; 98(8):2319-2325.

[21] Rummel MJ, Kafer G, Pfreundschuh M, et al. Fludarabine and epirubicin in the treatment of chronic lymphocytic leukaemia: a German multicenter phase II study. *Ann. Oncol.* 1999; 10(2):183-188.

[22] Frewin R, Turner D, Tighe M, Davies S, Rule S, Johnson S. Combination therapy with fludarabine and cyclophosphamide as salvage treatment in lymphoproliferative disorders. *Br. J. Haematol.* 1999; 104(3):612-613.

[23] Flinn IW, Byrd JC, Morrison C, et al. Fludarabine and cyclophosphamide with filgrastim support in patients with previously untreated indolent lymphoid malignancies. *Blood.* 2000; 96(1):71-75.

[24] Hallek M, Schmitt B, Wilhem M, et al. Fludarabine plus cyclophosphamide is an efficient treatment for advanced chronic lymphocytic leukemia (CLL): results of a phase II study of the German CLL Study Group. German Chronic Lymphocytic Leukaemia Study Group. *Br. J. Haematol.* 2001; 114(2):342-348.

[25] Schulz H, Winkler U, Staak JO, Engert A. The monoclonal antibodies Campath-1H and Rituximab in the Therapy of Chronic Lymphocytic Leukemia. *Onkologie.* 2000; 23(6):526-532.

[26] Kennedy B, Rawstron A, Carter C, et al. Campath-1H and fludarabine in combination are highly active in refractory chronic lymphocytic leukemia. *Blood.* 2002; 99(6):2245-2247.

[27] Byrd JC, Peterson BL, Morrison VA, et al. Randomized phase 2 study of fludarabine with concurrent versus sequential treatment with rituximab in symptomatic, untreated patients with B-cell chronic lymphocytic leukemia: results from Cancer and Leukemia Group B 9712 (CALGB 9712). *Blood.* 2003; 101(1):6-14.

[28] Schulz H, Klein SK, Rehwald U, et al. Phase 2 study of a combined immunochemotherapy using rituximab and fludarabine in patients with chronic lymphocytic leukemia. German CLL Study Group. *Blood.* 2002; 100(9):3115-3120.

[29] Byrd JC, Rai K, Peterson BL, et al. Addition of rituximab to fludarabine may prolong progression-free survival and overall survival in patients with previously untreated chronic lymphocytic leukemia: an updated retrospective comparative analysis of CALGB 9712 and CALGB 9011. *Blood.* 2005; 105(1):49-53.

[30] Dreger P, Stilgenbauer S, Benner A, et al. The prognostic impact of autologous stem cell transplantation in patients with chronic lymphocytic leukemia: a risk-matched analysis based on the VH gene mutational status. *Blood.* 2004; 103(7): 2850-2858.

[31] Gribben JG, Zahrieh D, Stephans K, et al. Autologous and allogeneic stem cell transplantations for poor-risk chronic lymphocytic leukemia. *Blood.* 2005; 106(13):4389-4396.

[32] Oscier DG, Gardiner AC, Mould SJ, et al. Multivariate analysis of prognostic factors in CLL: clinical stage, IGVH gene mutational status, and loss or mutation of the p53 gene are independent prognostic factors. *Blood.* 2002; 100(4):1177-1184.

[33] Wasil T, Rai KR, Albesiano E, et al. In B-CLL, time to first treatment is dependent on the IgVH gene mutations and expression of CD38 on leukemic cells. *J. Clin. Oncol.* 2005; 23(16S):573s (abstr 6555).

[34] Rosenwald A, Alizadeh AA, Widhopf G, et al. Relation of gene expression phenotype to immunoglobulin mutation genotype in B cell chronic lymphocytic leukemia. *J. Exp. Med.* 2001; 194(11):1639-1647.

[35] Crespo M, Bosch F, Villamor N, et al. ZAP-70 expression as a surrogate for immunoglobulin-variable-region mutations in chronic lymphocytic leukemia. *N. Engl. J. Med.* 2003; 348(18):1764-1775.

[36] Reinherz EL, Kung PC, Goldstein G, Levey RH, Schlossman SF. Discrete stages of intrathymic differentiation: analysis of normal thymocytes and leukemic lymphoblasts of T lineage. *Proc. Natl. Acad. Sci. USA.* 1980; 77(3):1588-1592.

[37] Mehta K, Shahid U, Malavasi F. Human CD38, a cell surface protein with multiple functions. *FASEB J.* 1996; 10(12):1408-1417.

[38] Dianzani U, Funaro A, DiFranco D, et al. Interaction between endothelium and CD4+CD45RA+ lymphocytes. Role of the human CD38 molecule. *J. Immunol.* 1994; 153(3):952-959.

[39] Kumagai M, Coustan-Smith E, Murray DJ, et al. Ligation of CD38 suppresses human B lymphopoiesis. *J. Exp. Med.* 1995; 181(3):1101-1110.

[40] Zupo S, Rugari E, Dono M, Taborelli G, Malavasi F, Ferrarini M. CD38 signaling by agonistic monoclonal antibody prevents apoptosis of human germinal center B cells. *Eur. J. Immunol.* 1994; 24(5):1218-1222.

[41] Omede P, Boccadoro M, Gallone G, et al. Multiple myeloma: increased circulating lymphocytes carrying plasma cell-associated antigens as an indicator of poor survival. *Blood.* 1990; 76(7):1375-1379.

[42] Keyhani A, Huh YO, Jendiroba D. Increased CD38 expression is associated with favourable prognosis in adult acute leukemia. *Leuk. Res.* 2000; 24(2):153-159.

[43] Hamblin TJ, Orchard JA, Ibbotson RE, et al. CD38 expression and immunoglobulin variable region mutations are independent prognostic variables in chronic lymphocytic leukemia, but CD38 expression may vary during the course of disease. *Blood.* 2002; 99(3):1023-1029.

[44] Jelinek DF, Tschumper RC, Geyer SM, et al. Analysis of clonal B-cell CD38 and immunoglobulin variable region sequence status in relation to clinical outcome for B-chronic lymphocytic leukaemia. *Br. J. Haematol.* 2001; 115(4):854-861.

[45] Krober A, Seiler T, Benner A, et al. V(H) mutation status, CD38 expression level, genomic aberrations, and survival in chronic lymphocytic leukemia. *Blood.* 2002; 100(4):1410-1416.

[46] Ghia P, Guida G, Stella S, et al. The pattern of CD38 expression defines a distinct subset of chronic lymphocytic leukemia (CLL) patients at risk of disease progression. *Blood.* 2003; 101(4):1262-1269.

[47] Thunberg U, Johnson A, Roos G, et al. CD38 expression is a poor predictor for VH gene mutational status and prognosis in chronic lymphocytic leukemia. *Blood.* 2001; 97(6):1892-1894.

[48] Matrai Z, Lin K, Dennis M, et al. CD38 expression and Ig VH gene mutation in B-cell chronic lymphocytic leukemia. *Blood.* 2001; 97(6):1902-1903.

[49] Vilpo J, Tobin G, Hulkkonen J, et al. Surface antigen expression and correlation with variable heavy-chain gene mutation status in chronic lymphocytic leukemia. *Eur. J. Haematol.* 2003; 70(1):53-59.

[50] Del Poeta G, Maurillo L, Venditti A, et al. Clinical significance of CD38 expression in chronic lymphocytic leukemia. *Blood.* 2001; 98(9):2633-2639.

[51] Mainou-Fowler T, Dignum H, Taylor PR, et al. Quantification improves the prognostic value of CD38 expression in B-cell chronic lymphocytic leukaemia. *Br. J. Haematol.* 2002; 118(3):755-761.

[52] Chevallier P, Penther D, Avet-Loiseau H, et al. CD38 expression and secondary 17p deletion are important prognostic factors in chronic lymphocytic leukaemia. *Br. J. Haematol.* 2002; 116(1):142-150.

[53] D'Arena G, Musto P, Cascavilla N, et al. CD38 expression correlates with adverse biological features and predicts poor clinical outcome in B-cell chronic lymphocytic leukemia. *Leuk. Lymphoma.* 2001; 42(1-2):109-114.

[54] Domingo-Domenech E, Domingo-Claros A, Gonzalez-Barca E, et al. CD38 expression in B-chronic lymphocytic leukemia: association with clinical presentation and outcome in 155 patients. *Haematologica.* 2002; 87(10):1021-1027.

[55] Durig J, Naschar M, Schmucker U, et al. CD38 expression is an important prognostic marker in chronic lymphocytic leukaemia. *Leukemia.* 2002; 16(1):30-35.

[56] Morabito F, Mangiola M, Oliva B, et al. Peripheral blood CD38 expression predicts survival in B-cell chronic lymphocytic leukemia. *Leuk. Res.* 2001; 25(11):927-932.

[57] Morabito F, Mangiola M, Stelitano C, Deaglio S, Callea V, Malavasi F. Peripheral blood CD38 expression predicts time to progression in B-cell chronic lymphocytic leukemia after first-line therapy with high-dose chlorambucil. *Haematologica.* 2002; 87(2):217-218.

[58] Ibrahim S, Jilani I, O'Brien S, et al. Clinical relevance of the expression of the CD31 ligand for CD38 in patients with B-cell chronic lymphocytic leukemia. *Cancer.* 2003; 97(8):1914-1919.

[59] Chang CC and Cleveland RP. Conversion of CD38 and/or myeloid-associated marker expression status during the course of B-CLL: association with a change to an aggressive clinical course. *Blood.* 2002; 100(3):1106.

[60] Stilgenbauer S, Lichter P, Dohner H. Genomic aberrations in B cell chronic lymphocytic leukemia. In: Cheson BD, ed. *Chronic lymphoid leukemias* (2nd ed.). New York: Marcel Dekker; 2001; 353-376.

[61] Krober A, Bloehdorn J, Hafner S, et al. Additional genetic high-risk features such as 11q deletion, 17p deletion, and V3-21 usage characterize discordance of ZAP-70 and VH mutation status in chronic lymphocytic leukemia. *J. Clin. Oncol.* 2006; 24(6):969-975.

[62] Byrd JC, Gribben JG, Peterson B, et al. Select high risk genetic features predict earlier progression following chemoimmunotherapy with fludaribine and rituximab in chronic lymphocytic leukemia (CLL): preliminary justification for risk-adapted therapy. *Blood.* 2004; 104(11):139a (abstr 476).

[63] Catovsky D, Richards S, Matutes E, et al. Response to therapy and survival in CLL is influenced by genetic markers: preliminary analysis from the LRF CLL4 trial. *Blood.* 2004; 104(11):8a (abstr 13).

[64] Vigorito AC, Azevedo WM, Marques JF, et al. A randomised, prospective comparison of allogeneic bone marrow and peripheral blood progenitor cell transplantation in the treatment of haematological malignancies. *Bone Marrow Transplant.* 1998; 22(12):1145-1151.

[65] Ringden O, Labopin M, Bacigalupo A, et al. Transplantation of peripheral blood stem cells as compared with bone marrow from HLA-identical siblings in adult patients with acute myeloid leukemia and acute lymphoblastic leukemia. *J. Clin. Oncol.* 2002; 20(24):4655-4664.

[66] Trisolini SM, Mauro FR, Capria S, et al. Follow-up of Chronic Lymphocytic Leukemia (CLL) patients who relapse after autologous stem cell transplantation (ASCT). *Blood.* 2004; 104(11):388b (abstr 5212).

[67] Pavletic Z, Bierman P, Vose J, et al. High incidence of relapse after autologous stem-cell transplantation for B-cell chronic lymphocytic leukemia or small lymphocytic lymphoma. *Ann. Oncol.* 1998; 9(9):1023-1026.

[68] Pavletic S, Zhou G, Sobocinski K, et al. Identical-Twin Transplants for B-Cell Chronic Lymphocytic Leukemia (B-CLL). *Blood.* 2004; 104(11):909a (abstr 3330).

[69] Dreger P, Michallet M, Schmitz N. Stem-cell transplantation for chronic lymphocytic leukemia: The 1999 perspective. *Ann. Oncol.* 2000; 11(Suppl 1):49-53.

[70] Michallet M, Van Biesen A, Bandini G, et al. Allotransplants and autotransplants in Chronic Lymphocytic Leukemia (CLL). *Bone Marrow Transplant.* 1999; 23(Suppl 1):S53 (abstract 170).

[71] Pavletic ZS, Arrowsmith ER, Bierman PJ, et al. Outcome of allogeneic stem cell transplantation for B cell chronic lymphocytic leukemia. *Bone Marrow Transplant.* 2000; 25(7):717-722.

[72] Khouri IF, Przepiorka D, van Besien K, et al. Allogeneic blood or marrow transplantation for chronic lymphocytic leukaemia: timing of transplantation and potential effect of fludarabine on acute graft-versus-host disease. *Br. J. Haematol.* 1997; 97(2):466-473.

[73] O'Reilly RJ. Allogenic bone marrow transplantation: current status and future directions. *Blood.* 1983; 62(5):941-964.

[74] OKunewick JP, Kociban DL, Buffo MJ. Comparative effects of various T cell subtypes on GVHD in a murine model for MHC-matched unrelated donor transplant. *Bone Marrow Transplant.* 1990; 5(3):145-152.

[75] Sprent J, Schaefer M, Gao EK, Korngold R. Role of T cell subsets in lethal graft-versus-host disease (GVHD) directed to class I versus class II H-2 differences. I. L3T4+ cells can either augment or retard GVHD elicited by Lyt-2+ cells in class I different hosts. *J. Exp. Med.* 1988; 167(2):556-569.

[76] Soiffer RJ, Alyea EP, Hochberg E, et al. Randomized trial of CD8+ T-cell depletion in the prevention of graft-versus-host disease associated with donor lymphocyte infusion. *Biol. Blood Marrow Transplant.* 2002; 8(11):625-632.

[77] Sehn LH, Alyea EP, Weller E, et al. Comparative outcomes of T-cell-depleted and non-T-cell-depleted allogeneic bone marrow transplantation for chronic myelogenous leukemia: impact of donor lymphocyte infusion. *J. Clin. Oncol.* 1999; 17(2):561-568.

[78] Korngold R, Leighton C, Manser T. Graft-versus-myeloid leukemia responses following syngeneic and allogeneic bone marrow transplantation. *Transplantation.* 1994; 58(3):278-287.

[79] Slavin S, Morecki S, Weiss L, Or R. Donor lymphocyte infusion: the use of alloreactive and tumor-reactive lymphocytes for immunotherapy of malignant and nonmalignant diseases in conjunction with allogeneic stem cell transplantation. *J. Hematother. Stem. Cell Res.* 2002; 11(2):265-276.

[80] Weiden PL, Flournoy N, Thomas ED, et al. Antileukemic effect of graft-versus-host disease in human recipients of allogeneic-marrow grafts. *N. Engl. J. Med.* 1979; 300(19):1068-1073.

[81] Sprent J, Surh CD, Agus D, Hurd M, Sutton S, Heath WR. Profound atrophy of the bone marrow reflecting major histocompatibility complex class II-restricted destruction of stem cells by CD4+ cells. *J. Exp. Med.* 1994; 180(1):307-317.

[82] Mohty M, Bagattini S, Chabannon C, et al. CD8+ T cell dose affects development of acute graft-vs-host disease following reduced-intensity conditioning allogeneic peripheral blood stem cell transplantation. *Exp. Hematol.* 2004; 32(11):1097-1102.

[83] Aggarwal S and Pittenger MF. Human mesenchymal stem cells modulate allogeneic immune cell responses. *Blood.* 2005; 105(4):1815-1822.

[84] Hill GR, Ferrara JL. The primary of the gastrointestinal tract as a target organ of acute graft-versus-host disease: rationale for the use of cytokine shields in allogeneic bone marrow transplantation. *Blood.* 2000; 95(9):2754-2759.

[85] Schetelig J, Thiede C, Bornhauser M, et al. Evidence of a graft-versus-leukemia effect in chronic lymphocytic leukemia after reduced-intensity conditioning and allogeneic stem-cell transplantation: the Cooperative German Transplant Study Group. *J. Clin. Oncol.* 2003; 21(14):2747-2753.

[86] Sorror ML, Maris MB, Sandmaier BM, et al. Hematopoietic Cell Transplantation After Nonmyeloablative Conditioning for Advanced Chronic Lymphocytic Leukemia. *J. Clin. Oncol.* 2005; 23(16):3819-3829.

[87] Rabinowe SN, Soiffer RJ, Gribben JG, et al. Autologous and allogeneic bone marrow transplantation for poor prognosis patients with B-cell chronic lymphocytic leukemia. *Blood.* 1993; 82(4):1366-1376.

[88] Gribben JG, Sequist L, Zahrieh D, et al. Outcome following autologous or allogeneic stem cell transplantation in Chronic Lymphocytic Leukemia. *Blood.* 2003; 102(11): 980a (abstr 3649).

[89] Dreger P, von Neuhoff N, Kuse R, et al. Early stem cell transplantation for chronic lymphocytic leukaemia: a chance for cure? *Br. J. Cancer.* 1998; 77(12): 2291-2297.

[90] Sutton L, Maloum K, Gonzalez H, et al. Autologous hematopoietic stem cell transplantation as salvage treatment for advanced B cell chronic lymphocytic leukemia. *Leukemia.* 1998; 12(11):1699-1707.

[91] Dreger P, Busch R, Stilgenbauer S, et al. Early autologous stem cell transplantation (SCT) in genetically poor-risk Chronic Lymphocytic Leukemia is feasible and effective: Results from a prospective multicenter study (GCLLSG CLL3 Protocol). *Blood.* 2004; 104(11):45a (abstr 146).

[92] Milligan DW, Fernandes S, Dasgupta R, et al. Results of the MRC pilot study show autografting for younger patients with chronic lymphocytic leukemia is safe and achieves a high percentage of molecular responses. *Blood.* 2005; 105(1):397-404.

[93] Dreger P, Viehmann K, von Neuhoff N, et al. A prospective study of positive/ negative ex vivo B-cell depletion in patients with chronic lymphocytic leukemia. *Exp. Hematol.* 2000; 28(10):1187-1196.

[94] Tedeschi A, Cairoli R, Montillo M, et al. Feasibility of autologous stem cell transplantation (ASCT) in Chronic Lymphocytic Leukemia (CLL) patients treated with sequential Fludarabine and Campath-1H. *Blood.* 2004; 104(11):520a (abstr 1879).

[95] Montillo M, Tedeschi A, Cairoli R, et al. Purging "In Vivo" with Alemtuzumab (Campath-1H) before autologous stem cell transplantation (ASCT) in Chronic Lymphocytic Leukemia (CLL). *Blood.* 2005; 106(11):820a (abstr 2927).

[96] Montillo M, Tedeschi A, Miqueleiz S, Veronese S, Cairoli R, Intropido L, et al. Alemtuzumab as consolidation after a response to fludarabine is effective in purging residual disease in patients with chronic lymphocytic leukemia. *J. Clin. Oncol.* 2006; 24(15): 2337-2342.

[97] Michallet M, Archimbaud E, Bandini G, et al. HLA-identical sibling bone marrow transplants for chronic lymphocytic leukaemia. *Ann. Intern. Med.* 1996; 124(3):311-315.

[98] Michallet M, Van Biezen A, Bandini G, et al. Allotransplants and autotransplants in CLL. *Bone Marrow Transplant.* 1999; 23(Suppl 1): S53.

[99] Doney KC, Chauncey T, Appelbaum FR; Seattle Bone Marrow Transplant Team. Allogeneic related donor hematopoietic stem cell transplantation for treatment of chronic lymphocytic leukemia. *Bone Marrow Transplant.* 2002; 29(10):817-823.

[100] Toze CL, Galal A, Barnett MJ, et al. Myeloablative allografting for chronic lymphocytic leukemia: evidence for a potent graft-versus-leukemia effect associated with graft-versus-host disease. *Bone Marrow Transplant.* 2005; 36(9):825-830.

[101] Dreger P, Brand R, Milligan D, et al. Reduced-intensity conditioning lowers treatment-related mortality of allogeneic stem cell transplantation for chronic lymphocytic leukemia: a population-matched analysis. *Leukemia.* 2005; 19(6):1029-1033.

[102] Ritgen M, Stilgenbauer S, Von Neuhoff N, et al. Graft-versus-leukemia activity may overcome therapeutic resistance of chronic lymphocytic leukemia with unmutated

immunoglobulin variable heavy-chain gene status: implications of minimal residual disease measurement with quantitative PCR. *Blood.* 2004; 104(8):2600-2602.

[103] Moreno C, Villamor N, Colomer D, et al. Allogeneic Stem-Cell Transplantation May Overcome the Adverse Prognosis of Unmutated *VH* Gene in Patients With Chronic Lymphocytic Leukemia. *J. Clin. Oncol.* 2005; 23(15):3433–3438.

[104] Caballero D, Garcia-Marco JA, Martino R, et al. Allogeneic transplant with reduced intensity conditioning regimens may overcome the poor prognosis of B-cell chronic lymphocytic leukemia with unmutated immunoglobulin variable heavy-chain gene and chromosomal abnormalities (11q- and 17p-). *Clin. Cancer Res.* 2005; 11(21):7757-7763.

[105] Dreger P, Brand R, Hansz J, et al. Treatment-related mortality and graft-versus-leukemia activity after allogeneic stem cell transplantation for chronic lymphocytic leukemia using intensity-reduced conditioning. *Leukemia.* 2003; 17(5):841-848.

[106] Delgado J, Thomson K, Russell N, et al. Results of alemtuzumab-based reduced-intensity allogeneic transplantation for chronic lymphocytic leukemia: a British Society of Blood and Marrow Transplantation Study. *Blood.* 2006; 107(4):1724-1730.

[107] Waldmann H, Polliak A, Hale G, et al. Elimination of graft-versus-host disease by in-vitro depletion of alloreactive lymphocytes with a monoclonal rat anti-human lymphocyte antibody (CAMPATH-1). *Lancet.* 1984; 2(8401):483-486.

[108] Kottaridis PD, Milligan DW, Chopra R, et al. In vivo CAMPATH-1H prevents graft-versus-host disease following nonmyeloablative stem cell transplantation. *Blood.* 2000; 96(7):2419-2425.

[109] Chakraverty R, Peggs K, Chopra R, et al. Limiting transplantation-related mortality following unrelated donor stem cell transplantation by using a nonmyeloablative conditioning regimen. *Blood.* 2002; 99(3): 1071-1078.

[110] Khouri IF, Albitar M, Saliba RM, et al. Low-dose alemtuzumab (Campath) in myeloablative allogeneic stem cell transplantation for CD52-positive malignancies: decreased incidence of acute graft-versus-host-disease with unique pharmacokinetics. *Bone Marrow Transplant.* 2004; 33(8):833-837.

[111] Chakrabarti S, Mackinnon S, Chopra R, et al. High incidence of cytomegalovirus infection after nonmyeloablative stem cell transplantation: potential role of Campath-1H in delaying immune reconstitution. *Blood.* 2002; 99(12):4357-4363.

[112] Willemze R, Suciu S, Mandelli F, et al. EORTC and GIMEMA Leukemia Groups. Autologous versus allogeneic stem cell transplantation in acute myeloid leukemia. *Ann. Hematol.* 2004; 83(Suppl 1):S134.

[113] Dreger P, Viehmann K, von Neuhoff N, et al. Autografting of highly purified peripheral blood progenitor cells following myeloablative therapy in patients with lymphoma: a prospective study of the long-term effects on tumor eradication, reconstitution of hematopoiesis and immune recovery. *Bone Marrow Transplant.* 1999; 24(2):153-161.

[114] Paulus U, Schmitz N, Viehmann K, et al. Combined positive/negative selection for highly-effective purging of PBPC grafts: Towards clinical application in patients with B-CLL. *Bone Marrow Transplant.* 1997; 20(5):415-420.

[115] Montillo M, Schinkoethe T, Elter T. Eradication of minimal residual disease with alemtuzumab in B-cell chronic lymphocytic leukemia (B-CLL) patients: the need for a standard method of detection and the potential impact of bone marrow clearance on disease outcome. *Cancer Invest.* 2005; 23(6): 488-496.

[116] Moreno C, Villamor N, Colomer D, et al. Clinical significance of minimal residual disease, as assessed by different techniques, after stem cell transplantation for chronic lymphocytic leukemia. *Blood.* 2006; 107(11):4563-4569.

[117] Ritgen M, Dreger P, Stilgenbauer S, et al. Minimal residual disease (MRD) kinetics after autologous stem cell transplantation for Chronic Lymphocytic Leukemia (CLL) correlate to IgH VH mutational status and predict post transplant outcome: Real time ASO-IgH PRC results of the GCLLSG CLL3 Protocol. *Blood.* 2004; 104(11):45a (abstr 147).

[118] Trneny M, Salkova J, Karban J, et al. Rituximab, Fludarabin and Cyclophosphamide followed by high dose therapy with autologous stem cell rescue leads to high molecular remission rate in Chronic Lymphocytic Leukemia patients but relapses are observed. *Blood.* 2004; 104(11):948a (abstr 3480).

In: Chronic Lymphocytic Leukemia Research Focus
Editor: Chadi Nabhan, pp. 77-111
ISBN: 978-1-60021-526-1

Chapter IV

What Is the Clinical and Biological Significance of the Bcl-2 Family in Apoptosis Regulation in B-Cell Chronic Lymphocytic Leukemia (CLL)?

Anurag Saxena and Oksana Moshynska
Department of Pathology and Laboratory Medicine, Saskatoon Health Region and University of Saskatchewan, Saskatoon, Saskatchewan, Canada.

Abstract

Chronic lymphocytic leukemia of B-cell lineage (CLL) is the most common leukemia in the Western world and is characterized by an accumulation of clonal B-cells with a unique immunophenotype (including CD5+, CD23+, CD10-, FMC 7-). The presence or absence of somatic hypermutations identifies two subgroups with different clinical behaviors. These two subgroups also have different gene expression signatures as identified by DNA microarray.

Prolonged cell survival is one of the fundamental attributes of CLL cells. Although many abnormalities in apoptosis regulation in CLL have been reported, the exact defects and underlying mechanisms are far from being completely delineated. Further, the clinical relevance of apoptosis regulatory molecules as prognostic or predictive variables is not established.

Here in we discuss the relevant literature on apoptosis regulation in CLL, extrinsic and intrinsic pathway abnormalities in CLL, our work on the Bcl-2 family regulators, and research directions in this area. After identifying aberrant expression of Bcl-2, Bax, and Mcl-1 in CLL we sequenced BAX and MCL-1 genes. A novel single nucleotide polymorphism (SNP), G125A, in the BAX promoter was found to be associated with reduced Bax protein expression and poor clinical outcome. Subsequently, we have provided evidence of direct effect of this SNP on gene expression by showing its association with reduced Bax mRNA and by Luciferase gene reporter assay. We also identified a short sequence insertion in the MCL-1 promoter and demonstrated its association with gene expression and clinical end-points and using a functional assay a cause-effect relationship between these promoter changes and gene expression. Using DNA microarray focused on 112 apoptosis regulatory genes, we have identified consistent aberrant expression of 10 molecules (from the Bcl-2 family, IAP proteins, PI3 kinase pathway and death receptors) in CLL compared to normal flow-sorted age and sex-matched B-cells and confirmed this data by RT-PCR.

Target-based treatment, as opposed to "carpet-bombing" chemotherapy requires strong justification from basic cancer biology. In order to improve response rates and develop effective treatments that might be able to replace combination chemotherapy, identification of genes/proteins that play a crucial role in the pathogenesis of CLL is required. Some of these genes/proteins may also help identify patients at risk of aggressive disease. In this article we discuss the biologic and clinical relevance of apoptosis regulation in CLL.

Introduction

CLL, a malignancy of CD5+ B-cells [1] is the most common leukemia in the Western world and is primarily a disease of the elderly. CLL is now considered a tumor arising from mature, antigenically experienced, immunologically competent B-cell lymphocytes with a very low but definable cell turnover rate [2]. There is an accumulation of mature-appearing B-cells with a characteristic immunophenotype [CD19+, CD5+, and CD23+, low or absent expression of CD22, FMC7, CD79b, and surface immunoglobulins (IgM and IgD)] [3].

The clinical course is variable probably reflecting a combination of diagnosis in the preclinical phase, different rates of disease progression and response to treatment. However, in general, one group of patients, approximately a third, has a relatively stable clinical course – patients live for >20-25 years and usually do not require treatment [4], while the other exhibits disease progression or is severe from the outset [5-7]. In some of these patients the disease may transform to a lymphoma [8] or prolymphocytic leukemia [9]. The two basic clinical categories (stable vs progressive disease) can be separated by IgV_H mutational status, which is also the currently accepted gold standard for prognosis [10-15]. Patients with mutated IgV_H genes have a median survival of 25 years compared with 8 years for patients with unmutated IgV_H genes [11].

Prognostic factors: Attempts to predict/ prognosticate the disease course take into account assessment of traditional risk factors [gender, lymphocyte count at diagnosis, lymphocyte morphology, lymphocyte doubling time, serum markers (LDH, beta-2 microglobulin, soluble CD23, M protein), pattern of bone marrow infiltration, clinical stage (Rai / Binet) data] [16, 17] and incorporation of newer parameters such as cell surface CD38 expression [18-21], chromosomal aberrations [14], and intracellular ZAP-70 [22]. We have recently reported on the prognostic importance of the promoter changes in BAX and MCL-1 genes [23-25].

The impact of IgV_H mutational status on prognosis is not without controversies. There is still controversy as to the percentage of mutations which best correlates with clinical outcome (between 98% and 95% homology to the germline gene) [21]. Since IgV_H mutational status is a relatively newer finding, is time consuming and cumbersome, attempts to find surrogate or better markers to predict/ prognosticate the disease are currently being sought. Although unmutated IgV_H status is associated with high CD38 expression, this is not an absolute phenomenon and CD38 expression can change over a period of time [26], therefore, CD38 is not a surrogate marker of IgV_H status. Although the preliminary data suggest that mRNA and protein expression of ZAP-70 (by flow cytometry) correlates with IgV_H gene mutation status [22, 27, 28], the jury is still out on ZAP-70's role as a surrogate marker of IgV_H mutation status. A quantitative RT-PCR for as few as three genes (ZAP-70, IM1286077 and activation-induced C-type lectin) has been suggested as a surrogate marker for IgV_H status [7, 13]. It has also been suggested recently that the use of particular IgV_H gene segments such as the $V_H3.21$ gene may confer a poor prognosis independent of IgV_H mutation status [29].

The impact of some prognostic factors depends upon the type of treatment received by the patients e.g. P53 abnormalities predict a poor response to alkylating agents, purine analogs and rituximab but not to high-dose steroids and alemtuzumab [30, 31]. The impact of IgV_H mutation status remains constant while the impact of others such as cytogenetic abnormalities and CD38 expression depends upon when these are acquired by the malignant clone [17]. Multivariate analysis is helpful in delineating the value of these prognostic factors that are closely linked. IgV_H mutation status (cut-off at 98% and 97% germline homology), P53 abnormalities and deletion of chromosome 11q have been identified to be independent prognostic factors in two recent reports using multivariate studies [15, 32].

DNA microarray insights: *DNA microarray studies have identified a signature pattern unique to CLL and different from normal B-cells and other lymphomas such as diffuse large B-cell lymphoma and follicular lymphoma [32-35]. DNA microarray studies have reported differences in the gene expression signatures of the two groups as separated by the IgV_H mutational status [13, 34, 36]. DNA microarray studies have also been immensely helpful in identifying CLL subsets by segregating the CLL based upon CD38 expression [12, 37, 38], 11q23 deletion [39], and in-vitro sensitivity to ionizing radiation [40, 41].*

Treatment response: Approximately, one third of CLL patients exhibit treatment resistance [16, 42] during the clinical course, although eventually all patients with this incurable disease will become unresponsive to treatment [6, 16, 42]. Indeed, the revised NCI criteria for diagnosis and treatment do not even attempt to define cure [43]. Newer treatments such as rituximab [44], alemtuzamab [45], gene therapy to exploit CD40 ligation [46], allogeneic transplantation [47] have improved complete response rates.

In order to improve response rates and develop effective treatments that might be able to replace combination chemotherapy, identification of genes/ proteins that play a crucial role in the pathogenesis of CLL is required. Some of these genes/ proteins may also help identify patients at risk of aggressive disease.

Defective Apoptosis Is One of the Two Fundamental Pathogenic Attributes of CLL

It has recently emerged that DNA repair dysfunction may also be a defining attribute of CLL, in addition to defective apoptosis, that has already been established as a hallmark of CLL [7]. Even though cell proliferation rates in CLL may have been underestimated up until now [48], resistance to apoptosis is still a fundamental attribute of this malignancy [49].

DNA repair dysfunction based genomic instability is believed to be; a) the basis of multiple cytogenetic abnormalities, as opposed to a unique defining karyotypic lesion in CLL [40] and b) a contributory factor to drug resistance [50]. CLL is characterized by prolonged survival of tumor cells [1, 10, 51]. A relatively apoptosis-resistant phenotype due to alterations in the apoptotic pathway is associated with disease progression and inadequate response to treatment [33, 42, 52-56]. Further, proteins associated with cell proliferation e.g; cyclin A and B, Ki67, p16 and thymidine kinase are downregulated or not expressed in CLL [32-34]. It is believed that a general defect in the apoptotic cell-death pathways in CLL cells is reinforced by other factors in CLL subsets. For example, radioresistance of a subset of leukemic cells is related to both the ubiquitin–proteasome and DNA repair systems, and that these events correspond to specific gene-expression patterns [7].

Why Are CLL Cells Resistant to Apoptotic Death?

A) Apoptosis Regulation; General Aspects

Apoptosis regulation, in general, has three aspects; stimuli for initiating apoptosis, signal transduction pathway culminating at effectors and regulators of the process [57, 58] (summarized in table 1).

Table 1. General aspects of apoptosis regulation in CLL

EXTRINSIC PATHWAY	
Membrane receptor signaling and microenvironment have an essential role, as CLL cells are extremely sensitive to apoptosis by various stimuli in vitro	[7].
Survival Signals	
Interleukins, colony stimulating factors, Interferons, integrins, CD40, CD6-mediated protection from IgM-induced apoptosis	[55, 180-183]
High expression of TNF-α and a TNF superfamily member B-Lymphocyte stimulator (Blys)	[184]
CD40-dependent transcriptional induction of surviving	[72]
Follicular dendritic cell induction of Mcl-1 via a CD44-dependent pathway	[185]
Apoptotic stimuli	
TRAIL-induced apoptosis in CLL. Fas expression is weak on CLL cells and its role is not entirely clear	[186]
INTRINSIC PATHWAY	
Bcl-2 family: Mitochondrial pathway regulators are shown to be of central importance in CLL	[82]
P53 mutations linked to disease progression and treatment resistance	[187]
Overexpression of *mdm-2* may be a poor prognostic factor in CLL	[188, 189]
Nitric-oxide synthase induction	[190]
Constitutive activation of PI3 kinase and consequent activation of AKT serine/threonine kinase pathway of survival signaling	[191]
COMMON PATHWAY	
Caspase 3 and nuclear lamin protease have been implicated in cell death in CLL	[192, 193]

There are two major pathways for apoptosis - the extrinsic or death-receptor-induced pathway and the intrinsic or mitochondria-apoptosome-mediated pathway - both lead to caspase activation and subsequent cleavage of specific substrates. Crosstalk between these two pathways exists; caspase 8 (upstream caspase of the extrinsic pathway) can cleave BID protein resulting in the release of cytochrome c from mitochondria – the intrinsic pathway [59]. The extrinsic pathway includes membrane receptors and their ligands such as FAS, TNF, TRAIL, and subsequent caspase activation. The intrinsic pathway includes internal apoptotic stimuli such as DNA damage, oxidant stress, mitochondrial release of apoptosis inducing factors and caspase activation. Caspases are activated in a cascade-like fashion; upstream caspases (caspases 8, 9, and 10) can activate effectors or downstream caspases (caspases 3, 6, and 7) leading to induction of apoptosis [60-63].

The regulators of the process include the Bcl-2 family proteins, inhibitors of apoptosis proteins (IAP, including XIAP, cIAPs), survivin, the PI3K/AKT/NFB pathway, P53, and heat shock proteins (HSPs); these interact with the process and can alter the outcome [64].

B) Apoptosis Defects in CLL

I) IAPs

IAPs regulate apoptosis by inhibiting specific caspases (3,7 and 9) [65, 66] and via caspase-independent mechanisms involving MAPK/JNK signal transduction pathways [67] and by their involvement in protein degradation [68]. At present, eight human IAPs are known to exist: c-IAP1, c-IAP2, NAIP, Survivin, XIAP, Bruce, ILP-2, and Livin [69]. IAP expression has been identified in a variety of human cancers (reviewed in [66]. In CLL cells, survivin [70-72], X-IAP [73, 74], c-IAP1 and c-IAP2 [74] have been detected. In WSU-CLL cell line resistance to bryostatin and 2-chloro-2-deoxyadenosine (2-CdA) induced apoptosis has been shown to be due to upregulation in IAP family members [75]; however, it should be noted that this cell line has been shown to be contaminated by ALL(acute lymphoblastic leukemia) cells [76]. Further, flavopiridol induce decreases in Mcl-1 and XIAP preceded apoptosis in CLL cells in vitro [70]. IAP: IAPs are regulated by several mechanisms. Negative regulators including XIAP-associated factor 1 (XAF1) and direct IAP binding protein with low pI (Smac/DIABLO) [77] and the membrane-associated protein OMI/HTRA2 [78] inhibit IAPs. Effector caspases can cleave IAPs, e.g; c-IAP1 cleavage produces a pro-apoptotic C-terminal fragment [79]. Post-translation modifications such as phosphorylation [80] can affect protein activity. IAPs' gene expression is tightly regulated at the transcription level [66]. NF-kB is implicated in transcriptional control [66]; the presence of a long sequence in the 5'-UTR called internal ribosomal entry site permits translation of XIAP under conditions of stress [81] and enable the cells to resist chemotherapy-induced apoptosis.

II) BCL-2 Family and Tumor Cell Survival in CLL

Our contribution to further understanding of mechanisms underlying altered expression of apoptosis regulatory proteins in CLL

The Bcl-2 family, consisting of both pro- and anti-apoptotic proteins, is a central regulator of apoptosis acting mainly on the mitochondrial pathway of cell death [57, 58, 82-84]. It consists of three subfamilies [84, 85]; anti-apoptotic Bcl-2 cohort (Bcl-2, Bcl-xL, Bcl-w, Mcl-1, NR-13, BHRF-1, LMW5HL), pro-apoptotic Bax subfamily (Bax, Bak and Bok) and pro-apoptotic BH3 subfamily (Bik, Blk, Hrk, BNIP3, Bim_L, Bad, Bid, EGL-1). Although the list of apoptosis defects in CLL is long, as highlighted in table 1, P53, ATM and Bcl-2 family members (Bcl-2, Bax and Mcl-1) have already been shown to be of clinical relevance (table 1) [15, 86, 87]. Bcl-x and Bfl-1 are also implicated in prolonged cell survival in CLL [52, 88, 89]. High Bcl-xL mRNA levels (16/23) and low to trace amounts of Bcl-xS mRNA (13/23) have been reported in CLL [90] in CLL samples using RT-PCR. Bfl-1 transcripts are present in CLL cells and the levels have been shown to be increased after stimulation with CD40 and sIgM [89]. So far, Bcl-2/Bax ratio [91-95] and Mcl-1 [52, 91] expression have been identified as key determinants of cell survival and proven to affect clinical course and treatment response. The important place in pathogenesis of CLL is devoted to Bcl-2 family proteins, especially to Bcl-2, Bax, and Mcl-1 proteins. Increased levels of Bcl-2/Bax protein ratio [54, 92, 94] and Mcl-1 protein [52] were found to be associated with poorer response to treatment and failure to achieve a complete response in patients with CLL. Bcl-2/Bax is associated with progressive disease [54]. Our previous work has also determined the relative importance of these two variables, Bcl-2/Bax ratio and Mcl-1 and shown that low Mcl-1 predicts complete response while low Bcl-2/Bax ratio is associated with at least partial response [91].

Microarray data has also confirmed the role of apoptosis regulatory genes in CLL [34]. There is a general defect in apoptosis pathway in CLL, however, there are subsets in which this defect is reinforced by other factors [7]. Our preliminary work including DNA microrarray has identified additional Bcl-2 family members likely to be of pathogenic and clinical importance; these are discussed in this chapter.

One needs to determine the mechanisms underlying altered expression of key proteins proven to be of pathogenic and clinical relevance in order to exploit them as targets of novel treatments. Using immunoblotting [91], RNase protection assay (RPA), DNA microarray and real-time quantitative PCR (RT-PCR) we have identified altered expression of certain genes including those of the Bcl-2 family (BCL-2, BAX, MCL-1, BCL-X, and BFL-1). RNA was extracted from fresh samples from three male CLL patients (age median – 61) and a control sample from a healthy individual, converted into cDNA, and hybridized to GeneArray HS-002 (Affymetrix, Chip SuperArray Bioscience Corporation, Frederick, MD, USA). Two patients were newly diagnosed patients with stage 0 and 1, and one patient had stage 1 of the disease and was treated more than 6 months prior. Using DNA microarray (figure 1) focused on 96 apoptosis regulatory genes (table 2), and RT-PCR confirmation of data on genes expressed at a different level (compared to normal flow-sorted age and sex-matched B-cells), we determined consistent aberrant expression (at least more than 1.5-fold expression levels compare to a control sample) of the following functional group genes in CLL: CARD family (1 gene), Death effector domain family (1 gene), TNF ligand family (3 genes), TNF receptor family (1 gene), IAP family (3 genes), p53 and ATM pathway family (3 genes), and Bcl-2 family (8 genes) (table 2).

Table 2. Functional groups in human 112 gene (including 96 apoptosis regulatory genes) array

Functional Gene Grouping	No. of genes	List of genes
TNF Ligand Family	14	TNF-α, TNF-β, OX40 Ligand, CD40 Ligand, LT-β,FASL, CD27 Ligand, CD30 Lignad, 4-1 BB Ligand,TRAIL, TRANCE, APO3L, APRIL, HVEM-L
TNF Receptor Family	15	TNFR1, TNFR2, OX40, LTBR, CD40, FAS, CD27, CD30, 4-1 BB, DR4, DR5, DCR1, DCR2, DR3, HVEM
Bcl-2 Family	14	CASP1, CASP2, CASP3,CASP4, CASP5, CASP6, CASP7,CASP8, CASP9, CASP10, CASP13, CASP14
IAP Family	6	NIAP, IAP2, IAP1, XIAP, Survivin, Bruce
TRAF Family	8	I-TRAF, TRAF1, TRAF2, CRAF1, TRAF4, TRAF5,TRAF6, TRIP
CARD Family	6	APAF-1, ASC, BCL-10, CARD4, NOP30, CARDIAC
Death Domain Family	5	CRADD, DAPK2, FADD, MYD88, RIPK1
Death Effector Domain Family	4	FLASH, CASPER, FADD, BAR
CIDE Domain Family	4	CIDEA, CIDEB, DFFA, DFFB
P53 and ATM Pathway	8	ATM, CHEK1, CHEK2, GADD45A, MDM2, P63, RPA3, p53
Controls and Housekeeping Genes	16	Actin-β (2), Cyclophilin A (PPIA) (4), GAPDH (2), pUC18 (3), RPL13A (2), Blank (3)

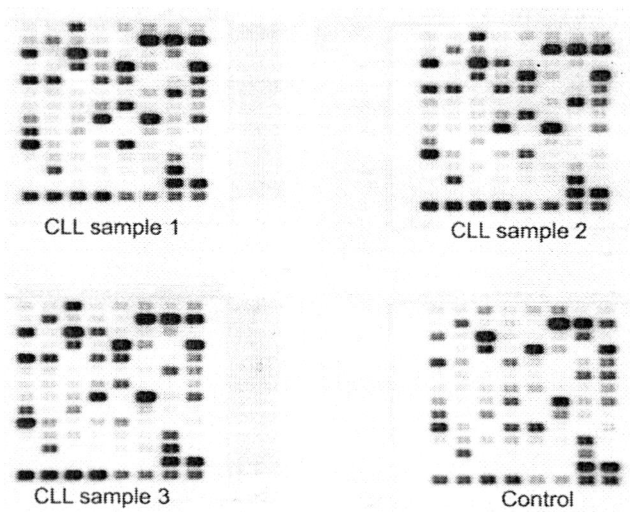

Figure 1. DNA microarray data of 112 including 96 apoptosis regulatory genes (Affymetrix Chip SuperArray Bioscience Corporation, Frederick, MD, USA) on three CLL samples and one control (purified B-cells = flow-sorted from a healthy age/sex matched volunteer). Lymphocyte and B-cell separation was done as described earlier [25]. Microarray was performed using HS-002 GeneChip (Affymetrix) as per manufacturers' instructions. Each dot represents one gene; data normalized to housekeeping gene PPIA (peptidylprolyl isomerase A = cyclophillin A).

This chapter is devoted to Bcl-2 family of apoptosis regulatory genes. The Bcl-2 family members that distinguished between the control and CLL samples were: BCL-2, BAX, MCL-1, BXL-X, BFL-1, BAD, BCL-W, and BLK (figure 1). Compared to the control sample all three CLL samples had higher levels of BCL-2 (2 times) and BFL-1 (1.7 times) and lower levels of BAX (3.1 times), with the BCL-2/ BAX ratio being the most clearly identifiable discriminator (6.9 times). The microarray data correlated with RT-PCR (by MultiGene-12 RT-PCR profiling kits, Affymetrix, as per manufacturers' instructions) and protein data (by immunobloting with primary antibodies against Bcl-2 (mouse monoclonal, Upstate Biotech, New York, NY), Bax (rabbit polyclonal (Pharmingen), Bfl-1 (A1, rabbit polyclonal, Santa-Cruz Biotechnology Inc, Santa Cruz, CA), and secondary antibodies conjugated to a horseradish peroxidase (goat anti-mouse IgG (for Bcl-2) and anti-rabbit IgG (for Bax and Bfl-1) antibodies (Bio-Rad), as described earlier [91]). All protein expression levels were normalized to β-Actin (mouse monoclonal, Sigma). The microarray results supported the suggested role of BCL-2, BAX [52, 91-95], MCL-1 [52, 91], and BFL-1 [96, 97], in CLL pathogenesis. BIK expression was similar in CLL and control samples, as per microarray studies (figure 1). These results are in agreement with [98], who reported absence of BIK expression, suggesting that the later is unlikely to play a major role in CLL.

We also performed RNAse protection assay (RPA) using the template set hApo-2c (Pharmingen) to study Bcl-2 family mRNA transcripts in 21 CLL samples and results are shown in figure 2. Higher BCL-W expression was noted in 24%, BCL-X – in 43%, BFL-1 – in 52%, BCL-2 – in 81%, MCL-1 – in 67%, and BCL-2/ BAX – in 38% CLL cases. These results are in agreement with [98], who reported high levels of BCL-2, BCL-W, and BCL-2/ BAX mRNA and involvement of BFL-1 in apoptosis, suggesting their importance in CLL.

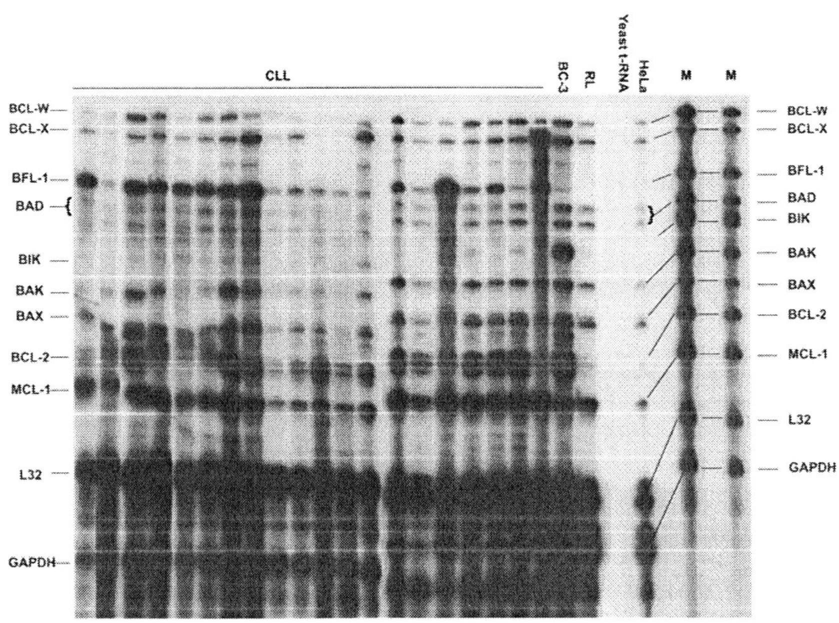

Figure 2. BCL-2 family mRNA expression in cell lines (BC-3, RL, and positive control RNA from HeLa cell line), and in 21 peripheral blood lymphocytes of chronic lymphocytic leukemia patients, analyzed by RNase protection assay. Total RNA was isolated and analyzed as described previously [25]. 10 μg of RNA sample was processed using *In Vitro* Transcription kit (Pharmingen, Mississauga, USA), hybridized with the hApo-2c (Pharmingen), human apoptosis multiprobe template set. RNAse protected probe following hybridization with yeast tRNA served as a negative control. The level of each mRNA was normalized against the housekeeping genes, L32 and GAPDH, and relative expression levels were determined.

Bcl-2

Bcl-2 protein found to be overexpressed in high percentage of CLL cases [99-101]. Hypomethylation of the promoter has been shown to underlie Bcl-2 overexpression in some CLL cases [102]. In rare bona-fide CLL cases (t14;18) has also been identified [103]. However, in many cases, the underlying mechanism remains unknown. We sequenced the open reading frame (ORF) and promoter of the BCL-2 gene (figure 3A, table 3) from ten CLL patients to identify any possible genetic alterations, which might be associated with or be the cause of altered gene expression and compared the results to GenBank sequences (AN: M13994 for ORF and AN: X51898 for 5'-region of the gene). In the ORF only one single nucleotide polymorpism (SNP) was found, A-to-G nucleotide replacement, it did not cause any amino acid change: Thr7Thr. We compared our findings to results obtained for the RL lymphoma cell line, which served as a positive control for Bcl-2 protein expression. This cell line has a translocation, t(14;18), and expresses high BCL-2 mRNA (figure 2) and protein levels. Five genetic alterations were found in RL cell line [104]. Two alterations resulted in amino acid changes, G-to-C (Asp31His), and C-to-T (Ala60Val). The other three, C-to-T (Leu23Leu), C-to-T (Ser50Ser) and A-to-G (Pro59Pro), did not change amino acid coding. One SNP at the amino acid position 23, Leu23Leu, was located in BH4 domain of the Bcl-2 protein and two others, Ala60Val and Pro59Pro, were in the non-conserved region. The absence of BCL-2 gene mutations in CLL suggests that other mechanisms, not genetic alterations are responsible for Bcl-2 overexpression in CLL.

Table 3. Primers used in this chapter for sequencing BCL-2, BAX, MCL-1, BFL-1, and BCL-X genes

Primers		Region	Fragment
BCL-2			
B2-PrF	5'-CAC ACA CAC ACA CAC ACT C-3'	5'-region	583bp
B2-PrR	5'-GCC TGT CCT CTT ACT TCA TTC-3'		
B2-F	5'-AGA GGT GCC GTT GGC CCC-3'	ORF	508bp
B2-R	5'-GAC CCC ACC GAA CTC AAA GAA G-3'		
BAX			
BAX-PrF [1]	5'-CGG GGT TAT CTC TTG GGC-3'	5'-region	441bp
BAX-PrR [2]	5'-GTG AGA GCC CCG CTG AAC-3'		
BPL-F	5'-CTT TTG CAC TTG CTA ATT CCT TC-3'	5'-region	158bp
BPL-R	5'-CAA TGA GCA TCT CCC GAT AAG T-3'		
BAX-1F[1]	5'-CGT TCA GCG GGG CTC TCA-3'	Exon1/Intron1	194bp
BAX-1R[1]	5'-CAG GCC GGT AGG AAG GAT-3'		
BAX-2.3F[1]	5'-CCC CTA GAA CCC AAG AGT C-3'	Exons2-3/Introns1-3	400bp
BAX-2.3R[1]	5'-GGC TGA GAG TCC TGT GTC C-3'		
BAX-4F[1]	5'-TCT CCT GCA GGA TGA TTG C-3'	Exon4/Intron4	209bp
BAX-4R[1]	5'-TCC CCA GGT CCT CAC AGA T-3'		
BAX-5F[1]	5'-CAC GCA GTG GGG ACA AGG TT-3'	Exon5/Intron5	192bp
BAX-5[1]	5'-GCG GTG GTG GGG GTG AGG AG-3'		
BAX-6F[1]	5'-CCC CTG GCC GAG TCA CTG AA-3'	Exon6	234bp
BAX-6R[1]	5'-AAT GCC CAT GTC CCC CAA TC-3'		

Table 3. (Continued).

Primers		Region	Fragment
MCL-1			
M-PrF	5'-TGG CAG TTT TAG CTT AGT CCA C-3'	5'-region	650bp
M-PrR	5'-ACT CCT TAC TGG AAG GAA GCG-3'		
MPS-F	5'-CGA GGT GCT CAT GGA AAG AG-3'	5'-region	218bp
MPS-R	5'-TTC CCA TAA AAG GGG AAA GG-3'		
M-1F	5'-TGT TGT TCC CAT CCA TCA CTG-3'	5'-region/ Exon1	992 bp
M-1R	5'-AGA TTC CCC GAC CAA CTC C-3'		
M-2F	5'-ACT GGC AAT GTT TGG CCT C-3'	Exon1	457bp
M-2R	5'-AGA TTC CCC GAC CAA CTC C-3'		
M-3F	5'-GGT AAT AAC ACC AGT ACG GAC-3'	Exon1/ Intron1	405bp
M-3R	5'-GAT GGG CGG AAA CAA TGA C-3'		
M-4F	5'-AGT TCT TCC ATG TAG AGG ACC-3'	Exon 3	397bp
M-4R	5'-TTC CAC CCT ACC ATC TTC AC-3'		
M-5F	5' GCA ACC TAG CCA GAA AAG CAA G 3'	Exon3/ 3'-region	470bp
M-5R	5' TGG AAC CAA GCC AAA GTA TAA CAG		
M-6F	5'-CTT TGG CTT GGT TCC ATG ATT C-3'	3'-region	452bp
M-6R	5'-TCC ATT CAT CAA CCT CTC AAT CC-3'		
M-7F	5'-CCT TTT GAA CTT TGC AAC TTC C-3'	3'-region	513bp
M-7R	5'-ACA AGG GAT CAA ATG TCT CTC C-3'		
M-8F	5'-GTC ATT GTT TCC CGC CCA TC-3'	Intron1/ Exon2	504bp
M-8R	5'-TCC TTC ATT CTA CTT CCA CTC C-3'		
M-9F	5'-AAG GAC AAA ACG GGA CTG GC-3'	Intron2	510bp
M-9R	5'-CCC TTA TTC TCA AAT CTC CAC CAC-3'		
M-10F	5'-TGA TGG CTT GGA AAA GCA G-3'	3'-region	723bp
M-10R	5'-AGC TTT CAA ATG ACC CTA GTT C-3'		
M-11F	5'-GGA ACT AGG GTC ATT TGA AAG C-3'	3'-region	764bp
M-11R	5'-CAC ACT ACA TTT GAC AAC CAA C-3'		
BFL-1			
BFLA-F	5'-GTG CTG CCA TGT ACA TTC GT-3'	5'-region	591bp
BFLA-R	5'-AGC ACT CTG GAC GTT TTG CT-3'		
BFLB-F	5'-GGA ATT GCT GGG TCT CTT GA-3'	5'-region	478bp
BFLB-R	5'-CGC ACT GCA GAT AGT CCT GA-3'		
BCL-X			
BX-F	5'-TCC TGA GCT TCG CAA TTC C-3'	5'-region/Exon1	592bp
BX-R	5'-GCT CAA CCA GTC CAT TGT CC-3'		
BCLXB-F	5'-CCT GGG CTG GTG CTT AAA TA-3'	5'-region/Exon1,2	819bp
BCLXB-R	5'-ATA GGG ATG GGC TCA ACC AG-3'		

[1] Primers adapted from Peng et al [194].

[2] All other primers were designed in our laboratory.

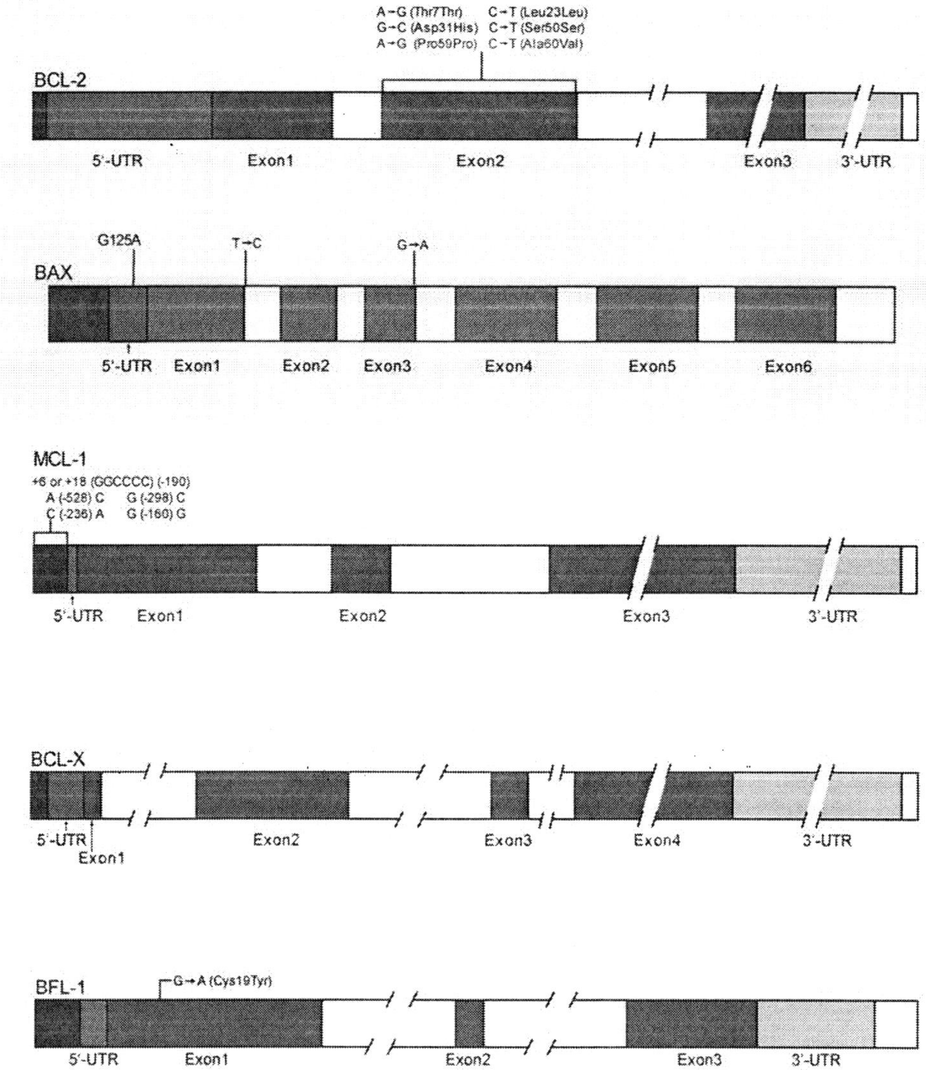

Figure 3. Schematic representation (not to scale) of BCL-2 family (BCL-2, BAX, MCL-1, BFL-1, and BCL-X) members gene structure with location of found genetic changes (arrows) in patients with chronic lymphocytic leukemia.

Bax

Bax has been shown to play a role in CLL pathogenesis [52, 105] and to be an important regulator of in vitro and in vivo response to chemotherapy in CLL [52, 95, 106]. Bax is under transcriptional regulation by P53; P53 mutations, a later event in CLL [107] may impair this downregulation and may contribute to altered Bax protein expression. Bax protein is capable of heterodimerizing with Bcl-2 [108], Mcl-1 [109], Bcl-X [110], and Bfl-1 [111] and Bcl-2/ Bax protein ratio is shown to be relevant in CLL [91-94, 112]. We sequenced the entire BAX gene (figure 3B) from DNA of 77 CLL patients using seven sets of primers (table 3), and compared the results to GenBank sequences (AN: U17193 for promoter, and from AF020353 to AF020362 for exons/ introns). We were the first to identify a novel polymorphism G125A

(125 nucleotides downstream from the transcription start) or G(-248)A (248 nucleotides upstream of translation start), in the 5'-untranslated region (5'-UTR) of the BAX promoter and demonstrate its association with reduced protein expression and treatment resistance [23, 24]. We also proposed the restriction enzyme analysis as a detection method for the new G125A, as a simple and cost-effective alternative to sequencing [23, 24]. This SNP was not in the P53 binding region and there were no mutations found in the open reading frame [23]. Further, a group from England [113] independently confirmed, on a larger group of CLL patients, the association of this SNP with reduced protein expression. They also found the association of SNP with reduced survival. Although, Skogsberg et al [114] did not agree with the influence of the SNP on survival but, interestingly, noted a trend towards shorter overall survival in the V_H mutated group for patients with GA and AA genotypes, compared to those bearing GG genotype. Overall, this SNP does not correlate with the V_H mutational status [114]. Moreover, we did not perform the correlation studies with the V_H mutational status, as there is no theoretical/ experimental evidence that the genetic alterations in the BCL-2 family members would correlate with the V_H mutational status [97]. Using luciferase gene reporter assay we have also shown that this SNP has a direct effect on gene expression [115].

In this chapter we provide the updated report on G125A status in the Saskatchewan population. Our reports on smaller groups of patients, 13 [116] and 35 [23] patients, were published earlier. Among 77 CLL patients screened up-to-date in our laboratory, 28 were women and 49 men, age range from 34 to 85 (median – 68). Twenty-five out of these 77 (32.5%) patients had G125A BAX SNP, while it was found in only seven out of 65 (10.8%) healthy age and sex matched (22 women and 43 men; age range from 33 to 91, median – 75) individuals. The frequency for G-to-A replacement at position 125 in healthy Saskatchewan population is estimated to be of 0.06, compared to the reported frequency of 0.13 in GenBank (GenBank, AN: AY217036). We found the G125A SNP frequency to be higher, 0.18, in CLL patients.

We also found two intronic nucleotide replacements [116]; these were T-to-C in intron 1 (43 nucleotides downstream from the exon 1), and G-to-A in intron 3. The latter one is located in one of the three TGGG repeats region and (A/T)GGG intronic repeats and may regulate alternative splicing [117]. We found this G-to-A replacement in 73% CLL patients, in most cases there was only one allele replaced. The frequency of allele G is believed to be of 0.5 in normal population (GenBank, AN: AY217036).

These findings strongly suggest the importance of Bax protein in CLL pathogenesis and its clinical relevance; genetic changes in BAX gene may alter the Bax protein expression in this type of cancer.

Mcl-1

Mcl-1 (myeloid cell leukemia-1) [118] protein is the most studied Bcl-2 family protein over the last two years. Mcl-1 promotes short-term increase of cell survival in different cell types [119]. The finding of immortalization of hematopoietic cells from MCL-1 transgenic mice, cultured in the presence if IL-3 [120], provide the proof of the importance of MCL-1 in carcinogenesis. Mcl-1 is known to be regulated both transcriptionally and post-transcriptionally [121, 122]. Mcl-1 regulation is reviewed in [123]; transcriptional regulation involves several pathways, MEK/ERK, MAP, PI3K/AKT, JAK/STAT. The increased expression of Mcl-1 in CLL may depend on certain growth factors, including interleukin 4 (IL-4), IL-3, IL-13 and granulocyte-macrophage colony-stimulating factor (GM-CSF) [121, 124-126]. The cytokine-dependent pathways e.g. GM-CSF [127] ultimately converge on to the transcriptional domains in the MCL-1 promoter [121, 128].

Mcl-1 can serve as a substitute for Bcl-2 and a reciprocal expression of these proteins was noted in immunobloting and RPA experiments in different cell types [52, 91], including in RL, BC-3, HeLa and some CLL cases in this chapter (figure 2). Mcl-1 can dimerize with other anti-apoptotic proteins, Bcl-2 and Bcl-x as supported by findings in yeast two-hybrid experiments [108]. Because Mcl-1 is able to heterodimerize with Bax [109], we sought to analyze the role of Mcl-1/ Bax protein ratio in CLL. There was no statistical significance between control (0.69 ± 0.25) and CLL (2.19 ± 5.88) groups (P=0.450), although the average ratio levels was 3 times higher in CLL group. Based on receiver operating characteristics (ROC) analysis, the cut-off point of 1.0 was considered to be a discriminator of high and low levels for Mcl-1/ Bax ratio. Twenty-five out of 58 (43.1%) patients had high Mcl-1/ Bax ratio, and among those four (6.9%) did not co-overexpressed other proteins, suggesting the possible role of these proteins ration in CLL pathogenesis.

Environment factors together with genetic changes are believed to lead to the immortalization of cells and cancer and Mcl-1 protein contributes to this process [123]. Genetic changes causing the induction of gene expression in MCL-1 gene and resulting in extended cell survival may contribute to tumorigenesis. We sequenced the entire MCL-1 gene (figure 3C) using twelve sets of primers (table 3), and compared the results to GenBank sequence (AN: AF198614). No changes were found in the open reading frame (ORF) of the gene. Four single nucleotide polymorphisms were found in the promoter area: G(-298)C, C(-236)A, A(-528)C, C(-160)G (figure 3C). They were described in [25], the role of these SNPs remains unclear; We were the first to report on a six-/ eighteen- short nucleotide sequence insertion in the MCL-1 promoter and its association with higher mRNA and protein and treatment resistance and survival [25]. Out of 76 CLL patients screened up-to-date in our laboratory, 19 (25%) had at least one allele with +6 or +18 insertion. While two out of seventy-three age and sex matched healthy individuals had one of the alleles replaced. This insertion was in a GC rich region that may contain binding sites for several transcription factors [129]. Deletions in the nearby region significantly decreases GM-CSF dependant activation of Mcl-1 protein expression [128]. On the other hand, multimerization in this region may lead to gene superactivation (dependent upon the distance form the TATA box); this might explain overexpression of the protein [130-132]. The cause-effect relationship between this promoter change and gene expression was studied by luciferase reporter gene analysis, eletrophoretic mobility shift assay (EMSA), and chromatin immunoprecipitation (ChIP) assay (manuscript submitted for publication). Unstimulated and especially phorbol 12-myristate 13-acetate (PMA)-stimulated activity of +6 and +18 promoters is significantly higher than the +0 promoter activity. We found also that the affected region does not bear Sp1/ Sp3 transcription factor sites. It would be interesting to exploit possible influence of the affected site on the located a few nucleotides downstream Sp1 site of the MCL-1 promoter.

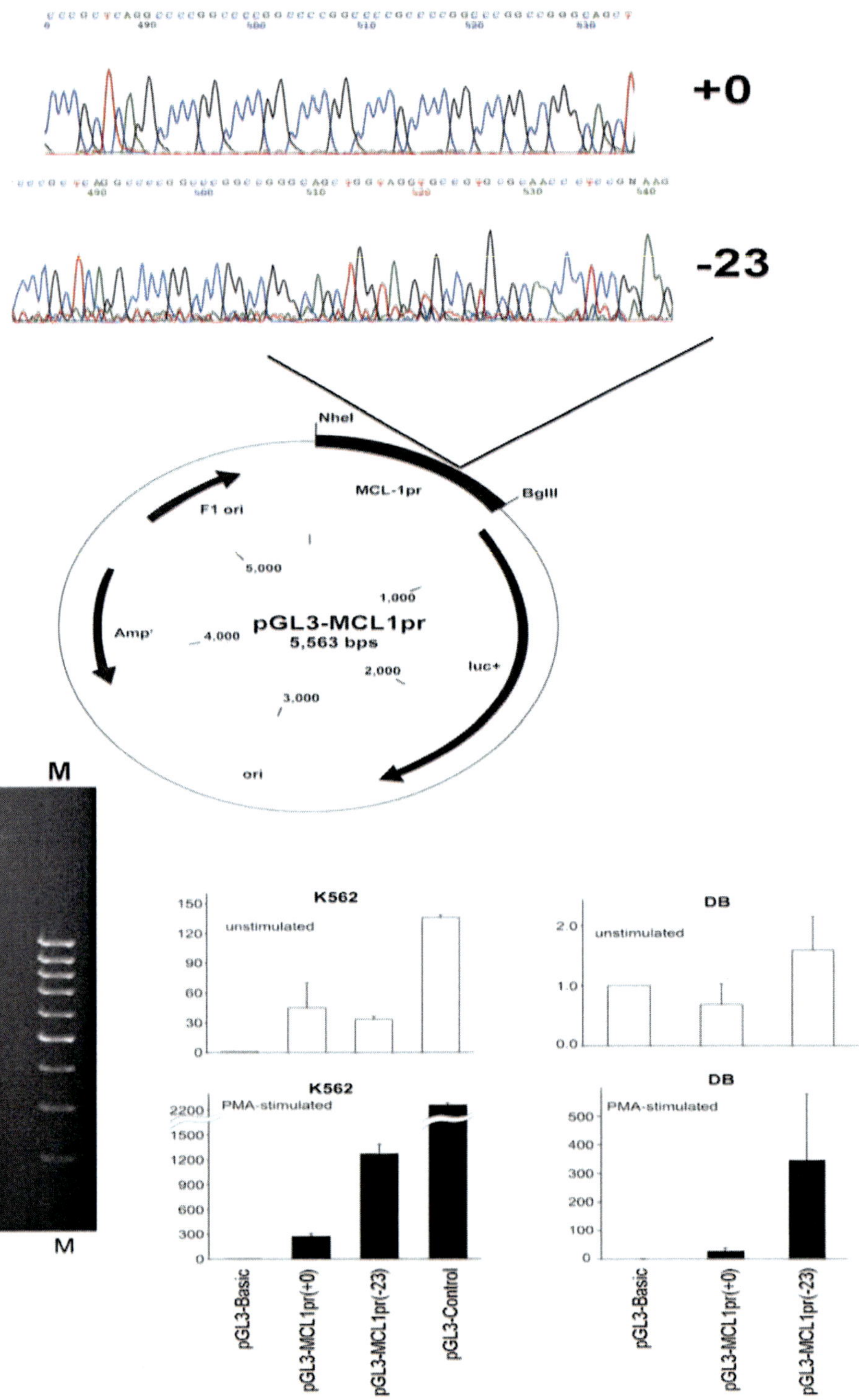

Figure 4. A: Schematic presentation of the pGL3-MCL1pr construct with the partial sequences of the +0 and –23 MCL-1 promoter segments, restriction enzymes sites of the MCL-1 promoter insertion (*NheI* and *BglII*); B: Agarose gel electrophoresis (1%) with experimental constructs digested with restriction enzymes, *NheI* and *BglII*: +0 = pGL3-MCL1pr(+0) – MCL-1 promoter with +0 insertion; -23

= pGL3-MCL1pr(-23) – MCL-1 promoter segment with deleted 23 nucleotides; M = marker, GeneRuler 100bp DNA Ladder (Fermentas); RVprimer3 (5'-CTAGCAAAATAGGCTGTCCC-3') and GLprimer2 (5'-CTTTATGTTTTTGGCGTCTTCCA-3') were used to verify the insertion of the MCL-1 promoter segment; C: Unstimulated and PMA-stimulated luciferase reporter gene activity after transfection of K562 and DB cells with the constructs. All luciferase experiments were performed in three-to-six independent transfection experiments, shown data represent mean values with error bars representing standard deviations. K562 (CCL-243) cell line was obtained from American Type Culture Collection, ATCC (VA, USA) and DB cell line (ACC 539) was from German collection of microorganisms and cell culture, DSMZ (Braunschweig, Germany). The K562 cells were grown in IMDM and DB cells in RPIM-1640 medium supplemented with FBS, at 37^{0}C under 5% CO_2. K562 and DB cells were transiently transfected with 5 μg of each plasmid using ExGen 500 *in vitro* Transfection Reagent (Fermentas, Burlington, ON), according to the manufacturer's instructions. The transfected cells were maintained for 16-18 hours 48 hours post-transfection. At this time cells were stimulated with complete fresh medium only (1.5 ml), or medium with 50 ng/ml phorbol 12-myristate 13-acetate (PMA, Sigma-Aldrich Canada Ltd, Oakville, ON), and cells were allowed to grow for another 24 hours. The cells were harvested twenty-four hours post-stimulation. All cells were co-transfected with pRL-SV40 used as transfection efficiency control (the plasmid's *Renilla* luciferase is enhanced by SV40 promoter). The luciferase activity (relative light units) was measured using Dual-Glo luciferase assay system (Promega, Madison, WI) according to the manufacturer's instructions in luminiscence counter (Packard Bioscience, Meriden, CT).

During the screening of CLL patients we found one case representing a novel allele that has not been reported previously, a deletion of 23 nucleotides (CGGCCCCGGCCCCGGCC-CCGCCC), the patient was heterozygous for this allele, +0/-23. This new allele, -23, is described for the first time in this chapter and represents a deletion of three (GGCCCC) and one (GCCCC) motifs. To study the possible effect of the deletion of 23 nucleotides in the MCL-1 promoter on the gene transcription, we constructed a reporter plasmid with (–23)-MCL-1 promoter segment linked to luciferase, and compared the luciferase acitivity of the +0 promoter reporter plasmid (figure 4), using standard cloning techniques and as described in [115], and primers PIM-F (5'-GGAATAG*GCTAGC*TGTGGCAGTTTTAG -3') and PIM-R (5'-CACAGT*AGATCT*TGAGTCCGATTAC -3'). Restriction enzyme analysis with *NheI* and *BglII* (figure 4B) and sequence analysis (figure 4A) provided evidence on the successful insertion of the MCL-1 promoter segments. *Firefly* luciferase activity in each individual transfection was normalized to *Renilla* luciferase activity and the results were compared to the expression of pGL3-Basic plasmid (accepted as equal to 1), and presented as fold activity (figure 4C). The luciferase activity for unstimulated K562 and DB cells did not differ significantly between +0 and –23 promoters (figure 4C). Then the cells were stimulated with phorbol 12-myristate 13-acetate (PMA). The PMA-stimulated luciferase activity of K562 and DB cells transfected with pGL3-MCLpr(-23) was 4.7 times (for K562, P=0.0001, t-test) and 12 times higher (for DB, P=0.033, t-test), compare to pGL3-MCLpr(+0) plasmid (figure 4C). We speculate that the reason for the upregulation of the luciferase activity is due to presence of possible negative regulatory element in the 23-nucleotide fragment in the MCL-1 promoter.

The obtained results on Mcl-1 suggest the important and independent role of Mcl-1 in CLL pathogenesis, the genetic changes contribute to the altered MCL-1 gene expression

Bcl-x

The Bcl-x protein belongs to the Bcl-2 family of proteins; alternative splicing generates long (or Bcl-xL) with anti-apoptotic and short (Bcl-xS) with pro-apoptotic function isoforms [133]. Post-translational modifications, such as phosphorylation at serine or threonine sites of Bcl-xL [134], affect its function. Cell surface receptor mediated signaling can affect Bcl-xL levels; CD40 ligand can upregulate Bcl-xL expression [135] while TNF-induced apoptosis is associated with reduced levels of Bcl-xL [136]. Bcl-x interacts with p53, binding of p53 to Bcl-x or Bcl-2 blocks these anti-apoptotic proteins resulting in apoptosis [137, 138]. Bcl-x and also Bcl-2 found to play a role in hematopoietic differentiation restricting lineage determination [139]. The Bcl-x/ c-Myc double transgenic mice have less than a six weeks survival rate (with a 50% lethality) due to developed lymphoproliferative diseases, clonal B-cell and plasma cell tumors [140]. In a murine model, the overexpression of Bcl-x promoted breast cancer xenograft tumor progression to the metastatic state [141]. Using Bcl-x antisense oligonucleotides in combination with chemotherapy reduces in half the levels of Bcl-x protein and cell proliferation and increases apoptosis in Caco-2 colon cancer cells [142]. The Bcl-x antisense oligonucleotides sensitizing effect has been demonstrated also in prostate cancer [143] and breast cancer resistant to doxorubicin cells, MCF-7/ADR [144]. CD40L prolongs CLL cell survival in vitro and interestingly enough, this process is accompanied by BCL-X mRNA levels increase and not Bcl-2, after 3-6 hours of cell culture [145]. Bcl-x prevents TRAIL-induced apoptosis in acute myelogenous leukemia HL-60 cells and abolishes the activation of caspases 3 and 7, and cleavage of Bid' BH3 domain [146]. Inhibition of Bcl-x protein by Bcl-x-specific small interfering RNA inhibits proliferation of 5-fluorouracil resistant human colon cancer cells [147], demonstrating its role in suppressing proliferation in these cells. Bcl-x has been found to be expressed in 72.4% colorectal cancer cells [148], 96.9% Barrett's esophagus [149], and in 63.6% hepatocellular carcinomas [150]. Bcl-x overexpression provides a survival mechanism protecting ovarian cancer [151], renal cell cancer [152], and prostate cancer PC3 and LNCaP cells [143] from apoptosis. Overexpressed Bcl-x overcomes the p53 action by inhibiting p53-induced apoptosis resulting in advance disease and poor prognosis in head and neck squamous cell carcinoma [153]. Hematopoietic cancers with overexpressed Bcl-x include non-Hodgkin's [154], Hodgkin's [155], and HIV-associated [156] lymphomas, and in acute myelogenous leukemia [157] and multiple myeloma [158]. Moreover, the worse overall and disease-free survival in patients with hepatocellular carcinoma expressing Bcl-x [150] and the association of Bcl-x protein expression with histological grade of colorectal cancer [148] suggest that Bcl-x protein may have prognostic significance in at least some tumors. In follicular lymphoma, Bcl-x overexpression is associated with extranodal involvement, an International Prognostic Index (IPI) higher group risk, and shorter overall survival [159].

Bcl-x is not found to be commonly expressed in CLL cells [52]. We noticed that the patients with high Bcl-xL and treatment resistance also had high Bcl-2/Bax ratio and Mcl-1, suggesting that these proteins may be involved in a redundant fashion, as is a feature of Bcl-2 family gene expression [160]. CD40 ligation increases expression of Bcl-xL in CLL cells protecting them from apoptosis in freshly isolated cells co-cultured with a follicular dendritic cell line [161] and the relative importance of Mcl-1 and Bcl-xL might be dependent upon cell-cell interactions in vivo. A study evaluating apoptosis induction by fludarabine in CLL cells demonstrated Bcl-xL (along with Bcl-2) was followed by high Bcl-x (along with Bax) in the early stages of apoptosis [162]. Though, most of CLL cells express high levels of Bcl-2 and/or Mcl-1, there are cases in which these proteins are downregulated.

We hypothesize that in those CLL cells with low Mcl-1 and Bcl-2/Bax ratio, Bcl-x anti-apoptotic serves as a substitute providing the cell protection. In this chapter we tested the hypothesis that Bcl-x protein expressed in a minority of CLL patients helps tumor cells to escape apoptosis in CLL patients with low Mcl-1 protein and low Bcl-2/Bax ratio, acting on its own, and in other cases it provides supportive redundancy to Mcl-1 and Bcl-2 proteins.

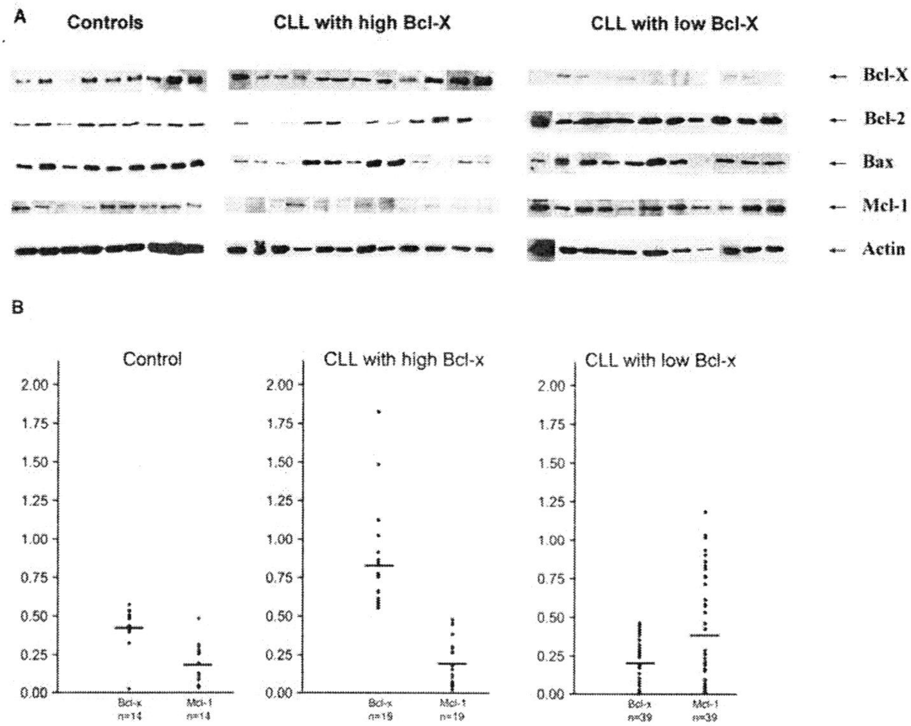

Figure 5. A: Immunoblotting data of Bcl-X, Bcl-2, Bax, Mcl-1, and Actin (the later used to normalize the data) protein levels in control group and CLL group with low Bcl-X protein levels and CLL group with high Bcl-X protein levels. Immunobloting was performed as described earlier [25, 91]), with primary antibodies against Bcl-x (mouse monoclonal, Dako), Bcl-2 (mouse monoclonal, Upstate Biotech, New York, NY), Bax (rabbit polyclonal (Pharmingen), Mcl-1 (mouse monoclonal, NeoMarkers, Fremont, CA), and β-Actin (mouse monoclonal, Sigma), and secondary antibodies conjugated to a horseradish peroxidase (goat anti-mouse IgG (for Bcl-2, Mcl-1, Bcl-x, and β-Actin), anti-rabbit (for Bax) IgG antibodies (Bio-Rad). B: Association of high Bcl-x protein levels with low Mcl-1 levels. Horizontal lines represent the mean values. Control group = age/ sex matched healthy volunteers; CLL groups with high or low Bcl-x protein levels and corresponding Mcl-1 protein levels (P=0.037, t-test).

For this, we evaluated Bcl-x, Bcl-2, Bax, Bcl-2/Bax ratio, and Mcl-1 protein levels by immunoblotting in 58 CLL patients and 13 controls (four purified B-cells and nine - peripheral blood of healthy individuals); proteins were normalized to actin, compared with clinical data using SPSS statistical software, and presented as mean values plus/ minus standard deviations. The results are shown in figure 5A. The bands were quantitated in densitometry units, relative to signals from actin, using Quantity One software and Gel Documantation system (Bio-Rad). The cut-off points were determined using curves of receiver operating characteristics (ROC) to distinguish protein levels of CLL patients from those levels of healthy individuals, as described previously [91].

There was no difference in the overall Bcl-x protein expression levels between CLL (0.41 $\pm$ 0.38) and control (0.43 $\pm$ 0.14) groups (P > 0.05) (figure 5, A and B). The same observation was made by Brajuskovic et al [163] who reported no difference in means between CLL and control cells. The expression levels of all proteins varied within a CLL group. The survival of the patients with high Bcl-x was not statistically different from those patients with low Bcl-x (P=0.54, Kaplan-Meier survival analysis); this requires further confirmation in a larger number of patients. Interestingly, among seven patients with high Bcl-x, for whom treatment response was available, six patients did not respond to treatment and one patient achieved partial response.

CLL patients were divided into two groups: with high Bcl-x (n=19; protein levels >0.5) and low-to-intermediate Bcl-x (n=39; protein levels: 0 to 0.49). Nineteen patients (32.8%) had higher (P=0.0001, t-test) Bcl-x (mean - 0.83) compared to those with low-to-moderate levels (mean – 0.21). None of the nineteen patients with high Bcl-x had high Mcl-1 levels while 24/39 patients with low Bcl-x had also low Mcl-1 (P=0.005, X^2-test) (figure 5B). Mcl-1 levels were significantly (P=0.037, t-test) lower (mean - 0.21) in the patients (n=39) with high Bcl-x compared to the Mcl-1 levels (mean – 0.39) in the patients (n=19) with Bcl-x. The reverse co-expressions were noted between Bcl-x levels on one side and Mcl-1 (P=0.005, X^2-test) or Bcl-2/ Bax ratio (P=0.015, X^2-test) on the other (figure 5). We believe that the Bcl-x may act on its own based on the fact that seven out of 19 (12.1% of all studied in this chapter CLL cases) patients with high Bcl-x levels studied in this chapter did not express high levels of other Bcl-2 family proteins. In other cases it was co-expressed with Bcl-2 (n=7), Bcl-2/ Bax (n=5), Mcl-1/ Bax (n=3), Bfl-1 (n=6), Bfl-1/ Bax (n=6).

It appears then that Bcl-x is an important cell regulator in CLL patients with low Mcl-1 and it is likely to have both pathogenic and clinical significance. The antiapoptotic role of Bcl-x is of both pathogenic and clinical significance in CLL either on its own or by providing supportive redundancy to Mcl-1 and Bcl-2. Bcl-x may act as an independent cell survival factor in a minority of patients.

As Bcl-X [110] is capable of heterodimerizing with Bax, we sought to study the Bcl-x/ Bax protein ratio in CLL. As expected, based on the study of other Bcl-2 family proteins in CLL, there was no statistically significant difference between Bcl-x/ Bax ratio levels between control (1.14 $\pm$ 0.52) and CLL (2.68 $\pm$ 5.43) groups (P=0.295, t-test), due to high variations in the later group. Although, in two out of 13 cases with high Bcl-x/ Bax, there was no other overexpressed proteins, suggesting a possible role in cell survival, at least in some case. Based on ROC analysis, Bcl-x/ Bax protein ratio was considered high when normalized levels were >2, and low ratio was considered to be between 0 and 1.99 (relative units). The reciprocity in expression levels with Mcl-1 was observed; none out of thirteen patients with high Bcl-x/ Bax ratio expressed high Mcl-1 levels (P=0.016, X^2–test) (see Mcl-1 section in this chapter).

To see whether the genetic changes are responsible for the overexpression of Bcl-x protein, we sought to sequence the BCL-X gene in five CLL patients (figure 2D, table 3). The sequences were compared to GenBank sequences (AN: D30746). No changes were found suggesting the role of other regulatory mechanisms/ or presence genetic changes in the areas not studied in this chapter. This requires a confirmation in a larger study. A limited number of publications have reported on BCL-X genetic structural changes in human disease; a) BCL-X promoter polymorphisms in multiple sclerosis [164] and b) a Ser-to-Gly mutation (codon 154) in a patient with diffuse large cell lymphoma [165]; but, there are no published data on any underlying genetic mechanisms responsible for Bcl-xL in CLL.

Among 58 CLL patients, we identified six (10.3%) cases where Bcl-X, Mcl-1, Bcl-2 proteins, and Bcl-2/Bax protein ratio were low suggesting the importance of other proteins. Based on the microarray, RT-PCR, and RPA data we sought of Bfl-1 protein as a potential regulator of cell survival in CLL.

Bfl-1

Bfl-1 is a member of Bcl-2 family and able to reduce the release of pro-apoptotic cytochrome c from mitochoindria. It is a direct target of NF-kappaB [166], Bfl-1 transcripts in CLL cells increase after stimulation with CD40 and sIgM [89]; the role of GM-CSF in the upregulation of Bfl-1 is controversial [167, 168]. Upregulation of Bfl-1 by extracellular signals such as CD40 [169], IgM [169], and phorbol esters and inflammatory cytokine TNF [111, 166, 170], has also been shown in many human cancer cell lines. Bfl-1 is overexpressed in stomach cancer [171] and in advanced stages of breast cancer [172]. Using RNAse protection assay (RPA), we have shown earlier in a smaller group of patients [96], and in this chapter, in a larger sample size, BFL-1 mRNA overexpression in CLL (figure 2). HeLa and BC-3 cell lines both lacked BFL-1 expression, while it was low in RL cell line. Recently, the reciprocity of BFL-1 and BCL-2 has been reported at RNA level, as demonstrated by microarray and RT-PCR analysis [97] and earlier by us using RNase protection assay [96]. We have demonstrated the reciprocity between gene expression of BFL-1 and BCL-2, and co-expression of BFL-1 with MCL-1, at the RNA level. The reciprocity of BFL-1 and BCL-2 found at the RNA level was lost at the protein level (this chapter). We studied Bfl-1 protein levels (mean – 0.51) in 47 CLL patients and did not find statistically significant difference from protein levels (mean – 0.51) expressed in control group (n=8), due to protein level variations in CLL group (figure 6). Interestingly, there was a trend towards significance of reciprocal expression of Bfl-1 and Bax protein levels (P=0.077, Fisher's exact test). This reciprocity reached significance once Mcl-1/ Bax ratio was considered. Higher Bfl-1 protein levels were associated with lower Mcl-1/ Bax ratio: 18 out of 19 (94.7%) cases with high Mcl-1/ Bax ratio had low Bfl-1 protein levels, while 13 out of 28 cases with low Mcl-1/ Bax had high Bfl-1 (P=0.003, Fisher's exact test). The group of patients (n=14) with high Bfl-1 levels (mean – 3.10) had lower Mcl-1/ Bax ratio (mean – 0.39), while the group (n=33) with low Bfl-1 levels (mean – 0.37) had higher Mcl-1/ Bax ratio (mean – 3.38 (P=0.0172, logistic regression analysis, figure 6), suggesting that only one of these two parameters is needed for cell survival. Moreover, three out of fourteen (5.2% of all CLL cases) patients with high Bfl-1 expressed did not have other highly expressed proteins from Bcl-2 family studied in this chapter. Three cases had also high Mcl-1, one – Bcl-2/ Bax ratio, four - Bcl-2, five – Bcl-x, one – Mcl-1/ Bax ratio.

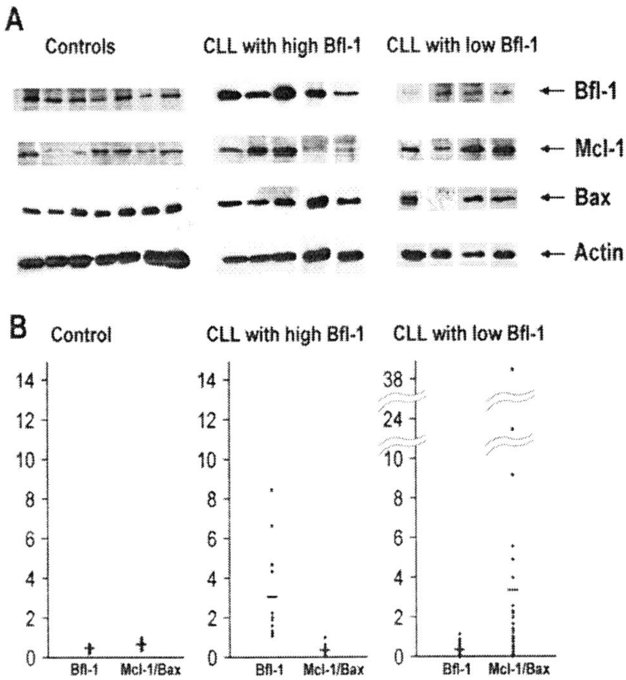

Figure 6. A: Immunoblotting data of Bfl-1, Mcl-1, Bax, and Actin (the later used to normalize the data) protein levels in control group and CLL group with low Bfl-1 protein levels and CLL group with high Bfl-1 protein levels. Immunobloting was performed as described earlier [25, 91]), with primary antibodies against Bfl-1 (A1, rabbit polyclonal, Santa-Cruz Biotechnology Inc, Santa Cruz, CA), Mcl-1 (mouse monoclonal, NeoMarkers, Fremont, CA), Bax (rabbit polyclonal (Pharmingen), β-Actin (mouse monoclonal, Sigma), and secondary antibodies conjugated to a horseradish peroxidase (goat anti-mouse IgG (for Mcl-1 and β-Actin) and anti-rabbit (for Bax and Bfl-1) IgG antibodies (Bio-Rad). B: Association of high Bfl-1 levels with low Mcl-1/ Bax ratio. Horizontal lines represent the mean values. Control group = age/ sex matched healthy volunteers; CLL groups with high or low Bfl-1 protein levels and corresponding Mcl-1/ Bax protein ratio levels (P=0.0172, logistic regression analysis).

We correlated Bfl-1 protein expression levels with treatment response. Among twenty-six patients received treatment, thirteen were non-responders (NR) and thirteen achieved some degree of response, partial (PR) or complete (CR), to standard chemotherapy (figure 7). Low Bfl-1 levels were associated with ability of CLL patients to achieve response to treatment: twelve out of thirteen patients (92.3%) who achieved some degree of response (CR + PR) had low levels of Bfl-1, while seven out of thirteen non-responders (NR) had high levels of Bfl-1 (P=0.030, Fisher's exact test) (figure 7). Average expression of Bfl-1 was higher in non-responders group (mean – 2.19), compared to patients who achieved response (mean – 0.59) (P=0.043, t-test). These data are in agreement with [97], who reported the association of BFL-1 mRNA levels with progressive disease and resistance to apoptosis in CLL. Moreover, using microarray analysis of 96 apoptosis regulatory genes, they showed that BFL-1 was the strongest discriminating gene between sensitive and resistant to treatment CLL groups.

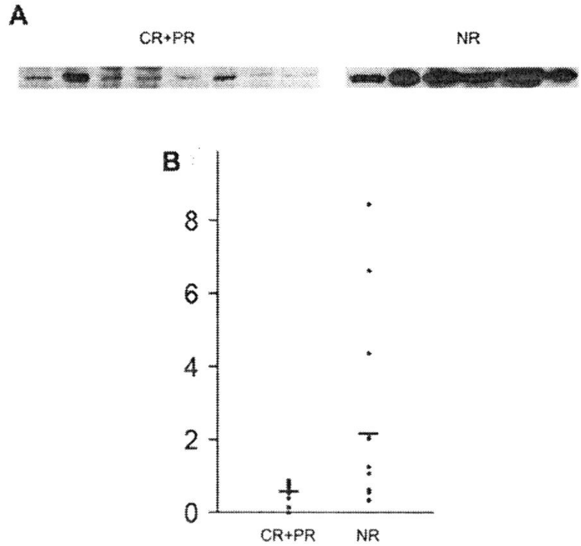

Figure 7. Association of Bfl-1 protein relative levels with treatment response in chronic lymphocytic leukemia patients. A: Immunobloting results; B: Logistic regression analysis of Bfl-1 protein levels in two groups of patients, resistant to treatment (NR = no response) and achieved some kind of response to standard treatment – complete or partial response (CR+PR) (P=0.030, Fisher's exact test).

BFL-1 gene is a direct transcription target of NF-kappa B [89, 111, 169, 173], its promoter consists of sites for NF-kappaB-dependant induction [166]. Moreover, the mutated glycine-rich N-terminal region of the Bfl-1 might be responsible for extended survival of cells in the absence of cell proliferation [174]. Bfl-1 interacts with Bid [175], Bad [176], and Bax [177], but is not able to interact with Bcl-2 [177] and Bcl-xL [177]. Based on this information, we hypothesize that structural changes in the Bfl-1 are present either within the glycine-rich N-terminal region or BH domains or in the promoter binding sites for NF-kappaB. Although mutational analysis of Bfl-1 provides a basis for the structure-function relationships of Bfl-1 gene segments [174], there are no published data on structural changes in the BFL-1 gene in CLL. Therefore, we sequenced the promoter and part of the ORF of the BFL-1 gene (figure 2E, table 3), and compared the sequences to ones available in GenBank (AN: AF479683, DQ081729, and NM_004049). One SNP was found, it was G-to-A nucleotide replacement changing an amino acid sequence at position 19 (Cys19Tyr). Fourteen out of 25 (56%) CLL patients had one allele replaced (G/A), all others were homozygous for allele G. We did not find an association of this SNP with Bfl-1 protein expression. Among patients with G/G haplotype, six had low and five expressed high Bfl-1 protein levels, while three were with low and eleven had high levels of expression in the G/A haplotype group (P=0.115, Fisher's exact test). The frequency of allele A in CLL patients was 0.28, and this is comparable to reported frequencies in normal populations in: East Asia – 0.50, West Africa – 0.22, and in Europe – 0.28 [178]. Moreover, the Cys19Tyr change is predicted to be benign (by PolyPhen prediction of functional effect of human SNPs program; http://genetics.bwh.harvard.edu/pph/index.html). Based on the above, it is unlikely this SNP is a cause of high Bfl-1 expression in CLL.

Other Apoptosis Regulatory Proteins

The other proteins/ genes form Bcl-2 family and from other apoptosis regulatory families that are not described in this chapter may contribute to the pathogenesis in CLL. By microarray analysis and confirmation by RT-PCR we found other genes with altered gene expression, the role of which has not yet been completely defined in CLL. These are shown in figure 8. Compared to a control, the following genes were found to be overexpressed: BCL-W (1.5 times), BLK (3 times), MDM2 (1.9 times), Chk1 (1.5 times), IAP1 (6.2 times) and BCL-10 (5.4 times). IAP1 and other IAP family genes have been reported to be overexpressed in CLL and postulated to be upregulated by tumor necrosis factor receptor-associated factor 1 (TRAF1) [74]. Suprisingly, when we analyzed the microarray data, TRAF members did not differ significantly from the control, and, thus, six times upregulation of IAP1 is unlikely to be explained by TRAF family upregulation in all cases. We were surprised to see overexpreesion of BCL-10 in CLL. BCL-10 is usually overexpressed in lymphomas [179] and no overexpression has been reported in CLL. Ours is the first report on overexpression of BCL-10 gene in CLL. The expression levels of the BCL-W gene were found to be altered in CLL samples confirmed by microarray (figure 1), RT-PCR (figure 8), and RPA (figure 2) assays. CLL cells expressed 2-4 times higher BCL-W compared to cord blood samples and it was found to be co-expressed with BCL-2 ($r = +0.814$). Higher BLK gene expression was confirmed by microarray (figure 1) and RT-PCR (figure 8). The results described in this chapter are in agreement with our previous report [96], [98]. No reports on Blk protein expression are available in the literature.

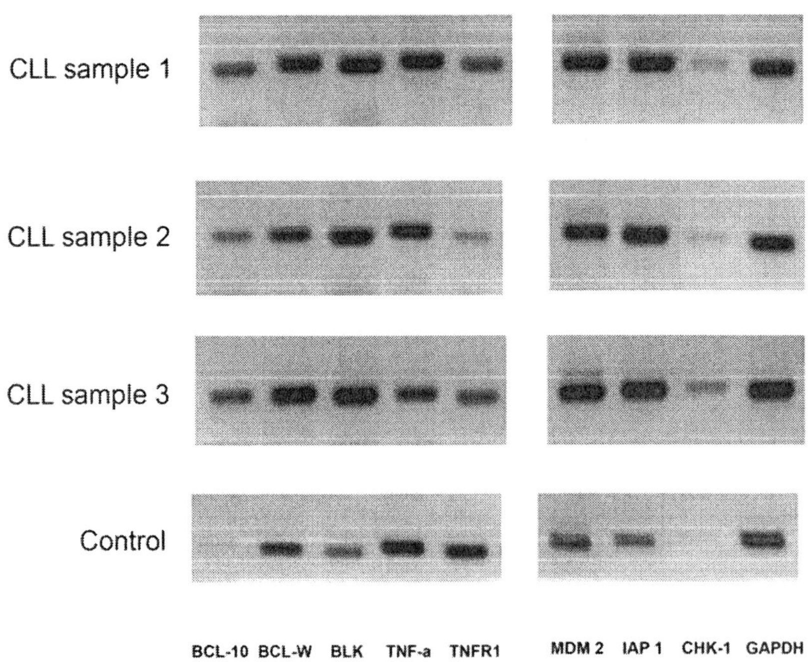

Figure 8. RT-PCR (confirmation of microarray data) of other genes not described in this chapter but showing altered levels of gene expression. RT-PCR was performed with the profiling kit PCR (MultiGene-12 RT-PCR profiling kits, Affymetrix, as per manufacturers' instructions). GAPDH was used to normalize the data for each gene.

Conclusion

CLL is the commonest leukemia in the Western world and usually affects adults and elderly. With the increasing geriatric population, the health care costs of managing patients with this incurable malignancy are likely to increase. Defective apoptosis is a fundamental attribute of CLL. Bcl-2 family proteins play a role in the pathogenesis and clinical behavior of CLL patients. Identification of key proteins that are of biological and clinical significance and delineating mechanisms underlying their altered expression is a necessary first step in order to exploit these genes/ proteins for developing specific target-directed treatments for this incurable malignancy. Further, polymorphisms / mutations linked with a particular clinical outcome would improve risk stratification and help identify patients for appropriate management strategies aimed at could optimizing patient care.

Acknowledgements

This work has been supported by a grant from Canadian Institutes of Health Research and Saskatchewan Health Research Foundation, Regional Partnership Program. We would also like to thank the following individuals: Dr. V. Misra and Dr. I. Moshynskyy for help with luciferase reporter assays; Dr. L. Qualtiere for long-term maintaining of cell lines for this project, Dr. J.F. DeCoteau and Dr. S. Scott for help with RNase protection assay, Dr. K. Bonham for help with electrophoretic mobility shift assays, Mr. S. Viswanathan and Ms. H. Neufeld for assisting with immunobloting experiments, and Mr. T. Reichert and Ms. M. Hesson for technical assistance with figures preparation.

References

[1] Boumsell L, Bernard A, Lepage V, Degos L, Lemerle J, Dausset J. Some chronic lymphocytic leukemia cells bearing surface immunoglobulins share determinants with T cells. *Eur. J. Immunol.* 1978;8:900-4.

[2] Chiorazzi N, Ferrarini M. B cell chronic lymphocytic leukemia: lessons learned from studies of the B cell antigen receptor. *Annu. Rev. Immunol.* 2003;21:841-94.

[3] Matutes E, Polliack A. Morphological and immunophenotypic features of chronic lymphocytic leukemia. *Rev. Clin. Exp. Hematol.* 2000;4:22-47.

[4] Dighiero G, Binet JL. When and how to treat chronic lymphocytic leukemia. *N. Engl. J. Med.* 2000;343:1799-801.

[5] Molica S, Levato D. What is changing in the natural history of chronic lymphocytic leukemia? *Haematologica.* 2001;86:8-12.

[6] Molica S. Progression and survival studies in early chronic lymphocytic leukemia. *Blood.* 1991;78:895-9.

[7] Guipaud O, Deriano L, Salin H, Vallat L, Sabatier L, Merle-Beral H, Delic J. B-cell chronic lymphocytic leukaemia: a polymorphic family unified by genomic features. *Lancet Oncol.* 2003;4:505-14.

[8] Giles FJ, O'Brien SM, Keating MJ. Chronic lymphocytic leukemia in (Richter's) transformation. *Semin. Oncol.* 1998;25:117-25.

[9] Hercher C, Robain M, Davi F, *et al.* A multicentric study of 41 cases of B-prolymphocytic leukemia: two evolutive forms. *Leuk. Lymphoma.* 2001;42:981-7.

[10] Caligaris-Cappio F, Hamblin TJ. B-cell chronic lymphocytic leukemia: a bird of a different feather. *J. Clin. Oncol.* 1999;17:399-408.

[11] Hamblin TJ, Davis Z, Gardiner A, Oscier DG, Stevenson FK. Unmutated Ig V(H) genes are associated with a more aggressive form of chronic lymphocytic leukemia. *Blood.* 1999;94:1848-54.

[12] Damle RN, Wasil T, Fais F, *et al.* Ig V gene mutation status and CD38 expression as novel prognostic indicators in chronic lymphocytic leukemia. *Blood.* 1999;94:1840-7.

[13] Rosenwald A, Alizadeh AA, Widhopf G, *et al.* Relation of gene expression phenotype to immunoglobulin mutation genotype in B cell chronic lymphocytic leukemia. *J. Exp. Med.* 2001;194:1639-47.

[14] Krober A, Seiler T, Benner A, *et al.* V(H) mutation status, CD38 expression level, genomic aberrations, and survival in chronic lymphocytic leukemia. *Blood.* 2002;100:1410-6.

[15] Oscier DG, Gardiner AC, Mould SJ, *et al.* Multivariate analysis of prognostic factors in CLL: clinical stage, IGVH gene mutational status, and loss or mutation of the p53 gene are independent prognostic factors. *Blood.* 2002;100:1177-84.

[16] Kipps TJ. Chronic lymphocytic leukemia. *Curr. Opin. Hematol.* 1997;4:268-76.

[17] Oscier D, Fegan C, Hillmen P, *et al.* Guidelines on the diagnosis and management of chronic lymphocytic leukaemia. *Br. J. Haematol.* 2004;125:294-317.

[18] Zupo S, Isnardi L, Megna M, *et al.* CD38 expression distinguishes two groups of B-cell chronic lymphocytic leukemias with different responses to anti-IgM antibodies and propensity to apoptosis. *Blood.* 1996;88:1365-74.

[19] Domingo-Domenech E, Domingo-Claros A, Gonzalez-Barca E, *et al.* CD38 expression in B-chronic lymphocytic leukemia: association with clinical presentation and outcome in 155 patients. *Haematologica.* 2002;87:1021-7.

[20] Ghia P, Guida G, Stella S, *et al.* The pattern of CD38 expression defines a distinct subset of chronic lymphocytic leukemia (CLL) patients at risk of disease progression. *Blood.* 2003;101:1262-9.

[21] Lin K, Sherrington PD, Dennis M, Matrai Z, Cawley JC, Pettitt AR. Relationship between p53 dysfunction, CD38 expression, and IgV(H) mutation in chronic lymphocytic leukemia. *Blood.* 2002;100:1404-9.

[22] Orchard JA, Ibbotson RE, Davis Z, *et al.* ZAP-70 expression and prognosis in chronic lymphocytic leukaemia. *Lancet.* 2004;363:105-11.

[23] Saxena A, Moshynska O, Sankaran K, Viswanathan S, Sheridan DP. Association of a novel single nucleotide polymorphism, G(-248)A, in the 5'-UTR of BAX gene in chronic lymphocytic leukemia with disease progression and treatment resistance. *Cancer Lett.* 2002;187:199-205.

[24] Moshynska O, Sankaran K, Saxena A. Molecular detection of the G(-248)A BAX promoter nucleotide change in B cell chronic lymphocytic leukaemia. *Mol. Pathol.* 2003;56:205-9.

[25] Moshynska O, Sankaran K, Pahwa P, Saxena A. Prognostic significance of a short sequence insertion in the MCL-1 promoter in chronic lymphocytic leukemia. *J. Natl. Cancer Inst.* 2004;96:673-82.

[26] Hamblin TJ, Orchard JA, Ibbotson RE, Davis Z, Thomas PW, Stevenson FK, Oscier DG. CD38 expression and immunoglobulin variable region mutations are independent prognostic variables in chronic lymphocytic leukemia, but CD38 expression may vary during the course of the disease. *Blood.* 2002;99:1023-9.

[27] Crespo M, Bosch F, Villamor N, *et al.* ZAP-70 expression as a surrogate for immunoglobulin-variable-region mutations in chronic lymphocytic leukemia. *N. Engl. J. Med.* 2003;348:1764-75.

[28] Wiestner A, Rosenwald A, Barry TS, *et al.* ZAP-70 expression identifies a chronic lymphocytic leukemia subtype with unmutated immunoglobulin genes, inferior clinical outcome, and distinct gene expression profile. *Blood.* 2003;101:4944-51.

[29] Tobin G, Thunberg U, Johnson A, *et al.* Somatically mutated Ig V(H)3-21 genes characterize a new subset of chronic lymphocytic leukemia. *Blood.* 2002;99:2262-4.

[30] Dohner H, Fischer K, Bentz M, *et al.* p53 gene deletion predicts for poor survival and non-response to therapy with purine analogs in chronic B-cell leukemias. *Blood.* 1995;85:1580-9.

[31] Byrd JC, Smith L, Hackbarth ML, Flinn IW, Young D, Proffitt JH, Heerema NA. Interphase cytogenetic abnormalities in chronic lymphocytic leukemia may predict response to rituximab. *Cancer Res.* 2003;63:36-8.

[32] Korz C, Pscherer A, Benner A, *et al.* Evidence for distinct pathomechanisms in B-cell chronic lymphocytic leukemia and mantle cell lymphoma by quantitative expression analysis of cell cycle and apoptosis-associated genes. *Blood.* 2002;99:4554-61.

[33] Alizadeh AA, Staudt LM. Genomic-scale gene expression profiling of normal and malignant immune cells. *Curr. Opin. Immunol.* 2000;12:219-25.

[34] Klein U, Tu Y, Stolovitzky GA, *et al.* Gene expression profiling of B cell chronic lymphocytic leukemia reveals a homogeneous phenotype related to memory B cells. *J. Exp. Med.* 2001;194:1625-38.

[35] Wang J, Coombes KR, Highsmith WE, Keating MJ, Abruzzo LV. Differences in gene expression between B-cell chronic lymphocytic leukemia and normal B cells: a meta-analysis of three microarray studies. *Bioinformatics.* 2004.

[36] Staudt LM. Gene expression profiling of lymphoid malignancies. *Annu. Rev. Med.* 2002;53:303-18.

[37] Ibrahim S, Keating M, Do KA, *et al.* CD38 expression as an important prognostic factor in B-cell chronic lymphocytic leukemia. *Blood.* 2001;98:181-6.

[38] Del Poeta G, Maurillo L, Venditti A, *et al.* Clinical significance of CD38 expression in chronic lymphocytic leukemia. *Blood.* 2001;98:2633-9.

[39] Aalto Y, El-Rifa W, Vilpo L, *et al.* Distinct gene expression profiling in chronic lymphocytic leukemia with 11q23 deletion. *Leukemia.* 2001;15:1721-8.

[40] Blaise R, Alapetite C, Masdehors P, Merle-Beral H, Roulin C, Delic J, Sabatier L. High levels of chromosome aberrations correlate with impaired in vitro radiation-induced apoptosis and DNA repair in human B-chronic lymphocytic leukaemia cells. *Int. J. Radiat. Biol.* 2002;78:671-9.

[41] Vallat L, Magdelenat H, Merle-Beral H, *et al.* The resistance of B-CLL cells to DNA damage-induced apoptosis defined by DNA microarrays. *Blood.* 2003;101:4598-606.

[42] Faguet GB. Chronic lymphocytic leukemia: an updated review. *J. Clin. Oncol.* 1994;12:1974-90.

[43] Cheson BD, Bennett JM, Grever M, Kay N, Keating MJ, O'Brien S, Rai KR. National Cancer Institute-sponsored Working Group guidelines for chronic lymphocytic leukemia: revised guidelines for diagnosis and treatment. *Blood.* 1996;87:4990-7.

[44] Keating MJ, O'Brien S, Albitar M. Emerging information on the use of rituximab in chronic lymphocytic leukemia. *Semin. Oncol.* 2002;29:70-4.

[45] Keating MJ, Flinn I, Jain V, *et al.* Therapeutic role of alemtuzumab (Campath-1H) in patients who have failed fludarabine: results of a large international study. *Blood.* 2002;99:3554-61.

[46] Kipps TJ. Advances in classification and therapy of indolent B-cell malignancies. *Semin. Oncol.* 2002;29:98-104.

[47] Keating MJ. Progress in CLL, chemotherapy, antibodies and transplantation. *Biomed. Pharmacother.* 2001;55:524-8.

[48] Keating MJ, Chiorazzi N, Messmer B, *et al.* Biology and treatment of chronic lymphocytic leukemia. *Hematology (Am. Soc. Hematol. Educ. Program)* 2003:153-75.

[49] Kolb JP, Kern C, Quiney C, Roman V, Billard C. Re-establishment of a normal apoptotic process as a therapeutic approach in B-CLL. *Curr. Drug Targets Cardiovasc. Haematol. Disord.* 2003;3:261-86.

[50] Panasci L, Paiement JP, Christodoulopoulos G, Belenkov A, Malapetsa A, Aloyz R. Chlorambucil drug resistance in chronic lymphocytic leukemia: the emerging role of DNA repair. *Clin. Cancer Res.* 2001;7:454-61.

[51] Caligaris-Cappio F. B-chronic lymphocytic leukemia: a malignancy of anti-self B cells. *Blood.* 1996;87:2615-20.

[52] Kitada S, Andersen J, Akar S, *et al.* Expression of apoptosis-regulating proteins in chronic lymphocytic leukemia: correlations with In vitro and In vivo chemoresponses. *Blood.* 1998;91:3379-89.

[53] Binet JL, Plunkett W, Robertson B, Merle-Beral H, Mentz F, Hoffbrand AV, Panayiotidis P. What does apoptosis mean in CLL? *Leuk. Lymphoma.* 1996;22 Suppl 2:47-52.

[54] Aguilar-Santelises M, Rottenberg ME, Lewin N, Mellstedt H, Jondal M. Bcl-2, Bax and p53 expression in B-CLL in relation to in vitro survival and clinical progression. *Int. J. Cancer.* 1996;69:114-9.

[55] Bannerji R, Byrd JC. Update on the biology of chronic lymphocytic leukemia. *Curr. Opin. Oncol.* 2000;12:22-9.

[56] Osorio LM, Aguilar-Santelises M. Apoptosis in B-chronic lymphocytic leukaemia. *Med. Oncol.* 1998;15:234-40.

[57] Golstein P. Cell death: TRAIL and its receptors. *Curr. Biol.* 1997;7:R750-3.

[58] Green DR, Reed JC. Mitochondria and apoptosis. *Science.* 1998;281:1309-12.

[59] Korsmeyer SJ, Wei MC, Saito M, Weiler S, Oh KJ, Schlesinger PH. Pro-apoptotic cascade activates BID, which oligomerizes BAK or BAX into pores that result in the release of cytochrome c. *Cell Death Differ.* 2000;7:1166-73.

[60] Jaattela M. Programmed cell death: many ways for cells to die decently. *Ann. Med.* 2002;34:480-8.

[61] Reed JC. Mechanisms of apoptosis. *Am. J. Pathol.* 2000;157:1415-30.

[62] Hu W KJ. Anticancer therapy targeting apoptotic pathways. *Lancet Oncol.* 2003;4:721-9.

[63] Reed J. Apoptosis-based therapies. *Nature Reviews.* 2002;1:111-21.

[64] Yang YL LX. The IAP family: endogenous caspase inhibitors with multiple biological activities. *Cell Research.* 2000;10:169-77.

[65] Deveraux QL, Stennicke HR, Salvesen GS, Reed JC. Endogenous inhibitors of caspases. *J. Clin. Immunol.* 1999;19:388-98.

[66] Nachmias B, Ashhab Y, Ben-Yehuda D. The inhibitor of apoptosis protein family (IAPs): an emerging therapeutic target in cancer. *Semin. Cancer Biol.* 2004;14:231-43.

[67] Sanna MG, da Silva Correia J, Ducrey O, *et al.* IAP suppression of apoptosis involves distinct mechanisms: the TAK1/JNK1 signaling cascade and caspase inhibition. *Mol. Cell Biol.* 2002;22:1754-66.

[68] Suzuki Y, Nakabayashi Y, Takahashi R. Ubiquitin-protein ligase activity of X-linked inhibitor of apoptosis protein promotes proteasomal degradation of caspase-3 and enhances its anti-apoptotic effect in Fas-induced cell death. *Proc. Natl. Acad. Sci. U S A.* 2001;98:8662-7.

[69] Liston P, Fong WG, Korneluk RG. The inhibitors of apoptosis: there is more to life than Bcl2. *Oncogene.* 2003;22:8568-80.

[70] Kitada S, Zapata JM, Andreeff M, Reed JC. Protein kinase inhibitors flavopiridol and 7-hydroxy-staurosporine down-regulate antiapoptosis proteins in B-cell chronic lymphocytic leukemia. *Blood.* 2000;96:393-7.

[71] Nakagawa Y, Yamaguchi S, Hasegawa M, *et al.* Differential expression of survivin in bone marrow cells from patients with acute lymphocytic leukemia and chronic lymphocytic leukemia. *Leuk. Res.* 2004;28:487-94.

[72] Granziero L, Ghia P, Circosta P, *et al.* Survivin is expressed on CD40 stimulation and interfaces proliferation and apoptosis in B-cell chronic lymphocytic leukemia. *Blood.* 2001;97:2777-83.

[73] Schliep S, Decker T, Schneller F, Wagner H, Hacker G. Functional evaluation of the role of inhibitor of apoptosis proteins in chronic lymphocytic leukemia. *Exp. Hematol.* 2004;32:556-62.

[74] Munzert G, Kirchner D, Stobbe H, Bergmann L, Schmid RM, Dohner H, Heimpel H. Tumor necrosis factor receptor-associated factor 1 gene overexpression in B-cell chronic lymphocytic leukemia: analysis of NF-kappa B/Rel-regulated inhibitors of apoptosis. *Blood.* 2002;100:3749-56.

[75] Wall NR, Beck FW, Al-Katib AM, Mohammad RM. Treatment-induced expression of anti-apoptotic proteins in WSU-CLL, a human chronic lymphocytic leukemia cell line. *J. Drug Target.* 2001;9:329-39.

[76] Drexler HG, Quentmeier H, Dirks WG, Uphoff CC, MacLeod RA. DNA profiling and cytogenetic analysis of cell line WSU-CLL reveal cross-contamination with cell line REH (pre B-ALL). *Leukemia.* 2002;16:1868-70.

[77] Holcik M, Gibson H, Korneluk RG. XIAP: apoptotic brake and promising therapeutic target. *Apoptosis.* 2001;6:253-61.
[78] Verhagen AM, Silke J, Ekert PG, *et al.* HtrA2 promotes cell death through its serine protease activity and its ability to antagonize inhibitor of apoptosis proteins. *J. Biol. Chem.* 2002;277:445-54.
[79] Clem RJ, Sheu TT, Richter BW, He WW, Thornberry NA, Duckett CS, Hardwick JM. c-IAP1 is cleaved by caspases to produce a proapoptotic C-terminal fragment. *J. Biol. Chem.* 2001;276:7602-8.
[80] Dan HC, Sun M, Kaneko S, *et al.* Akt phosphorylation and stabilization of X-linked inhibitor of apoptosis protein (XIAP). *J. Biol. Chem.* 2004;279:5405-12.
[81] Holcik M, Sonenberg N, Korneluk RG. Internal ribosome initiation of translation and the control of cell death. *Trends Genet.* 2000;16:469-73.
[82] Cory S, Adams JM. Matters of life and death: programmed cell death at Cold Spring Harbor. *Biochim. Biophys. Acta.* 1998;1377:R25-44.
[83] Adams JM, Cory S. The Bcl-2 protein family: arbiters of cell survival. *Science.* 1998;281:1322-6.
[84] Gross A, McDonnell JM, Korsmeyer SJ. BCL-2 family members and the mitochondria in apoptosis. *Genes Dev.* 1999;13:1899-911.
[85] Zamzami N, Brenner C, Marzo I, Susin SA, Kroemer G. Subcellular and submitochondrial mode of action of Bcl-2-like oncoproteins. *Oncogene.* 1998;16:2265-82.
[86] Stankovic T, Stewart GS, Byrd P, Fegan C, Moss PA, Taylor AM. ATM mutations in sporadic lymphoid tumours. *Leuk. Lymphoma.* 2002;43:1563-71.
[87] Zent CS, Kay NE. Advances in the understanding of biology and prognosis in chronic lymphocytic leukemia. *Curr. Oncol. Rep.* 2004;6:348-54.
[88] Christodoulopoulos G, Fotouhi N, Krajewski S, Reed JC, Alaoui-Jamali M, Panasci L. Relationship between nitrogen mustard drug resistance in B-cell chronic lymphocytic leukemia (B-CLL) and protein expression of Bcl-2, Bax, Bcl-X and p53. *Cancer Lett.* 1997;121:59-67.
[89] Bernal A, Pastore RD, Asgary Z, Keller SA, Cesarman E, Liou HC, Schattner EJ. Survival of leukemic B cells promoted by engagement of the antigen receptor. *Blood.* 2001;98:3050-7.
[90] Gottardi D, Alfarano A, De Leo AM, *et al.* In leukaemic CD5+ B cells the expression of BCL-2 gene family is shifted toward protection from apoptosis. *Br. J. Haematol.* 1996;94:612-8.
[91] Saxena A, Viswanathan S, Moshynska O, Tandon P, Sankaran K, Sheridan DP. Mcl-1 and Bcl-2/Bax ratio are associated with treatment response but not with Rai stage in B-cell chronic lymphocytic leukemia. *Am. J. Hematol.* 2004;75:22-33.
[92] Pepper C, Hoy T, Bentley DP. Bcl-2/Bax ratios in chronic lymphocytic leukaemia and their correlation with in vitro apoptosis and clinical resistance. *Br. J. Cancer.* 1997;76:935-8.
[93] Thomas A, El Rouby S, Reed JC, Krajewski S, Silber R, Potmesil M, Newcomb EW. Drug-induced apoptosis in B-cell chronic lymphocytic leukemia: relationship between

p53 gene mutation and bcl-2/bax proteins in drug resistance. *Oncogene.* 1996;12:1055-62.

[94] Pepper C, Bentley P, Hoy T. Regulation of clinical chemoresistance by bcl-2 and bax oncoproteins in B-cell chronic lymphocytic leukaemia. *Br. J. Haematol.* 1996;95:513-7.

[95] Pepper C, Hoy T, Bentley P. Elevated Bcl-2/Bax are a consistent feature of apoptosis resistance in B-cell chronic lymphocytic leukaemia and are correlated with in vivo chemoresistance. *Leuk. Lymphoma.* 1998;28:355-61.

[96] Moshynska O, Saxena, A. Bfl-1 and bcl-w, along with bcl-2/bax ratio and mcl-1 may determine the longevity of cells in B-cell chronic lymphocytic leukemia (CLL). *Histopathology.* 2002;41:15 (Ab. 0.03.309.

[97] Morales AA, Olsson A, Celsing F, Osterborg A, Jondal M, Osorio LM. High expression of bfl-1 contributes to the apoptosis resistant phenotype in B-cell chronic lymphocytic leukemia. *Int. J. Cancer.* 2005;113:730-7.

[98] Sanz L, Garcia-Marco JA, Casanova B, de La Fuente MT, Garcia-Gila M, Garcia-Pardo A, Silva A. Bcl-2 family gene modulation during spontaneous apoptosis of B-chronic lymphocytic leukemia cells. *Biochem. Biophys. Res. Commun.* 2004;315:562-7.

[99] Robertson LE, Plunkett W, McConnell K, Keating MJ, McDonnell TJ. Bcl-2 expression in chronic lymphocytic leukemia and its correlation with the induction of apoptosis and clinical outcome. *Leukemia.* 1996;10:456-9.

[100] Schena M, Larsson LG, Gottardi D, Gaidano G, Carlsson M, Nilsson K, Caligaris-Cappio F. Growth- and differentiation-associated expression of bcl-2 in B-chronic lymphocytic leukemia cells. *Blood.* 1992;79:2981-9.

[101] Caligaris-Cappio F, Ghia P, Gottardi D, Parvis G, Gregoretti MG, Nilsson K, Schena M. BCL-2 in B-chronic lymphocytic leukemia. *Curr. Top Microbiol. Immunol.* 1992;182:279-86.

[102] Hanada M, Delia D, Aiello A, Stadtmauer E, Reed JC. bcl-2 gene hypomethylation and high-level expression in B-cell chronic lymphocytic leukemia. *Blood.* 1993;82:1820-8.

[103] Raghoebier S, van Krieken JH, Kluin-Nelemans JC, *et al.* Oncogene rearrangements in chronic B-cell leukemia. *Blood.* 1991;77:1560-4.

[104] Moshynska O, Saxena, A. MCL-1 and BCL-2 polymorphisms/ mutations in B-cell lymphoproliferative disorders. *FASEB J.* 2002;16:A1199, Ab185.13.

[105] Pepper C, Thomas A, Hoy T, Cotter F, Bentley P. Antisense-mediated suppression of Bcl-2 highlights its pivotal role in failed apoptosis in B-cell chronic lymphocytic leukaemia. *Br. J. Haematol.* 1999;107:611-5.

[106] Pepper C, Thomas A, Hoy T, Bentley P. Chlorambucil resistance in B-cell chronic lymphocytic leukaemia is mediated through failed Bax induction and selection of high Bcl-2-expressing subclones. *Br. J. Haematol.* 1999;104:581-8.

[107] Johnston JB, Daeninck P, Verburg L, *et al.* P53, MDM-2, BAX and BCL-2 and drug resistance in chronic lymphocytic leukemia. *Leuk. Lymphoma.* 1997;26:435-49.

[108] Sato T, Hanada M, Bodrug S, *et al.* Interactions among members of the Bcl-2 protein family analyzed with a yeast two-hybrid system. *Proc. Natl. Acad. Sci. U S A.* 1994;91:9238-42.

[109] Bodrug S, Aime-Sempe, C, Sato, T, Krajewski, S, Hanada, M, Reed, JC. Biochemical and functional comparisons of Mcl-1 and Bcl-2 proteins: evidence for a novel mechanism of regulating Bcl-2 family protein function. *Cell Death Differ.* 1995;2:173-82.

[110] Jeong SY, Gaume B, Lee YJ, Hsu YT, Ryu SW, Yoon SH, Youle RJ. Bcl-x(L) sequesters its C-terminal membrane anchor in soluble, cytosolic homodimers. *Embo J.* 2004;23:2146-55.

[111] Cheng Q, Lee HH, Li Y, Parks TP, Cheng G. Upregulation of Bcl-x and Bfl-1 as a potential mechanism of chemoresistance, which can be overcome by NF-kappaB inhibition. *Oncogene.* 2000;19:4936-40.

[112] Korsmeyer SJ, Shutter JR, Veis DJ, Merry DE, Oltvai ZN. Bcl-2/Bax: a rheostat that regulates an anti-oxidant pathway and cell death. *Semin. Cancer Biol.* 1993;4:327-32.

[113] Starczynski J, Pepper C, Pratt G, *et al.* Common polymorphism G(-248)A in the promoter region of the bax gene results in significantly shorter survival in patients with chronic lymphocytic Leukemia once treatment is initiated. *J. Clin. Oncol.* 2005;23:1514-21.

[114] Skogsberg S, Tobin G, Krober A, *et al.* The G(-248)A polymorphism in the promoter region of the Bax gene does not correlate with prognostic markers or overall survival in chronic lymphocytic leukemia. *Leukemia.* 2006;20:77-81.

[115] Moshynska O, Moshynskyy I, Misra V, Saxena A. G125A single-nucleotide polymorphism in the human BAX promoter affects gene expression. *Oncogene.* 2005;24:2042-9.

[116] Moshynska O, Sheridan, DP, Saxena A. Novel BAX mutations in chronic lymphocytic leukemia (B-CLL). *FASEB J.* 2001;15:A589, Ab473.22.

[117] Wendel B, Hoehe MR. The human mu opioid receptor gene: 5' regulatory and intronic sequences. *J. Mol. Med.* 1998;76:525-32.

[118] Kozopas KM, Yang T, Buchan HL, Zhou P, Craig RW. MCL1, a gene expressed in programmed myeloid cell differentiation, has sequence similarity to BCL2. *Proc. Natl. Acad. Sci. U S A.* 1993;90:3516-20.

[119] Zhou P, Qian L, Kozopas KM, Craig RW. Mcl-1, a Bcl-2 family member, delays the death of hematopoietic cells under a variety of apoptosis-inducing conditions. *Blood.* 1997;89:630-43.

[120] Zhou P, Qian L, Bieszczad CK, Noelle R, Binder M, Levy NB, Craig RW. Mcl-1 in transgenic mice promotes survival in a spectrum of hematopoietic cell types and immortalization in the myeloid lineage. *Blood.* 1998;92:3226-39.

[121] Chao JR, Wang JM, Lee SF, *et al.* mcl-1 is an immediate-early gene activated by the granulocyte-macrophage colony-stimulating factor (GM-CSF) signaling pathway and is one component of the GM-CSF viability response. *Mol. Cell Biol.* 1998;18:4883-98.

[122] Bingle CD, Craig RW, Swales BM, Singleton V, Zhou P, Whyte MK. Exon skipping in Mcl-1 results in a bcl-2 homology domain 3 only gene product that promotes cell death. *J. Biol. Chem.* 2000;275:22136-46.

[123] Craig RW. MCL1 provides a window on the role of the BCL2 family in cell proliferation, differentiation and tumorigenesis. *Leukemia.* 2002;16:444-54.

[124] Lomo J, Smeland EB, Krajewski S, Reed JC, Blomhoff HK. Expression of the Bcl-2 homologue Mcl-1 correlates with survival of peripheral blood B lymphocytes. *Cancer Res.* 1996;56:40-3.

[125] Klampfer L, Zhang J, Nimer SD. GM-CSF rescues TF-1 cells from growth factor withdrawal-induced, but not differentiation-induced apoptosis: the role of BCL-2 and MCL-1. *Cytokine.* 1999;11:849-55.

[126] Moulding DA, Quayle JA, Hart CA, Edwards SW. Mcl-1 expression in human neutrophils: regulation by cytokines and correlation with cell survival. *Blood.* 1998;92:2495-502.

[127] Corcione A, Pistoia V. B-cell-derived granulocyte-colony stimulating factor (G-CSF). *Methods.* 1997;11:143-7.

[128] Akgul C, Turner PC, White MR, Edwards SW. Functional analysis of the human MCL-1 gene. *Cell Mol. Life Sci.* 2000;57:684-91.

[129] Heinemeyer T, Wingender E, Reuter I, *et al.* Databases on transcriptional regulation: TRANSFAC, TRRD and COMPEL. *Nucleic Acids Res.* 1998;26:362-7.

[130] Cross SH, Bird AP. CpG islands and genes. *Curr. Opin. Genet. Dev.* 1995;5:309-14.

[131] Segal R, Berk AJ. Promoter activity and distance constraints of one versus two Sp1 binding sites. *J. Biol. Chem.* 1991;266:20406-11.

[132] Segal JA, Barnett JL, Crawford DL. Functional analyses of natural variation in Sp1 binding sites of a TATA-less promoter. *J. Mol. Evol.* 1999;49:736-49.

[133] Boise LH, Gonzalez-Garcia M, Postema CE, *et al.* bcl-x, a bcl-2-related gene that functions as a dominant regulator of apoptotic cell death. *Cell.* 1993;74:597-608.

[134] Poruchynsky MS, Wang EE, Rudin CM, Blagosklonny MV, Fojo T. Bcl-xL is phosphorylated in malignant cells following microtubule disruption. *Cancer Res.* 1998;58:3331-8.

[135] Zarnegar B, He JQ, Oganesyan G, Hoffmann A, Baltimore D, Cheng G. Unique CD40-mediated biological program in B cell activation requires both type 1 and type 2 NF-kappaB activation pathways. *Proc. Natl. Acad. Sci. U S A.* 2004;101:8108-13.

[136] Grethe S, Ares MP, Andersson T, Porn-Ares MI. p38 MAPK mediates TNF-induced apoptosis in endothelial cells via phosphorylation and downregulation of Bcl-x(L). *Exp. Cell Res.* 2004;298:632-42.

[137] Mihara M, Erster S, Zaika A, Petrenko O, Chittenden T, Pancoska P, Moll UM. p53 has a direct apoptogenic role at the mitochondria. *Mol. Cell.* 2003;11:577-90.

[138] Petros AM, Gunasekera A, Xu N, Olejniczak ET, Fesik SW. Defining the p53 DNA-binding domain/Bcl-x(L)-binding interface using NMR. *FEBS Lett.* 2004;559:171-4.

[139] Haughn L, Hawley RG, Morrison DK, von Boehmer H, Hockenbery DM. BCL-2 and BCL-XL restrict lineage choice during hematopoietic differentiation. *J. Biol. Chem.* 2003;278:25158-65.

[140] Linden M, Kirchhof N, Carlson C, Van Ness B. Targeted overexpression of Bcl-XL in B-lymphoid cells results in lymphoproliferative disease and plasma cell malignancies. *Blood.* 2004;103:2779-86.

[141] Fernandez Y, Gu B, Martinez A, Torregrosa A, Sierra A. Inhibition of apoptosis in human breast cancer cells: role in tumor progression to the metastatic state. *Int. J. Cancer.* 2002;101:317-26.

[142] Wacheck V, Selzer E, Gunsberg P, *et al.* Bcl-x(L) antisense oligonucleotides radiosensitise colon cancer cells. *Br. J. Cancer.* 2003;89:1352-7.

[143] Yang CC, Lin HP, Chen CS, Yang YT, Tseng PH, Rangnekar VM, Chen CS. Bcl-xL mediates a survival mechanism independent of the phosphoinositide 3-kinase/Akt pathway in prostate cancer cells. *J. Biol. Chem.* 2003;278:25872-8.

[144] Mercatante DR, Mohler JL, Kole R. Cellular response to an antisense-mediated shift of Bcl-x pre-mRNA splicing and antineoplastic agents. *J. Biol. Chem.* 2002;277:49374-82.

[145] Younes A, Snell V, Consoli U, *et al.* Elevated levels of biologically active soluble CD40 ligand in the serum of patients with chronic lymphocytic leukaemia. *Br. J. Haematol.* 1998;100:135-41.

[146] Lamothe B, Aggarwal BB. Ectopic expression of Bcl-2 and Bcl-xL inhibits apoptosis induced by TNF-related apoptosis-inducing ligand (TRAIL) through suppression of caspases-8, 7, and 3 and BID cleavage in human acute myelogenous leukemia cell line HL-60. *J. Interferon Cytokine Res.* 2002;22:269-79.

[147] Zhu H, Guo W, Zhang L, *et al.* Bcl-XL small interfering RNA suppresses the proliferation of 5-fluorouracil-resistant human colon cancer cells. *Mol. Cancer Ther.* 2005;4:451-6.

[148] Kanczuga-Koda L, Sulkowski S, Koda M, Skrzydlewska E, Sulkowska M. Connexin 26 correlates with Bcl-xL and Bax proteins expression in colorectal cancer. *World J. Gastroenterol.* 2005;11:1544-8.

[149] Barwijuk-Machala M, Reszec J, Kemona A, Sobaniec-Lotowska M. C-erb-B2 and Bcl-xl protein expression in Barrett's oesophagus in correlation with morphological parameters. *Rocz. Akad. Med. Bialymst.* 2004;49 Suppl 1:40-2.

[150] Watanabe J, Kushihata F, Honda K, *et al.* Prognostic significance of Bcl-xL in human hepatocellular carcinoma. *Surgery.* 2004;135:604-12.

[151] Yang X, Zheng F, Xing H, *et al.* Resistance to chemotherapy-induced apoptosis via decreased caspase-3 activity and overexpression of antiapoptotic proteins in ovarian cancer. *J. Cancer Res. Clin. Oncol.* 2004;130:423-8.

[152] Gobe G, Rubin M, Williams G, Sawczuk I, Buttyan R. Apoptosis and expression of Bcl-2, Bcl-XL, and Bax in renal cell carcinomas. *Cancer Invest.* 2002;20:324-32.

[153] Schoop RA, Kooistra K, Baatenburg De Jong RJ, Noteborn MH. Bcl-xL inhibits p53- but not apoptin-induced apoptosis in head and neck squamous cell carcinoma cell line. *Int. J. Cancer.* 2004;109:38-42.

[154] Hermann M, Scholman HJ, Marafioti T, Stein H, Schriever F. Differential expression of apoptosis, Bcl-x and c-Myc in normal and malignant lymphoid tissues. *Eur. J. Haematol.* 1997;59:20-30.

[155] Schlaifer D, March M, Krajewski S, *et al.* High expression of the bcl-x gene in Reed-Sternberg cells of Hodgkin's disease. *Blood.* 1995;85:2671-4.

[156] Schlaifer D, Krajewski S, Galoin S, *et al.* Immunodetection of apoptosis-regulating proteins in lymphomas from patients with and without human immunodeficiency virus infection. *Am. J. Pathol.* 1996;149:177-85.

[157] Pallis M, Zhu YM, Russell NH. Bcl-x(L) is heterogenously expressed by acute myeloblastic leukaemia cells and is associated with autonomous growth in vitro and with P-glycoprotein expression. *Leukemia.* 1997;11:945-9.

[158] Tu Y, Renner S, Xu F, *et al.* BCL-X expression in multiple myeloma: possible indicator of chemoresistance. *Cancer Res.* 1998;58:256-62.

[159] Zhao WL, Daneshpouy ME, Mounier N, *et al.* Prognostic significance of bcl-xL gene expression and apoptotic cell counts in follicular lymphoma. *Blood.* 2004;103:695-7.

[160] Sorenson CM. Bcl-2 family members and disease. *Biochim. Biophys. Acta* 2004;1644:169-77.

[161] Pedersen IM, Kitada S, Leoni LM, *et al.* Protection of CLL B cells by a follicular dendritic cell line is dependent on induction of Mcl-1. *Blood.* 2002;100:1795-801.

[162] Clodi K, Kliche KO, Zhao S, *et al.* Cell-surface exposure of phosphatidylserine correlates with the stage of fludarabine-induced apoptosis in chronic lymphocytic leukemia and expression of apoptosis-regulating genes. *Cytometry.* 2000;40:19-25.

[163] Brajuskovic G, Vukosavic S, Dimitrijevic J, *et al.* [Expression of the Bcl-2 family of proteins in peripheral blood B-lymphocytes in patients with chronic lymphocytic leukemia]. *Vojnosanit. Pregl.* 2004;61:41-6.

[164] Kuhlmann T, Glas M, zum Bruch C, Mueller W, Weber A, Zipp F, Bruck W. Investigation of bax, bcl-2, bcl-x and p53 gene polymorphisms in multiple sclerosis. *J. Neuroimmunol.* 2002;129:154-60.

[165] Yamaguchi H, Inokuchi K, Tarusawa M, Dan K. Mutation of bcl-x gene in non-Hodgkin's lymphoma. *Am. J. Hematol.* 2002;69:74-6.

[166] Zong WX, Edelstein LC, Chen C, Bash J, Gelinas C. The prosurvival Bcl-2 homolog Bfl-1/A1 is a direct transcriptional target of NF-kappaB that blocks TNFalpha-induced apoptosis. *Genes Dev.* 1999;13:382-7.

[167] Lin EY OA, Berger MS. Characterization of A1, a novel hemopoietic-specific early-response gene with sequence similarity to bcl-2. *J. Immunol.* 1993;151:1979-88.

[168] Epling-Burnette PK ZB, Bai F, Jiang K, Bailey RD, Garcia R, Jove R, Djeu JY, Loughran TP Jr, Wei S. Cooperative regulation of Mcl-1 by Janus kinase/stat and phosphatidylinositol 3-kinase contribute to granulocyte-macrophage colony-stimulating factor-delayed apoptosis in human neutrophils. *J. Immunol.* 2001;166:7486-95.

[169] Lee HH, Dadgostar H, Cheng Q, Shu J, Cheng G. NF-kappaB-mediated up-regulation of Bcl-x and Bfl-1/A1 is required for CD40 survival signaling in B lymphocytes. *Proc. Natl. Acad. Sci. U S A.* 1999;96:9136-41.

[170] Karsan A, Yee E, Harlan JM. Endothelial cell death induced by tumor necrosis factor-alpha is inhibited by the Bcl-2 family member, A1. *J. Biol. Chem.* 1996;271:27201-4.

[171] Choi SS, Park IC, Yun JW, Sung YC, Hong SI, Shin HS. A novel Bcl-2 related gene, Bfl-1, is overexpressed in stomach cancer and preferentially expressed in bone marrow. *Oncogene.* 1995;11:1693-8.

[172] Yoon HS, Hong SH, Kang HJ, Ko BK, Ahn SH, Huh JR. Bfl-1 gene expression in breast cancer: its relationship with other prognostic factors. *J. Korean Med. Sci.* 2003;18:225-30.

[173] Wang CY, Guttridge DC, Mayo MW, Baldwin AS, Jr. NF-kappaB induces expression of the Bcl-2 homologue A1/Bfl-1 to preferentially suppress chemotherapy-induced apoptosis. *Mol. Cell Biol.* 1999;19:5923-9.

[174] D'Sa-Eipper C, Chinnadurai G. Functional dissection of Bfl-1, a Bcl-2 homolog: anti-apoptosis, oncogene-cooperation and cell proliferation activities. *Oncogene.* 1998;16:3105-14.

[175] Werner AB, de Vries E, Tait SW, Bontjer I, Borst J. Bcl-2 family member Bfl-1/A1 sequesters truncated bid to inhibit is collaboration with pro-apoptotic Bak or Bax. *J. Biol. Chem.* 2002;277:22781-8.

[176] Bae J, Hsu SY, Leo CP, Zell K, Hsueh AJ. Underphosphorylated BAD interacts with diverse antiapoptotic Bcl-2 family proteins to regulate apoptosis. *Apoptosis.* 2001;6:319-30.

[177] Zhang H, Xu Q, Krajewski S, *et al.* BAR: An apoptosis regulator at the intersection of caspases and Bcl-2 family proteins. *Proc. Natl. Acad. Sci. U S A.* 2000;97:2597-602.

[178] Haga H, Yamada R, Ohnishi Y, Nakamura Y, Tanaka T. Gene-based SNP discovery as part of the Japanese Millennium Genome Project: identification of 190,562 genetic variations in the human genome. Single-nucleotide polymorphism. *J. Hum. Genet.* 2002;47:605-10.

[179] Capello D, Gaidano G. Molecular pathophysiology of indolent lymphoma. *Haematologica.* 2000;85:195-201.

[180] McConkey DJ, Aguilar-Santelises M, Hartzell P, Eriksson I, Mellstedt H, Orrenius S, Jondal M. Induction of DNA fragmentation in chronic B-lymphocytic leukemia cells. *J. Immunol.* 1991;146:1072-6.

[181] Osorio LM, De Santiago A, Aguilar-Santelises M, Mellstedt H, Jondal M. CD6 ligation modulates the Bcl-2/Bax ratio and protects chronic lymphocytic leukemia B cells from apoptosis induced by anti-IgM. *Blood.* 1997;89:2833-41.

[182] Osorio LM, Jondal M, Aguilar-Santelises M. Regulation of B-CLL apoptosis through membrane receptors and Bcl-2 family proteins. *Leuk. Lymphoma.* 1998;30:247-56.

[183] Romano MF, Lamberti A, Tassone P, *et al.* Triggering of CD40 antigen inhibits fludarabine-induced apoptosis in B chronic lymphocytic leukemia cells. *Blood.* 1998;92:990-5.

[184] Novak AJ, Bram RJ, Kay NE, Jelinek DF. Aberrant expression of B-lymphocyte stimulator by B chronic lymphocytic leukemia cells: a mechanism for survival. *Blood.* 2002;100:2973-9.

[185] Pedersen IM, Buhl AM, Klausen P, Geisler CH, Jurlander J. The chimeric anti-CD20 antibody rituximab induces apoptosis in B-cell chronic lymphocytic leukemia cells through a p38 mitogen activated protein-kinase-dependent mechanism. *Blood.* 2002;99:1314-9.

[186] Johnston JB, Kabore AF, Strutinsky J, *et al.* Role of the TRAIL/APO2-L death receptors in chlorambucil- and fludarabine-induced apoptosis in chronic lymphocytic leukemia. *Oncogene.* 2003;22:8356-69.

[187] Cordone I, Masi S, Mauro FR, *et al.* p53 expression in B-cell chronic lymphocytic leukemia: a marker of disease progression and poor prognosis. *Blood.* 1998;91:4342-9.

[188] Watanabe T, Hotta T, Ichikawa A, *et al.* The MDM2 oncogene overexpression in chronic lymphocytic leukemia and low-grade lymphoma of B-cell origin. *Blood.* 1994;84:3158-65.

[189] Watanabe T, Ichikawa A, Saito H, Hotta T. Overexpression of the MDM2 oncogene in leukemia and lymphoma. *Leuk. Lymphoma.* 1996;21:391-7, color plates XVI following 5.

[190] Zhao H, Dugas N, Mathiot C, Delmer A, Dugas B, Sigaux F, Kolb JP. B-cell chronic lymphocytic leukemia cells express a functional inducible nitric oxide synthase displaying anti-apoptotic activity. *Blood.* 1998;92:1031-43.

[191] Ferrajoli A, Keating MJ, Manshouri T, *et al.* The clinical significance of tumor necrosis factor-alpha plasma level in patients having chronic lymphocytic leukemia. *Blood.* 2002;100:1215-9.

[192] Chandra J, Gilbreath J, Freireich EJ, Kliche KO, Andreeff M, Keating M, McConkey DJ. Protease activation is required for glucocorticoid-induced apoptosis in chronic lymphocytic leukemic lymphocytes. *Blood.* 1997;90:3673-81.

[193] Krajewski S, Gascoyne RD, Zapata JM, *et al.* Immunolocalization of the ICE/Ced-3-family protease, CPP32 (Caspase-3), in non-Hodgkin's lymphomas, chronic lymphocytic leukemias, and reactive lymph nodes. *Blood.* 1997;89:3817-25.

[194] Peng H, Aiello A, Packham G, Isaacson PG, Pan L. Infrequent bax gene mutations in B-cell lymphomas. *J. Pathol.* 1998;186:378-82.

In: Chronic Lymphocytic Leukemia Research Focus
Editor: Chadi Nabhan, pp. 113-127
ISBN: 978-1-60021-526-1

Chapter V

Mode of Action of Rituximab in Chronic Lymphocytic Leukemia; Activation of Tis11b, an Inducer of mRNA Instability and Induction of Apoptosis

Maria Baou[1,2], Andrew P. Jewell[*3], Kwee L. Yong[4] and John Murphy[2]

[1] School of Life Sciences, Kingston University, Surrey KT1 2EE, UK.
[2] Immunology Department, King's College London WC2R 2LS, UK.
[3] Faculty of Health and Social Care Sciences, Kingston University and St George's University of London, SW17 0RE, UK.
[4] Haematology Department, University College London, WC1E 6HX.

Abstract

Tis11b/Berg36 is a member of a family of proteins involved in post-transcriptional gene regulation. Members of this family bind to AU rich elements in certain mRNA species and increase mRNA stability. mRNA that contain AU rich elements and can be regulated in this way include molecules involved in the regulation of apoptosis eg bcl-2 and several cytokines eg TNF, IL-2. We have recently shown that Tis11b/Berg36 is involved in the regulation of apoptosis following Rituximab treatment of CLL cells suggesting a pro-apoptotic role of this gene in CLL cells. Thus B-CLL cells were stimulated with IL-4, anti-CD40, anti-CD40+IL-4 and PMA. It was found that IL-4, CD40 and their combination significantly inhibited spontaneous apoptosis while PMA was able to inhibit spontaneous apoptosis in some but not all patients tested and the overall effect did not reached statistical significance.. Unexpectedly Tis11b/Berg36 mRNA remained unchanged following IL-4 treatment, but expression was induced following anti-CD40 or PMA treatment. Because it was found that regulation of

[*] Correspondence: A. P. Jewell; Faculty of Health and Social Care Sciences, Kingston University and St George's University of London, London SW17 0RE; Telephone: 0208 547 2000 x62468; Fax: 0208 547 8800; Email: ajewell@hscs.sgul.ac.uk

Tis11b/Berg36 mRNA following the latter two treatments was under the control of NF-κB pathway and not p38 (as found for Rituximab treatment) it was hypothesized that this gene may be involved in the regulation of cell cycle or differentiation of B-CLL cells. Indeed it was found that stimulation of B-CLL cells with either PMA or anti-CD40 resulted in increase in sIgM as a marker of B-CLL differentiation. CLL cells were also found to express high basal levels of two other members of this family, Tis11 and Tis11d. From these stimuli, IL-4 or anti-CD40 were found to downregulate the basal expression of Tis11 mRNA. These data suggest that members of the Tis11b/Berg36 family are involved in the regulation of apoptosis and differentiation in B-CLL cells

Rituximab

Rituximab is a monoclonal chimeric mouse/human IgG κ1 antibody raised against CD20. CD20 is only expressed in the B lineage and specifically on mature B cells (stem cells, pre-B and plasma cells do not express CD20). The gene for CD20 belongs to the MS4 family and is located on chromosome 11q12-13.1. The protein product exists as 3 different isoforms of 33, 35 and 37 KDa that result from multiple phosphorylation on serine and threonine residues (Janas *et al* 2005). CD20 has emerged as an ideal target for immunotherapy because it is expressed on >90% of B cell malignancies such as Non-Hodgkin's Lymphoma (NHL), and is also not internalised or shed after antibody binding. Rituximab binds between residues K142 and Y184 in the extracellular loop and its binding depends on the presence of a disulfide bond between C167-C183 (Ernst *et al* 2005).

Initially several antibodies were raised against the CD20 Receptor such as 1F5, B1, 11B8, AT80, KT20, 2F2, 7D8 as well as Rituximab (Chan *et al* 2003, Teeling *et al* 2004). Rituximab has been "humanised" with human constant regions replacing those of the mouse. This increases the half life of the antibody (making it more stable *in vivo)* and the effectiveness to activate the complement cascade and mediate cell lysis . In general all the anti-CD20 antibodies are subdivided into Type I (Rituximab and most of the anti-CD20 antibodies) which are potent inducers of Complement Dependent Cytotoxicity (CDC) but weak inducers of apoptosis and Type II (B1 and 11B8) which are potent inducers of apoptosis but weak activators of complement (Teeling *et al* 2004). Both Type I and Type II antibodies are equally efficient in mediating Antibody Dependent Cell Cytotoxicity (ADCC) (Teeling *et al* 2004).

Rituximab binds with high affinity to CD20 ($5x10^{-9}$ mol/L) (Maloney *et al* 1997) and been used as a thearapeutic agent in several patients including low grade Non-Hogkin's Lymphoma (NHL), Chronic Lymphocytic Leukaemia (CLL) and Folicullar Lymphoma (FL). Parameters that were measured after this treatment were the overall response rate (ORR), complete remission (CR) and time to progression (TTP) which were in general better after multiple administrations of Rituximab compared to a single administration of the antibody (Davis *et al* 2000, Ghielmini *et al* 2004). Rituximab treatment will induce cell death to all CD20 positive cells (normal or malignant) but normal B cells will re emerge after few months due to constant haematopoiesis from the bone marrow. Recovery of normal B cells has been shown to correlate with the levels of SDF-1 (Stromal Derived Factor 1) in serum of previously treated patients (Dunleavy *et al* 2005). One study involving follicular lymphoma grade I (low grade NHL) showed that the overall response rate was 72% with 36% achieving CR and that the median TTP was 2.2 years (Witzing *et al* 2005). When FL patients were tested it was found that the ORR was 52% with 8% achieving CR. Interestingly in the same study it was shown that 56% of Blood samples and 39% of Bone Marrow samples became t(14;18) negative (Bcl2 negative) after Rituximab treatment (Ghielmini *et al* 2004). Finally in CLL the standard dose of Rituximab (375 mg/m^2) as well as increased doses of 500, 650 , 825 , 1000 , 1500 , 2250 mg/m^2 were tested. The same study included patients with Mantle Cell Lymphoma (MCL) and B Pro Lymphocytic Leukaemia(B-PLL) (O' Brien *et al* 2001). They have found that the ORR was 36% when the standard dose was used which was increased to 75% by Rituximab at 2250 mg/m^2 (O' Brien *et al* 2001). In CLL patients the response rates were better in Rai stage I and II than in Rai stages III and IV (O'Brien *et al* 2001). Later on, Rituximab was combined with standard chemotherapeutic agents such as Fludarabine and CHOP and much better response rates were achieved. In combination with Fludarabine in low grade NHL, 70-83% of patients showed complete remission (Cruczman *et al* 2005) while in combination with CHOP (Cyclophosphamide, Doxorubicin, Vincristine, Prednisone) in low grade and follicular NHL the ORR was 100% with 58% being CR (Cruczman *et al* 2004). Interestingly in the latter study, 7 patients that achieved complete remission became Bcl2 negative after this treatment and 3/7 patients were still Bcl2 negative for up to 99 months (more than 8 years) after the end of treatment (Cruczman *et al* 2004).

Apart from B cell malignancies Rituximab has been used recently to treat autoimmune disorders such as Rheumatoid Arthritis (RA) and Systemic Lupus Erythematosous (SLE) (Chambers and Isenberg 2005).

Mode of Action of Rituximab

The efficacy of Rituximab in inducing cell death in B-cells, appears to depend on three mechanisms; Complement Dependent Cytotoxicity (CDC); Antibody Dependent Cell Cytotoxicity (ADCC); Induction of Apoptosis.

A) Complement Dependent Cytotoxicity (CDC)

There are several studies that have highlighted the importance of complement in Rituximab mediated cell lysis in B cell lines and primary malignant B cells such as CLL. In one such study it was shown that several B cell lines and CLL cells showed significant higher cell death in the presence of complement (Chow *et al* 2002) while an earlier study has shown that this complement mediated cell lysis is accompanied by loss in $\Delta\Psi m$ (drop in mitochondrial membrane potential) and production of Reactive Oxygen Species (ROS) (Bellosillo *et al* 2001). Rituximab activates the classical complement pathway as it was shown that binding of the antibody induces deposition of C1q, C3b and C6 to the plasma membrane (Cragg and Glennie 2004).

Thus it is not surprising the fact that efficiency of Rituximab in mediating CDC depends on the presence of complement inhibitors such as CD55, CD35 and CD46 molecules. In B cell lines such as Karpas 422, DHL4 (FL) and BJAB (Burkitt's Lymphoma), CDC induced by Rituximab correlated with the levels of CD55 expression (Golay *et al* 2000) and inclusion of anti-CD55 antibodies alone or combined with anti-CD59 antibodies further increased Rituximab mediated cell death (Golay *et al* 2000). A similar finding has been reported by Belosillo *et al* who showed that the levels of CD55 expression determined CDC mediated lysis in CLL and again use of anti-CD55 and anti-CD59 synergistically increased the efficiency of Rituximab treatment (Bellosillo *et al* 2001). In contrast to these findings, another study in CLL cells showed that there is no such correlation but instead reported that Rituximab treatment may lead to increase in expression of CD55 and CD59 which in turn may result in poor response to the antibody due to selection of a certain subclone (Bannerji *et al* 2003). CD55 binds to C3b and C4b and induces their decay while CD59 binds to C8 and C9 and prevents the formation of the membrane attack complex (Bellosillo *et al* 2001) preventing complement activation at the early or late stages respectively.

B) Antibody Dependent Cell Cytotoxicity (ADCC)

This mechanism includes cell killing mediated by effector cells such as Natural Killer (NK) cells which are activated by binding to the Fcγ portion of Rituximab. Fcγ Receptors such as CD64 (FcγRI), CD16 (FcγRIIIA) , CD32 (FcγRIIB) are found on monocytes, macrophages, granulocytes, NK and B cells (CD32). The importance of effector cells in mediating cell death after Rituximab treatment has been shown in B cell lines and primary B cells. One such study showed that treatment of 3FL cell lines and BJAB with Rituximab in the presence of purified PBMC (Peripheral Blood Mononuclear Cells) enriched for CD16+ cells further increased cell lysis by 10-20% (Golay *et al* 2000). Another study was conducted in normal B cells and showed that purified CD56+ NK cells as well as CD14+ Monocytes were important for ADCC mediated by Rituximab when added at a concentration 10x the concentration of normal B cells (Voso et al 2002). Finally Voso *et al* showed that mononuclear cells enhanced cell killing by Rituximab in B-CLL cells when used at 10:1 ratio and this effect was further increased by addition of GM-CSF which induces differentiation of CD14+ monocytes to macrophages (Voso *et al* 2002).

C) Induction of Apoptosis

Rituximab can also induce apoptosis directly but requires cross-linking with a secondary antibody for *in vitro* studies (van der Kolk *et al* 2002, Pedersen *et al* 2002, Shan *et al* 1998). Induction of apoptosis by Rituximab was evident as early as 4-8h in Ramos cells and was accompanied by activation of caspase 9 and caspase 3 as well as PARP cleavage (van der Kolk *et al* 2002). Interestingly though when the activation of caspases was blocked with a pan- caspase inhibitor (zVAD-fmk), there was a partial inhibition of Rituximab induced apoptosis (van der Kolk *et al* 2002) suggesting that Rituximab may induce apoptosis through an unknown caspase independent mechanism. Similarly another study showed apoptosis induction but without any change in Bcl2 and Bax protein levels (Shan *et al* 2000). In contrast to these studies that were carried out in Burkitt's Lymphoma cell lines, Rituximab alone (no cross-linking) was able to downregulate Bcl-2 protein in two NHL cell lines derived from AIDS patients with Burkitt's Lymphoma (Alas et al 2001) and this was correlated with inhibition of IL-10 secretion and downregulation of p38 and STAT3 pathways (Alas *et al* 2001 , Vega *et al* 2004).

On the other hand, in CLL it has been shown that cross-linked Rituximab induces apoptosis at 24h and among the signalling pathways activated (ERK, p38 and JNK) only the p38 MAPK pathway was involved in induction of apoptosis (Pedersen *et al* 2002). Interestingly when a specific inhibitor of the p38 pathway called SB203580 was utilised the effect of Rituximab was only partially abrogated (Pedersen *et al* 2002) suggesting that other signalling events/pathways may be also involved. Apart from these *in vitro* studies there is evidence of apoptosis induction *in vivo* by Rituximab, as it was shown that there is activation of caspase 9 and caspase 3 as well as PARP cleavage 24h after the initiation of Rituximab treatment in CLL patients (Byrd *et al* 2002). This was accompanied by downregulation of Mcl1 and XIAP anti-apoptotic proteins (Byrd *et al* 2002).

Signal Transduction Pathways Activated by Rituximab

Rituximab binding induces redistribution of CD20 into lipid raft microdomains (rich in cholesterol, sphingomyelin and other lipids), activation of Src kinases and subsequent activation of signal transduction pathways such as PLCγ , ERK, p38, JNK etc. It has been shown that treatment of Ramos with 1F5 induced phosphorylation of Src kinases such as Lyn (p53/56), Lck (p56), Fyn (p59) whereas there was no activation of Blk (Deans *et al* 1993). Furthermore, activation of PLCγ1, PLCγ2 and calcium mobilisation has been demonstrated when the anti-CD20 antibody was crosslinked with a secondary antibody (Deans *et al* 1993). Other studies confirmed these findings and use of specific Src kinase inhibitors such as PP1 (Shan *et al* 2000) and PP2 (Hofmeister *et al* 2000) abolished induction of apoptosis and caspase 3 activation following Rituximab treatment in Ramos cells (Hofmeister *et al* 2000). In terms of calcium mobilisation it has been shown that release of intracellular calcium and extracellular calcium influx are both equally important for induction of apoptosis by anti-CD20 (Shan *et al* 2000). Localisation of Src kinases such as Lyn into Triton-X insoluble membrane compartments (lipid rafts) led to the hypothesis that CD20 may be redistributed into lipid rafts after binding to its ligand or an anti-CD20 antibody. This was shown initially in Raji cell line in which treatment with 1F5 and B1 antibodies resulted in 80-95% translocation of CD20 to a Triton -X insoluble fraction (Deans *et al* 1998). Similarly it was shown that Rituximab treatment results in redistribution of CD20 into Triton-X insoluble lipid rafts in Ramos cells which depends on the presence of cholesterol. This redistribution is absolutely critical for calcium influx (Janas *et al* 2005). Caspase 3 and 7 activities were reduced to background levels when the cells were treated with Rituximab in the presence of a cholesterol inhibitor suggesting that induction of apoptosis by Rituximab is dependent on redistribution into lipid rafts. Because there was no direct association of caspase 3 with lipid rafts it was proposed that other proteins link redistribution of CD20 into lipid rafts and activation of caspase 3 (Javas *et al* 2005). Activation of Src kinases or further downstream events such as activation of p38 (Pedersen et al 2002) may mediate this link since inhibition of these events inhibit apoptosis induction by Rituximab.

The Tis11 Family and Post-Transcriptional Gene Regulation

An alternative mechanism for the induction of apoptosis by Rituximab may be mediated through indirect mechanisms, such as post-transcriptional gene regulation by Tis11 family members.

The Tis11 (TPA inducible sequences 11) family consists of 3 members namely Tis11 (ZFP36, TTP, Nup475, GOS24), Tis11b (Berg36, ERF-1, ZFP36L1, BRF-1) and Tis11d (ZFP36L2, ERF-2, BRF-2). Very recently a fourth family member has been described in rodents, ZFP36L3, which was only expressed in mouse placenta but was not detected in human placenta or other human tissues (Blackshear *et al* 2005). Human Tis11 is located on chromosome 19q13.1, has been isolated and cloned by several groups after stimulation of cells with serum, PMA or insulin (Varnun *et al* 1991, DuBois *et al* 1990). Tis11b is located on chromosome 14q22-24 and was isolated from CLL cells after stimulation with PMA (Murphy and Norton 1993), from a variety of human cDNA libraries probed with the rat cMG1 (Bustin *et al* 1994, Varnum *et al* 1991) or as a gene that reverses the phenotype of mutant H1080 cells which are unable to degrade IL-3 mRNA (Stoecklin *et al* 2002). Finally, Tis11d is located on 2p22.3-p21 and has been isolated from human cDNA libraries probed with the mouse Tis11d mRNA (Nie *et al* 1995 , Varnum *et al* 1991) .

All 3 genes in the Tis11 family consist of 2 tandemly repeated zinc finger motifs (YKTEL) through which they bind to AU Rich elements (AREs) in MRNA and promote deadenylation, decapping and finally degradation of the mRNA body by either the exosome (3'-5' degradation) or XRN1 exonuclease (5'-3' degradation). Thus they are considered as destabilisation AU Binding proteins (AUBPs). Apart from Tis11 family other proteins are classified as AUBPs and these include the HuR family which is considered as stabilisation agents and the AUF1 family that are considered mainly as destabilisation agents (Wilusz *et al* 2001)

AREs are found at the 3' end of several proteins including transcription factors, cell cycle regulators, apoptosis regulators and cytokines . They are subdivided into 3 classes. Class I is characterised by the WAUUUAW motif and a U-rich region and is found in c-fos and c-myc onogenes, Class III is characterised by an U-rich sequence found in c-jun. Class II AREs are found in the mRNA for many cytokines. Class IIA contains 5 repeats of the nonameric sequence AUUUA and is found for example in TNF-α and GM-CSF, ClassIIB contains 4 AUUUA nonamers found for example in IFN-α , Class IIC contains 3 AUUUA nonamers and is found in VEGF and IL-2, Class IID contains 2 AUUUA nonamers and is found in FGF2 and finally Class IIE contains 1 AUUUA motif and is found in u-PA Receptor (Wilusz *et al* 2001).

One of the first indication that Tis11 functions as AUBPs came from Carballo *et al* which showed that Tis11 overexpression does not inhibit TNF-α transcription but it reduced the half life of the mRNA by binding to AREs in the 3' end (Carballo *et al* 1996). It should be noted that AREs are found at the 3' end of each of the three Tis11 genes suggesting that they may regulate themselves by a negative feedback loop (Bustin *et al* 1994, Ino *et al* 1995).

Overexpression of human Tis11 in HEK293 (embryonic kidney cells) caused a significant reduction of the levels of TNF-α mRNA (Lai *et al* 2000). Similarly in another study showed that Tis11 family are involved in the mRNA decay of GM-CSF and IL-3 (Lai *et al* 2001).

Proteins that have been shown to be targets of the Tis11 family include TNF-α in many cell types (Sauer *et al* 2006), GM-CSF (Lai *et al* 2001) , IL-3 (Lai *et al* 2001) , IL-2 (Ogilvie *et al* 2005), IL-6 (Sauer *et al* 2006) while VEGF was shown to be specifically a target of Tis11b (Ciais *et al* 2004). Most of these targets are cytokines and have been identified through overexpression of Tis11 family members in certain cell lines. It has been recently reported that Bcl2 contains a Class II ARE motif in it's 3'UTR with at least one UUAUUUAUU nonamer which is conserved between different species (humans, mice, chicken) and that this sequence is actively controlling the stability of Bcl2 (Schiavone *et al* 2000).

Because these genes have all been reported to be involved in the pathophysiology of CLL, we investigated the involvement if the Tis11 family in Rituximab mediated apoptosis.

Results

Induction of Apoptosis by Rituximab

Rituximab was utilised to induce apoptosis in B-CLL cells. Freshly isolated B-CLL cells were treated with 20 μg/ml Rituximab either alone or in combination with a secondary antibody (F(ab)2 of goat anti-human IgG specific for Fcγ) at 60 μg/ml. Apoptosis was monitored at 48 hours using Propidium Iodide (PI) exclusion.

At 48 hours Rituximab increased the percentage of apoptotic cells from 35.0% ± 8.0% (mean ± sem) in unstimulated cells to 63.0% ± 8.0% (mean ± sem). This effect was found to be highly significant since $p \leq 0.001$. Rituximab alone again had no effect (31.98% ± 8.20%, mean ± sem) (figure 1).

Rituximab Treatment and Induction of the Tis 11 Genes

Because previous work has shown that Tis11b/Berg36 was induced by immobilised Rituximab in a Burkitt's Lymphoma cell line (BL60-2) it was interesting to test whether XRituximab would induce the same gene in CLL cells. In order to do so, CLL cells were left untreated or were treated with cross-linked Rituximab for 3 hours and expression of Tis11b mRNA was tested by Northern Blot. It was found that in unstimulated cells Tis11b mRNA basal levels were low or undetectable and treatment with cross-linked Rituximab for 3h strongly induced expression of this gene (figure 2).

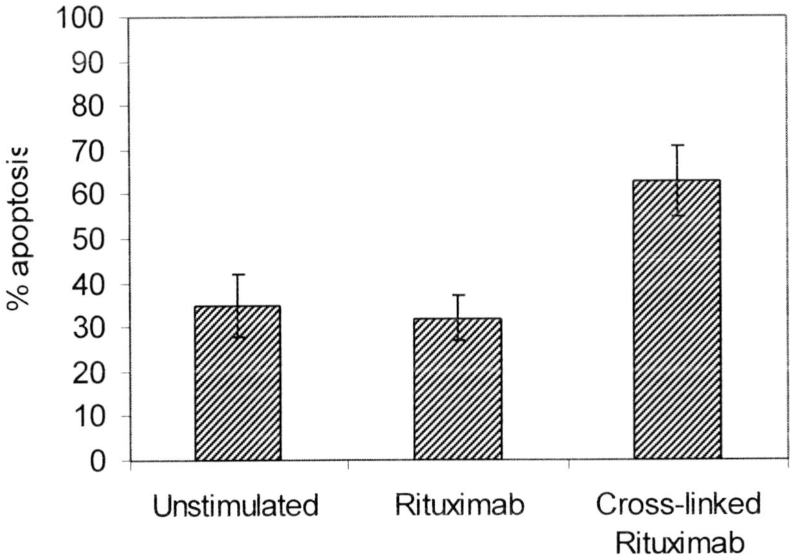

Figure 1. Freshly isolated CLL cells were unstimulated or were treated with Rituximab at 20 μg/ml in the presence of 3x F(ab)2 goat anti-human IgG antibody and % of apoptotic cells were measured by Propidium Iodide exclusion 48 hours. Values are mean ± sem, n=9.

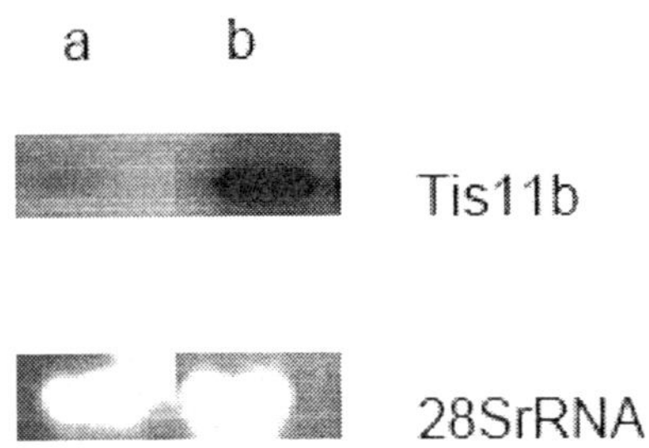

Figure 2. B-CLL have been left untreated (a) or were stimulated with Rituximab crosslinked with a secondary Ab for 3h (b). At the end of the culture period RNA was extracted, run on Formaldehyde/Agarose gels and transferred onto nylon membranes for Northern Blot analysis.The membranes were hybridised for 18h with the cDNA Probe for Tis11b/ Berg36 (3.5 kb) which was labelled with ^{32}P.

Tis11 was induced in some patients at 3h but remained unchanged in some others and this depended strongly on the basal levels of Tis11. In the cases that Tis11 was strongly expressed in unstimulated cells there was no induction of the gene by either Rituximab alone or with cross-linked Rituximab (data not shown). Tis11d was induced strongly at 3h by crosslinked Rituximab in some but not all patients tested.

Regulation of Tis11b Genes

In order to study the pathways involved in induction of Tis11b, B-CLL cells were cultured for 3 hours with cross-linked Rituximab in the presence or absence of inhibitors specific for p38 (SB203580) , JNK (SP600125) and ERK (U0126) pathways. Inhibition of the p38 pathway resulted in partial or complete abrogation of Tis11b mRNA levels following treatment with cross-linked Rituximab. Inhibition of the ERK or JNK pathways did not affect induction of this gene (figure 3). Similarly, Inhibition of the p38 pathway also abrogated Tis11 mRNA.

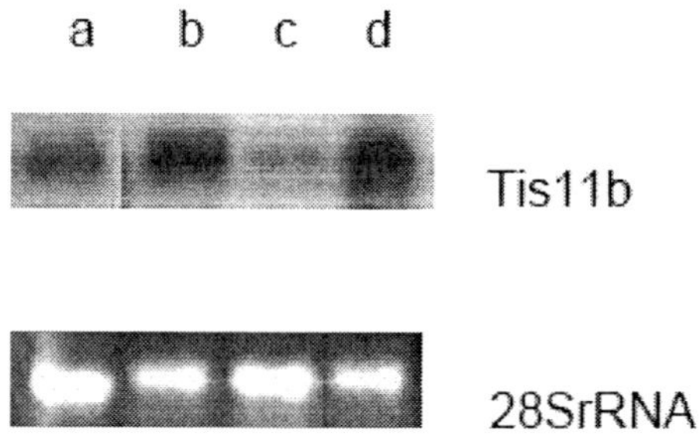

Figure 3. B-CLL cells have been left untreated (a) or were stimulated with crosslinked Rituximab for 3h (b) in the presence or absence of specific inhibitors for the (b) p38 (SB203580), and (c) JNK (SP600125) pathways. At the end of the culture period RNA was extracted, run on Formaldehyde/Agarose gels and transferred onto nylon membranes for Northern Blot analysis.The membranes were hybridised for 18h with the cDNA Probe for Tis11b/Berg36 which was labelled with ^{32}P.

siRNA Inhibition of Tis11b and Effect on Induction of Apoptosis

Since Tis11b is strongly induced by Rituximab it was interesting to test whether inhibition of this gene would affect apoptosis induction by Rituximab. In order to do so we utilised siRNA technology using a commercially available double stranded RNA oligo (siRNA) which specifically binds to Tis11b/Berg36 mRNA. The sequence of the Tis11b/Berg36 siRNA is : CAA GAU GCU CAA CUA UAG UdT for the sense strand and ACU AUA GUU GAG CAU CUU GdT for the antisense. In this experiment, cross-linked Rituximab increased the percentage of apoptotic cells from 29.7%±4.7% (mean ± sem) for the unstimulated cells to 44.7%±5.4% (mean ± sem) ($p \leq 0.001$) . When CLL cells treated with cross-linked Rituximab in the presence of Tis11b siRNA the percentage of apoptotic cells was reduced to 33.5%±4.6% (mean ± sem) ($p \leq 0.05$). Use of a control siRNA did not significantly affect the percentage apoptosis.

Conclusion

Previous work has shown that 2 strong inducers of apoptosis in Burkitt's Lymphoma cell lines, namely anti-IgM and immobilised Rituximab, induced Tis11b/ mRNA with similar kinetics (Mathas *et al* 2000) suggesting that Tis11b/Berg36 is involved in the regulation of apoptosis in Burkitt's Lymphoma B cells. It was interesting to test whether Rituximab could induce Tis11b in CLL cells as well since it has been used *in vivo* to treat CLL patients and also these cells are profoundly resistant to apoptosis.

Before testing induction of Tis11b following Rituximab treatment, it was necessary to evaluate the response of the patients participating in this study to this treatment. It was found that Rituximab alone was unable to induce apoptosis at 48 hours after the initiation of the culture period but was effective in doing so in the presence of 3x (60 μg/ml) F(ab)2 anti human IgG antibody as has been previously described (Pedersen *et al* 2002,van der Kolk *et al* 2002).

Northern Blot analysis was performed for Tis11b and for the other 2 Tis11 family members (Tis11 and Tis11d) following Rituximab treatment for 3 hours. It was found that Tis11b mRNA is clearly induced at 3 hours by cross-linked Rituximab and that this induction is regulated partially through the p38 pathway since inhibition of this pathway partially and in some cases completely abrogated induction of the gene.

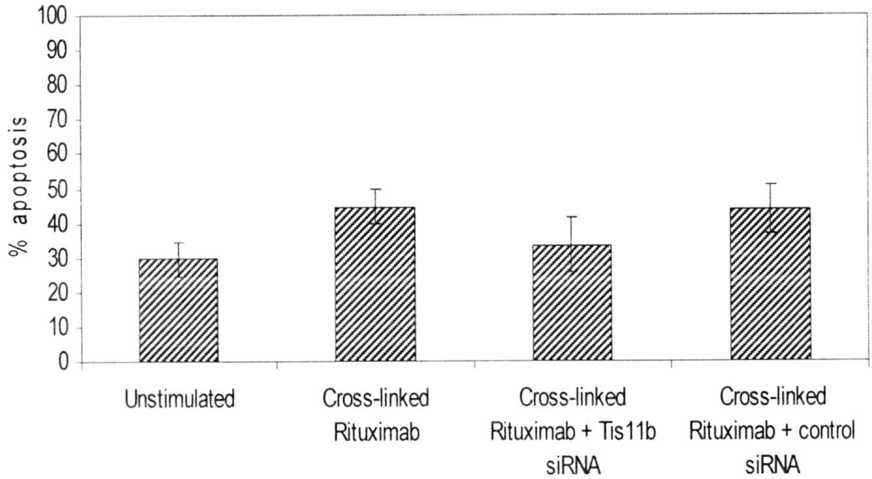

Figure 4. B-CLL cells from 7 CLL patients were left untreated (unstimulated) or were treated with cross-linked Rituximab at 20 μg/ml in the presence or absence of Tis11b siRNA transfected with the use of Hyperfect .The percentage of apoptotic cells was measured by Propidium Iodide staining. As a negative control a commercially available siRNA with no similarity to a mammalian gene was also used. Values are mean ± sem, n=7.

For the other 2 family members only Tis11d was induced by cross-linked Rituximab in some but not all patients tested but the magnitude of induction was lower than Tis11b/Berg36. On the other hand Tis11 seems to be weakly (and in some cases not at all) induced by Rituximab treatment. Variable induction of Tis11 in response to a stimuli (LPS) has been shown for human Peripheral Blood Lymphocytes as well as Neutrophils and Monocytes (Fairhurst et al 2003). In this study 4 normal donors were studied and showed significant variation in peak levels in response to LPS in lymphocytes with two donors showing no response, one donor showing a modest increase and only one donor showing significant induction of Tis11 protein (Fairhurst et al 2003).The reasons for such variation are currently unknown.

We then examined whether inhibition of Tis11b/Berg36 expression by siRNA would affect apoptosis induction by cross-linked Rituximab.

It was found that Tis11/Berg36 inhibition significantly reduced apoptosis induction by cross-linked Rituximab and this effect was specific to Tis11b/Berg36 inhibition since control siRNA had no effect on apoptosis induction when transfected under identical conditions.

In conclusion, this study proposes a novel mechanism of apoptosis induction by Rituximab, via post-transcriptional gene regulation mediated by members of the Tis11 family. Work is currently underway to identify genes which are targeted in this way.

References

Alas S., Emmanouilides C., Bonavida B. 2001 Inhibition of interleukin 1o by Rituximab results in downregulation of Bcl-2 and sensitization of B-cell Non Hodgkin's Lymphoma to apoptosis. *Clinical Cancer Research.* 2001; 7: 709-713

Bannerji R., Kitada S., Flinn I.W., Pearson M. , Young D. , Reed J.C. , Byrd J.C. Apoptotic-Regulatory and Complement expression in Chronic Lymphocytic Leukaemia: Relationship to in vivo Rituximab resistance. *The Journal of Clinical Oncology.* 2003; 21: 1466-1471

Bellosillo B. , Villamor N. , López-Guilermo A. , Marcé S. , Esteve S. , Campo E. , Colomer D. , Montserrat E. Complement-mediated cell death induced by rituximab in B- cell lymphoproliferative disorders is mediated in vitro by a caspase-independent mechanism involving the generation of reactive oxygen species. *Blood.* 2001; 98: 2771-2777

Bernal A. , Pastore R.D. , Asgary Z. , Keller S.A. , Cesarman E. , Liou H-C , Schattner E.J. Survival of leukemic B cells promoted by engagement of antigen receptor. *Blood.* 2001; 98: 3050-3057

Bezombes C. , Grazide S. , Garret C. , Fabre C. , Quillet-Mary A. , Műller S. , Jaffrézou J-P. , Laurent G. Rituximab antiproliferative effect in B-lymphoma cells is associated with acid-spingomyelinase activation in raft microdomains *Blood.* 2004; 104: 1166-1173

Blackshear P.J. , Phillips R.S. , Ghosh S. , Ramos S.V. , Richfield E.K. , Lai W.S. *Zfl36l3*, a rodent X chromosome gene encoding a placenta- specific member of the tristetraprolin family of CCCH tandem zinc finger proteins. *Biology of Reproduction.* 2005; 73: 297-307

Bustin S.A. and McKay I.A. The product of the primary response gene *BRF1* inhibits interaction between 14-3-3 proteins and cRaf-1 in the yeast trihybrid system *DNA and Cell Biology.* 1999; 18: 653-661

Byrd J.C , Smith L. , Hackbarth M.L. , Flinn I.W. , Young D. , Proffit J.H. , Heerema N.A. Interphase Cytogenetic Abnormalities in Chronic Lymphocytic Leukemia may predict response to Rituximab. *Cancer Research.* 2003; 63: 36-38

Byrd J. , Kitada S. , Flinn I.W. , Aron J.L. , Pearson M. , Lucas D. , Reed J.C. The mechanism of tumor cell clearance by Rituximab in vivo in patients with B- cell chronic lymphocytic leukaemia: evidence of caspase activation and apoptosis induction. *Blood.* 2002; 99:1038-1043

Carballo E. , Lai W.S. , Blackshear P.J. Feedback inhibition of macrophage Tumor Necrosis Factor-α production by Tristetraprolin *Science.* 1996; 281: 1001-1005

Chambers SA , Isenberg D. Anti-B cell therapy (Rituximab) in the treatment of autoimmune diseases. *Lupus.* 2005; 14: 210-214

Chan C.H. , Hughes D., French R.R, Tutt A.L., Walshe C.A., Teeling J.L., Glennie M.J., Cragg M.S. CD20-induced Lymphoma cell death is independent of both caspases and its redistribution into Triton X-100 insoluble membrane rafts. *Cancer Research.* 2003; 63,: 5480-5489

Chow K.U , Sommerland W.D. , Boehrer S. , Schneider B. , Seipelt G. , Rummel M.J. , Hoelzer D. , Mitrou P.S. , Weidmann E. Anti-CD20 antibody (IDEC-C2B8,Rituximab) enhances efficacy of cytotoxic drugs on neoplastic lymphocytes in *vitro* :role of cytokines,complement ,and caspases". *Haematologica.* 2002; 87: 33-43

Ciais D, Cherradi N, Bailly S, Grenier E, Berra E, Pouyssegur J, Lamarre J, Feige JJ. Destabilization of vascular endothelial growth factor mRNA by the zinc-finger protein TIS11b. *Oncogene.* 2004; 23: 8673-80

Cragg M.S. , Glennie M.J. Antibody specificity controls in vivo effector mechanisms of anti-CD20 reagents. *Blood.* 2004; 103: 2738-2743

Cruczman M.S. , Koryzna A. , Mohr A. , Stewart C. , Donohue K., Blumenson L. , Bernstein Z.P. , McCarthy P. , Alam A. , Hernandez-Ilizaliturri F. , Skipper M. , Brown K. , Chanan-Khan A. , Klippenstein D. , Loud P. , Rock M.K. , Benyunes M. , Grillo-Lopez A.J , Bernstein S.H. Rituximab in combination with Fludarabine chemotherapy in Low-grade or Follicular Lymphoma. *The Journal of Clinical Oncology.* 2005; 23: 694-704

Cruczman M.S. , Weaver R. , Alkuzweny B. , Berlfein J. , Grillo-Lopez A.J Pronged clinical and Molecular remission in patients with low grade or Follicular Non Hodgkin's Lymphoma treated with Rituximab plus CHOP chemotherapy : 9 years follow up. *The Journal of Clinical Oncology.* 2004; 22: 4711-4716

Davis T.A. , Grillo-Lopez A.J. , White C.A. , McLaughlin P. , Czuczman M.S. , Link B.K. , Maloney D.G. , Weaver R.L. , Rosenberg J. , Levy R. Rituximab anti-CD20 monoclonal antibody therapy in Non-Hodgkin's Lymphoma :Safety and efficacy of re-treatment. *The Journal of Clinical Oncology.* 2000; 18: 3135-3143

Deans J.P , Robbins S.M. , Polyak M.J. , Savage J.A. Rapid redistribution of CD20 to a low density detergent insoluble membrane compartment. *The Journal of Biological Chemistry.* 1998; 273: 344-348

Deans J.P. , Schieven G.L. , Shu G.L. , Valentine M. A. , Gilliland L.A., Aruffo A. , Clark E.A. , Ledbetter J.A. Association of Tyrosine and Serine Kinases with the B cell surface antigen CD20. *The Journal of Immunology.* 1993; 151: 4494-4504

DuBois R.N. , McLane M.W. , Ryder K. , Lau L.F. , Nathans D. A growth factor-inducible nuclear protein with novel Cysteine/Histidine repetive sequence. *The Journal of Biological Chemistry.* 1990; 265: 19185-19191

Dunleavy K. , Hakim F. , Kim H.K. , Janik J.E. , Grant N. , Nakayama T. , White T. , Wright G. , Kwak L. , Gress R. , Tosato G. , Wilson W.H. B-cell recovery following Rituximab-based therapy is associated with perturbations in stromal derived factor-1 and granulocyte hemostasi. *Blood.* 2005;106: 795-802

Ernst J. A. Li H. , Kim H.S. Nakamura G.R. , Daniel G. Yansura D.G. , Richard L. Vandlen R.L. Isolation and Characterization of the B-Cell Marker CD20. *Biochemistry.* 2005; 44: 15150-15158

Fairhurst A.M , Connolly J.E. , Hintz K.A. , Goulding N.J. , Rassias A.J. , Yeager M.P. , Rigby W. , Wallace P.K. Regulation and localisation of endogenous human tristetraprolin. *Arthritis Research and Therapy*. 2003; 5: R214-R225

Ghielmini M. , Schmitz S-F.H. , Cogliatti S.B. , Pichert G. , Hummerjohann J. , Waltzer U. , Fey M.F. , Betticher D.C. , Martinelli G. , Peccatori F. , Hess U. , Zucca E. , Stupp R. , Kovacsovics T. , Helg C. , Lohri A. , Bargetzi M. , Vorobiof D. , Cerny T. Prolonged treatment with rituximab in patients with follicular lymphoma significantly increases event-free survival and response duration compared with the standard weekly X 4 schedule. *Blood*. 2004; 103: 4416-4423

Golay J. , Zaffaroni L. , Vaccari T. , Borleri G.M. , Bernasconi S. , Tedesco F., Rambaldi A. , Introna M. Biologic response of B lymphoma cells to anti-CD20 monoclonal antibody Rituximab in vitro :CD55 and CD59 regulate complement –mediated cell lysis. *Blood*. 2004; 95: 3900-3908

Hainsworth J.D. , Burris III H.A. , Morrissey L.H. , Litchy S. , Scullin Jr D.C. , Bearden III J.D. , Richards P. , Greco F.A. Rituximab monoclonal antibody as initial systemic therapy for patients with low-grade non-Hodgkin lymphoma. *Blood*. 2000; 95: 3052-3056

Hofmeister J.K. , Cooney D. , Coggeshall K.M. Clustered CD20 induced Apoptosis :Src – Family Kinase, the proximal regulator of Tyrosine phosphorylation, Calcium Influx, and Caspase-3 –dependent apoptosis. *Blood Cells, Molecules, and Diseases*. 2000; 26: 133-143

Ino T. , Yasui H. , Hirano M. , Kurosawa Y.Identification of a member of the Tis11 early response gene family at the insertion point of a DNA fragment containing a gene for the T-cell receptor β chain in an acute T- cell Leukemia. *Oncogene*. 1995; 11: 2705-2710

Janas E. , Priest R. , Wilde J.I. , White J.H , Malhotra R. Rituxan (anti-CD20 antibody)-induced translocation of CD20 into lipid rafts for calcium influx and apoptosis. *Clinical and Experimental Immunology*. 2005; 139: 439-446

Kröber A. , Seiler T. , Benner A. , Bullinger L. , Brückle E. , Lichter E. , Döhner H., Stilgenbauer S. *VH* mutation status , CD38 expression level, genomic aberrations and survival in chronic lymphocytic leukaemia. *Blood*. 2002; 100: 1410-1416

Lai W.S, Blackshear P.J. Interactions of CCCH zinc finger proteins with mRNA: tristetraprolin-mediated AU-rich element-dependent mRNA degradation can occur in the absence of a poly(A) tail *The Journal of Biological Chemistry*. 2001; 276: 23144-54

Mathas S. , Rickers A. , Bommert K. , Dürken B. ,Mapara M.Y. Anti-CD20- and B-cellReceptor-mediated Apoptosis:Evidence for Shared Intracellular Signalling pathways. *Cancer Research*. 2000; 60: 7170-7176

Maloney D.G. , Grillo-Lopez A.J. , White C.A. , Bodkin D. , Schilder R.J. , Neidhart J.A. , Janakariman N. , Foon K.A. , Liles T-M. , Dalaire B.K. , Wey K. , Royston I. , Davis T. , Levy R. IDEC-C2B8 (Rituximab) anti-CD20 monoclonal antibody therapy in patients with Relapsed Low-Grade Non-Hodgkin's Lymphoma. *Blood*. 1997; 90: 2188-2195

Murphy J.J. , Norton J.D. Phorbol ester induction of early response gene expression in lymphocytic leukaemia and normal human B cells. *Leukemia Research*. 1993; 17: 657-662

Nédellec S. , Renaudineau Y. , Bordron A. , Berthou C. , Porakishvili N. , Lydyard P.M. , Pers J-O. , Youinou P. B cell response to surface IgM cross-linking identifies different prognostic groups of B-Chronic Lymphocytic Leukemia patients. *The Journal of Immunology*. 2005; 174: 3749-3756

Nie X-F. , Maclean K.N. , Kumar V. , McKay I.A. , Bustin S.A ERF-2 ,the human homologue of the murine Tis11d early response gene. *Gene*. 1995; 152: 285-286

Ning Z.Q. , Norton J.D. , Li J. , Murphy J.J. Distinct mechanisms for rescue from apoptosis in Ramos human B cells by signalling through CD40 and interleukin-4 receptor:role for inhibition of an early response gene,Berg36. *European Journal of Immunology*. 1996; 26: 2356-2363

O'Brien S.M. , Kantarjian H. , Thomas D.A. , Giles F.J. , Freireich E.J. , Cortes J., Lerner S. , Keating M.J. Rituximab dose-escalation trial in chronic lymphocytic leukaemia. *Journal of Clinical Oncology*. 2001; 19: 2165-2170

Ogilvie R.L. , Abelson M. , Hau H.H. , Vlasova I. , Blackshear P.J. , Bohjanen P.R.Tristetraprolin down-regulates *IL-2* gene expression through AU-Rich Element-mediated mRNA decay. *The Journal of Immunology*. 2005;174: 953-961

Pedersen I.M. , Buhi A.M. , Klausen P. , Geisler C.H. , Jurlander J. The chimeric anti-CD20 antibody rituximab induces apoptosis in B-cell chronic lymphocytic leukemia cells through a p38 mitogen activated protein-kinase-dependent pathway. *Blood*. 2002; 99: 1314-1319

Pickartz T. , Ringel F. , Wedde M. , Renz H. , Klein A. , von Neuhoff N. , Dreger P. , Kreuzer K.A. , Schmidt C.A. , Srock S. , Schoeler D. , Schiever F. Selection of B cell chronic lymphocytic leukemia cell variants by therapy with anti-CD20- monoclonal antibody Rituximab. *Experimental Hematology*. 2001; 29: 1410-1416

Renaudineau Y. , Nédellec S. , Berthou C. , Lydyard PM , Youinou P. , Pers J-O. Role of B-cell antigen receptor –associated molecules and lipid rafts in CD5-induced apoptosis of B-CLL cells. *Leukemia*. 2005; 19: , 223-229

Sauer I, Schaljo B, Vogl C, Gattermeier I, Kolbe T, Müller M, Blackshear PJ, and Kovarik P Interferons limit inflammatory responses by induction of tristetraprolin. *Blood*. 2006; 107: 4790 - 4797.

Schiavone N. , Rosini P. , Quattrone A. , Donnini M. , Lapucci A. , Citti L. , Bevilacqua A. , Nicolin A. , Capaccioli S. A conserved AU-rich element in the 3' untranslated region of *bcl2* mRNA is endowed with destabilizing function that is involved in *bcl2* down-regulation during apoptosis. *The FASEB Journal*. 2000; 14: 174-184

Shan D. , Ledbetter J.A. , Press O.W. Signalling events involved in anti-CD20-induced apoptosis of malignant human B cells. *Cancer Immunology Immunotherapy*. 2000; 48: 673-683

Shan D. , Ledbetter J.A. , Press O.W. Apoptosis of malignant human B cells by ligation of CD20 with monoclonal antibodies. *Blood*. 1998; 91: 1644-1652

Stoecklin G. , Colombi M. , Raineri I. , Leuenberger S. , Mallaun M. , Schmidlin M. , Gross B. , Lu M. , Kitamura T. , Moroni C. Functional cloning of BRF1 , a regulator of ARE-dependent mRNA turnover. *The EMBO Journal*. 2002; 21: 4709-4718

Teeling J.L. , French R.R. , Cragg M.S , van den Brakel , Pluyter M. , Huang H. , Chan C. , Parren P.W.H.I. , Hack C.E , Dechant M. , Valerius T. , van de Winkel J.G.J. , Glennie

M.J. Characterisation of new human CD20 monoclonal antibodies with potent cytolytic activity against non-Hodgkin Lymphomas. *Blood.* 2004; 104: 1793-1800

van der Kolk L.E. , Evers L.M. , Omene C. , Lens S.M.A. , Lederman S. , van Lier R.A.W. , van Oers M.H.J , Eldering E. CD20- induced B cell death can bypass mitochondria and caspase activation. *Leukemia.* 2002; 16: 1735-1744

Varnum B.C. , Ma Q. , Chi T. , Fletcher B. , Herschman H.R. The TIS11 primary response gene is a member of a gene family that encodes proteins with a highly conserved sequence containing an unusual Cys-His repeat. *Molecular and Cellular Biology.* 1991; 11:1754-1758

Vega M.I. , Huerta-Yepaz S. , Garban H. , Jazirehi A. , Emmanouilides C. , Bonavida B. Rituximab inhibits p38 MAPK activity in 2F7 B NHL and decreases IL-10 transcription: Pivotal role of p38 MAPK in drug resistance. *Oncogene.* 2004; 23: 3530-3540

Voso M.T. , Pantel G. , Rutella S. , Weis M. , D'Alo F. , Urbano R. , Leone G. , Haas , Hohaus S. Rituximab reduces the number of peripheral blood B-cells *in vitro* mainly by effector cell-mediated mechanisms. *Haematologica.* 2002; 87: 918-925

Wilusz C.J., Wormington M., Peltz S.W. The Cap-to-Tail guide to mRNA turnover. *Nature Reviews in Molecular Cell Biology.* 2001; 2: 237-246

Witzig T.E. , Vukov A.M. , Habermann T.M. , Geyer S. , Kurtin P.J. , Friedenberg W.R. , White W.L. , Chalchal H.J. , Flynn P.J. , Fitch T.R. , Welker D.A. Rituximab therapy for patients with newly diagnosed advanced stage, follicular grade I Non-Hodgin's Lymphoma: a phase II trial in the north central cancer treatment group. *Journal of Clinical Oncology.* 2005; 23:1103-1108

Zupo S. , Isnardi L. , Megna M. , Massara R. , Malavasi F. , Dono M. , Cosulich E. , Ferrarini M. CD38 expression distinguishes two groups of B-cell chronic lymphocytic leukemias with different responses to anti-IgM antibodies and propensity to apoptosis. *Bloo,* 1996; 88: 1365-1374.

In: Chronic Lymphocytic Leukemia Research Focus
Editor: Chadi Nabhan, pp. 129-141
ISBN: 978-1-60021-526-1

Chapter VI

Genetic Factors Impacting the Biology of B Cell Chronic Lymphocytic Leukemia(B-CLL) - Review

Vasantha Brito-Babapulle
Haemato-Oncolology, Institute of Cancer Research,
Royal Marsaden Hospital, Sutton, Surrey, UK

Absract

B-cell chronic lymphocytic leukaemia results from clonal expansion of cells expressing CD19, CD5 and CD23 with reduced levels of membrane IgM, IgD, and CD79b.A primary genetic abnormality characteristic of CLL has not been identified.Yet the monoclonal nature of the B-lymphocytes imply that an inducing lesion must exist in the progenitor clone. The clinical course in CLL is extremely diverse.Despite the heterogeneity the most consistent feature in CLL is the low membrane B cell receptor (BCR) expression. Three areas of genetic factors influencing the biology and prognosis of CLL will be reviewed.

VH mutations. CLL with unmutated immunoglobulin variable heavy chain (VH) gene displays worse prognosis compared to VH mutated CLL. V1-69 is the most common unmutated gene in CLL while the rearrangement of the V3-21 gene has been associated with poor clinical outcome irrespective of the VH mutation status. Biased use of VH genes may reflect antigenic selection in the development of CLL.

The nature of the cell of origin will be discussed.

Genetic imbalances. Losses and gains of DNA have been identified in CLL by fluorescence in situ hybridisation (FISH) and are associated with disease progression..Three consistent recurring abnormalities studied by FISH have been used extensively as prognostic markers. Deletion of 13q14 found in 50% of CLL and when present on its own is considered a good prognostic marker. Deletions of 11q22-23 (20%) is associated with disease progression while 17p deletion (10-15%) identifies the poorest prognostic group. The use of comparative genomic hybridisation (CGH) and array based CGH has allowed identification of other recurrent genetic imbalances.Further epigenetic mechanisms such as methylation can silence genes in the absence of a deletion and inactivating mutation in CLL. The genes of relevance in these regions will be discussed.

Genetic predisposition . The study of inherited susceptibility and identification of susceptibility genes could provide insights into pathogenic mechanism and potential targets for therapeutic intervention. Although familial susceptibility to CLL exists, with the age of onset of CLL being younger in CLL-families than in the general population., to date no gene has been shown to unambiguously confer susceptibility to CLL when mutated in the germ line. The investigation of apparently healthy relatives within CLL families with sublinical levels of CLL clones is an additional approach to increase the numbers of affected individuals to investigate genetic predisposion to CLL.

Introduction

Chronic lymphocytic leukaemia (CLL) is the most common form of leukaemia in the west and predominantly affects elderly individuals but can also occur in younger individuals.It follows an extremely variable clinical course ranging from months to decades. Several studies have attempted to understand the causes that result in stable or progressive CLL [1,2]. Although many new insights into the understanding of the disease are currently available the emerging picture is complex and challenges the concept that CLL cells are immune-incompetent cells which accumulate purely as a consequence of a defective apoptotic meachanism [3]. Despite the monoclonal nature of the cells a primary inducing genetic abnormality or a balanced diagnostic translocation as in other B-cell malignancies has not been identified in CLL. B-CLL is the leukaemic counter part of small cell lymphocytic lymphoma and results from clonal expansion of CD5 + B cells which have reduced levels of membrane IgM, IgD and CD79b , a phenotype of mature activated lymphocytes [4,5]. Genetic factors such as Immunoglobulin (Ig) variable heavy chain (V_H) gene mutational profiles in CLL and certain molecular cytogenetic abnormalities have been important prognostic indicators of disease progression , resistance to therapy and overall survival [6,7,8] . Epigenetic changes to gene transcription without DNA sequence alterations can also contribute to the development of CLL [9]. Further the recently discovered small functional noncoding RNAs called microRNAs have uncovered a hitherto unknown form of gene regulation and may have a role in the pathogenesis of CLL [10].

Immunoglobulin V Gene Mutations

In normal B cells the degree of somatic mutation in the immunoglobulin gene is indicative of the stage of differentiation with unmutated and mutated B cells being pregerminal and germinal centre (GC) B cells respectively. As the follicular B lymphocyte matures from a naïve pre germinal B-cell it rearranges its immunoglobulin V gene segments to create a code for an immunoglobulin molecule which serves as a B-cell receptor for antigen.When an antigen engages the receptor the cell moves to the GC of lymphoid follicles and divides rapidly and the rearranged V gene segments undergo somatic hypermutation. These mutations provide decendent cells with enhanced antigen binding affinity and proliferate in the presence of antigen.This selection pathway usually requires the help of T-cells[11]. CLL - B cells display a phenotype similar to mantle cell with the expression of CD5 antigen and hence was considered to be derived from naïve B-cells like mantle cell lymphoma. However the naïve B-cell has unmutated V genes while ~ 50% CLL have mutated V genes (>2% somatic mutation)[6,11,12]. In CLL with mutated Ig genes the proliferating cell may have transited through germinal centres (GC) the site of hyper mutation while CLL with unmutated Ig genes the malignant B cell could derive from a pregerminal centre naïve B-cell. On the other hand, the molecule activation induced cytidine deaminase (AID) required for somatic mutation and isotype switching , usually present in germinal centre B cells [13] is upregulated at the mRNA level in unmutated CLL generating isotype switched transcripts and protein together with mutation in the pre-switch mu transcripts [14,15].This factor again challenges the notion that unmutated CLL is derived from naïve B-cells .

Although several arguments have been put forward to support the idea that mutated B-CLL derives from memory B-cells exiting the GC, most post GC cells are IgG positive while mutated CLL B cells express IgM and IgD[15]. Another theory is that extra follicular Marginal Zone B (MZB) cells are stimulated by T-cell independent antigens with the production of poly specific antibodies and remain unmutated or subsequently acquire IgV mutations.[16,17,18] There is evidence that these cells can undergo IgV gene mutations outside the germinal centre [19,20]. The absence of diagnostic balanced translocations characteristic of lymphoid malignancies resulting from errors in VDJ or isotype switch recombination in the GC is also considered to be a further argument for the origin of CLL from a MZB cell [21]

Despite its heterogeneity the most consistent feature in CLL is the low membrane B cell receptor (BCR) expression.The receptor consists of surface Ig associated with a CD79a / CD79b heterodimer and plays a key role in signal transduction .In CLL it is associated with defective tyrosine kinase activity despite normal intra cellular transcription of the BCR components [22,23,24].The reason for BCR under expression in CLL remains elusive although evidence exists for the association of glycosylation and folding defects in the mu and CD79a chains of the BCR with low BCR expression[25]

CLL developing from a cell with unmutated V_H gene shows a much worse prognosis than that derived from a cell with mutated V_H genes.[1,2,5,6].Although overall BCR expression in CLL is low unmutated CLL has a better response to stimulation via the BCR pathway than CLL with mutated V genes. CLL lacking V_H mutations express larger numbers of BCR than CLL with mutations and could respond more strongly to stimulation via the BCR [22]. Further the cross linking of BCR to ZAP-70 a receptor associated tyrosine kinase and which shows increased expression in V_H unmutated CLL [26] could result in increased activation after BCR stimulation [27].The increased surface signal response in unmutated CLL has been suggested as mechanism accounting for the clinical differences between V_H mutated and unmutated CLL

An unusual finding in CLL is the over representation of certain V_H genes. Biased use of V_H 1-69 gene is observed in a large proportion of unmutated V_H gene rearrangements. Further V_H1-69+ CLL use preferential rearrangement of certain D (D3-3 or D2-2), J_H (J_H 6) and CDR3 which are longer than average and distinct from those of normal V_H 1-69+ B cells

[28]. Intriguingly, CLLs utilizing a mutated V_H 3-21 have poor overall survival irrespective of mutational status and many of these use J_H6 with almost complete loss of the D segment gene and also use light chain coded by $V_{lambda}2$-14/$J_{lambda}3$genes . However, those using V3-72 though rare are clinically benign [29,30,31]. Thus CLLs with a virtually identical antibody combining site imply that the clonal progenitor B cells arise in response to the same antigen [32]. These findings suggest a possible antigen component in the development of CLL with the hypothesis that specific V_HDJ_H combinations could generate Ig molecules with affinity to similar antigenic epitopes.[33]

Gene Expression

The overexpression of BCL2 the well known antiapoptotic gene in CLL supported the previously held notion that CLL cells were non proliferating cells which accumulated as a result of defective apoptosis[3].This assumption however is currently challenged as p27kip1 a negative regulator of cell cycle progression is found to have an elevated expression in CLL cells [34].Thus deregulation of the cell cycle also contributes to accumulation. Beside evidence for proliferation comes from the finding that the telomere in cells which have undergone division is shorter[35] and CLL B cells have a shorter telomere length than normal CD5+ B cells [36].Unmutated CLL have a significantly shorter telomere length with increased telomerase activity than mutated CLL [36,37] suggesting a more active proliferation in the unmutated CLL.

Despite a clear difference in prognosis and clinical outcome between V_H mutated and unmutated CLL , and although some genes showed a difference in expression , microarray studies displayed a surprisingly similar gene expression profile signature in the two forms of CLL. ZAP70 was found to show the most significant difference between the mutated and unmutated CLL [26]. This correlation however is not absolute and some discordant cases have been described. Although ZAP 70 expression is strongly associated with unmutated CLL it can also be present in some mutated CLL and the median time from diagnosis to treatment is more significant when ZAP70 expressing CLL is compared to ZAP 70 negative CLL than unmutated versus mutated CLL[38].Comparison of gene expression using the Affymetrix Gene Chip between indolent and progressive CLL on a relatively small sample of 21 patients, supervised learning procedures revealed groups of genes with combined powers to discriminate progressive from stable disease.The gene for protein phosphatase 2 regulatory subunit B'(B56) gamma isoform (PPP2R5C) and the gene for retinoblastoma like 2 (RBL2)(p130) were among the best discriminators. Both these were down regulated in progressive CLL. These genes are important regulators of both the cell cycle pathway and mitochondria/ cytochrome c apoptotic pathway [39,40,41]. A larger series is required to confirm these findings. Distinct differences in expression patterns of genes involved in BCR signalling , cell cycling and apoptosis control have been described between those using the V1-69 or V3-21 and V_H unmutated and mutated not using these genes[42,43]. Both proliferation and accumulation appear to be involved in the increase in cell numbers in CLL with ongoing antigenic stimulation playing an important role.

Genetic Imbalances

Molecular cytogenetic aberrations identified by fluorescence in-situ hybridisations using a panel of probes for deletions in 11q22-23 ,13q14.3 and 17p13.1 together with trisomy for 12 have been frequently used in a hierachical model to predict disease progression in CLL[7]. Deletions of 13q14.3 is found in 50-60% of CLL and when present on its own has been considered a marker for good prognosis. Trisomy 12 has often been associated with atypical morphology. Deletion at 11q22-23 was originally identified in patients presenting with extensive nodal disease and rapid disease progression [8] while treatment failure with alkylating agents and fludarabine and short survival times are associated with deletion of 17p13.1 . Both these abnormalities are currently the best molecular cytogenetic prognostic markers in CLL [7,5]. However epigenetic phenomena such as DNA methylation can also produce genetic imbalance and inactivation of tumour suppressor genes by silencing gene transcription without changes to DNA sequences[9,44]

11q22-23 , 17p13.1 and 13q14.3 Deletions

The 11q deletions are identified in CLL by FISH using a probe for the ataxia telangiectasia mutated (ATM) gene which maps to 11q22-23 . CGH and array based CGH studies reveal that these deletions are far more extensive than the segment of DNA containing the.gene[45,46]. ATM protein is the principal activator of the p53 protein in the response to DNA double strand breaks[47,48].Recent studies have shown heterogeneity within patients with 11q deletion. Mutations of ATM have been identified in patients with CLL but the mutations only show partial correlation with 11q deletions [49]. Compared to patients with wild type ATM/TP53 genes those with ATM mutations had a statistically significant reduction in overall and treatment free survival. [49]. The presence of ATM mutations can act upstream of the p53 gene with the loss of p53 function and hence identify patients refractory to treatment [50] . Patients with deletions who have a remaining wild type allele for ATM preserve their DNA damage response[49]. Thus other genes on 11q22-23 such as the micro RNA 34b/34c [51] and their role in disease progression together with gene dosage effects should be considered. ATM mutations are present in both IGVH mutated and unmutated subgroups but are associated significantly with unmutated CLL.

17p deletions have been identified in ~15% of CLL with the probe for TP53 which is localized to 17p13.1 .TP53 is associated with growth arrest , DNA repair and apoptosis. Cytogenetic studies and array based CGH have revealed that a majority of these deletions result from a loss of the major part of 17p with a break point cluster at 17p11.2 [45,46,52] the site which contains low copy repeats [53] and is also a region which contains breakpoints for several abnormalities associated with congenital disorders.

Patients with 17p deletion showing poor survival and resistance to treatment usually have inactivation of TP53 by mutation [45,46,54] and is probably a later event in disease progression. However LOH at 17p13.3 is common in several malignancies including lymphomas and leukemias and many genes associated with malignancies have been identified in this region.[55, 56] while 17p13.2 harbours a gene deleted in medulloblastoma [57].. A

CLL patient with only a 13q14.3 deletion tested with the FISH panel and with no TP53 mutation or deletion at 17p13.1 had a deletion at 17p13.3 identified by array based CGH. The patient had progressed to stage C at diagnosis but showed complete remission on treatment with chlorambucil. Thus progression related genes are likely to be present on 17p13.3 while resistance to treatment being predominantley associated with TP53 inactivation by mutation and deletion of the wild type allele at 17p13.1. Multiple targets on 17p13.1, 17p13.2 and 17p13.3 can be affected by loss of DNA in one homologue which extends from 17 p11.2. - 17pter . Array based CGH has also identified deletions of 18p, 20p and 22q amongst cases with 17p deletion and not in those without , and in some cases these deletions occurred at a higher level than the 17p deletion [45,46].These findings suggest that they arose prior to the TP53 deletion and could be conducive to the development of 17p deletion and TP53 mutation.

Both hemizygous and homozygous deletions at 13q14.3 are present in 50-60% of CLL.The commonly deleted region excludes the RB1 gene and none of the 11 known genes within the 790kb region of loss in this region are inactivated by mutation.[58,59].However the genes for small non coding RNAs or microRNAs miR-15a and miR-16-1which are located in the smallest region of deletion in 13q14.3 are deleted /downregulated in CLL cells.[60].These microRNAs negatively regulate BCL2 at a post transcriptional level[61].Thus down regulation of these microRNAs can contribute to increased expression of BCL2 and they can also be natural antisense interactors to BCL2 which could be used in therapy for BCL2 over expressing CLL. Besides a unique microRNA signature which distinguishes V_H mutated and V_H unmutated CLL has been described [62].The same microRNA signature has been associated with the presence or absence of disease progression.[62]

DNA-Methylation

An enzymatic reaction that results in the addition of a CH3 (methyl group) to the C-5 position of cytosine is called DNA methylation . The enzyme DNA methyl transferase targets small stretches of CpG dinucleotides forming CpG islands in the promoter region or the 1st exon of genes and can silence transcription in the absence of an inactivating mutation. Generally deleted regions in CLL contain many known and unknown genes and establishing the tumour suppressor role of these can be difficult. Besides methylation is a reversible process and genes regulated by this process are attractive targets for therapeutic intervention. Thus exploring epigenetic modification of DNA in which DNA sequence alterations are not involved is a another approach worth exploring to identify genes involved in this fatal disease.

There is evidence that TP53 promoter methylation may also be associated with poor prognosis in CLL [63].The reduction of TP53 gene dosage without an inactivating mutation has been shown to be sufficient to promote tumorigenesis in p53 heterozygous mice[64] and also in hepatocellular carcinoma [65].Thus haploinsufficiency resulting from down regulation in p53 expression by deletion or by promoter methylation in CLL could facilitate expansion of tumoral cells. The epigenetic marker Twist-2 was identified in CLL with significant promoter methylation in the V_H mutated CLL compared to unmutated CLL [66].Interestingly Twist-2 regulates the p19ARF/MDM2/p53 pathway and has antiapoptotic activity[66] .Genome wide methylation analysis using a two dimentional gel electrophoresis and methylation specific restriction enzymes , restriction landmark genome scanning (RLGS) has been performed on CLL revealing several hypermethylated targets which included genes for transcription factors, signalling, apoptosis, drug metabolism and cell adhesion [68,69] which emphasises that multiple pathways are affected by hypermethylation in CLL. Gene specific hypomethylation and upregulation of antiapoptic gene BCL-2 and survival gene TCL1 have also been described in CLL [69,70,71]

Genetic Predisposition

A subset of CLL results from increased genetic susceptibility and over 80 families have been reported with familial clustering of CLL.[72].This inherited predisposition contributing to the development of haematological malignancies was originally hypothesised by Videbaek. The tracing of 222 decendents of the original 57 family members of pedigree 14 described by Videbaek has revealed to date 10 with CLL, 1 with T cell lymphoma and 17 with non haematological cancers[73]. Population and family studies show a seven fold increased risk of CLL and two fold increased risk of Hodgkin lymphoma amongst first degree relatives of patients with CLL [74]. One intriguing feature seen in the expression of CLL in many families reported is the phenomenon of anticipation or the earlier onset (~10-20 years) of the disease[75,76] with a more aggressive clinical course in successive generations [77]. However to date no gene has been found to be mutated in the germline that unambiguously confers a risk of developing CLL.It is probable that inherited susceptibility results from a combination of both high and low risk alleles [78]. A key problem for undertaking linkage analysis is the rarity of multiple- case families with CLL available for study . Clues to early events in the development of CLL can come from the investigation of apparently healthy relatives within CLL families with subclinical levels of CLL clones and is another approach to increase the numbers of affected individuals to investigate genetic predisposion to CLL [78].Monoclonal B-lymphocytes with indolent B CLL characteristics are present in 3.5% of adults with normal blood counts[80, 81]

Conclusion

An inducing primary genetic lesion has not been identified in B-CLL with the most consistent feature being reduced surface BCR expression . Both accumulation and proliferation are involved in the expansion of the CLL clone.

The cell of origin of CLL remains an enigma.The presence of AID in V_H mutated and unmutated CLL however, is compatible with both being derived from an activated antigen experienced B-cell. CLL with the presence of unmutated V_H genes and the use of V_H3-21 irrespective of mutational status are associated with a much worse prognosis. Unmutated CLL shows an increased surface signal response and TWIST-2 a regulator of the p53 pathway is significantly hyper methylated in V_H mutated compared to unmutated CLL . These factors may contribute to the unfavourable prognosis of the V_H unmutated CLL. Further the significant telomere shortening in V_H unmutated CLL is compatible with increased proliferation within this group. The biased presence of specific VDJ combinations for heavy and light chain genes in CLL is suggestive of a role for a common antigen/infectious agent in the development of CLL.

Despite the observed heterogeneity , deletions of 11q22-23 and 17p13 with mutations of ATM gene and TP53 genes are significant poor prognostic features in CLL. Although these occur predominantly in the V_H unmutated CLL they can also occur in V_H mutated CLL. The study of methylation pattern in early stage CLL is important for identifying early and inducing genetic changes in CLL together with comparisons between the methylation status of stable and progressive CLL. Genetic predisposition with anticipation in CLL is well recognised and the study of monoclonal B-lymphocytes with B-CLL characteristics in healthy individuals should be an interesting approach to understand early events in CLL

References

[1] Chiorazzi N, Rai KR , Ferrarini M Chronic Lymphocytic Leukemia . *N. Engl. J. Med.* 2005 352:804 -815

[2] Hamblin TJ. Chronic lymphocytic leukaemia: clinical translations of biological features. *Curr. Top Microbiol. Immunol.* 2005;294:165-85.

[3] Dameshek W. Chronic lymphocytic leukaemia –an accumulative disease of immunologically incompetent lymphocytes. *Blood.* 1967 29:Suppl 566-584

[4] Moreau EJ, Matutes E, A'Hern RP et al. Improvement of the chronic lymphocyctic leukemia scoring system with monoclonal antibody SN8 (CD79b). *Am. J. Clin. Pathol.* 1997; 108:378-82

[5] Binet J-L, Caligaris-Cappio F, Catovsky D et al. Perspectives on the use of new diagnostic tools in the treatment of chronic lymphocytic leukaemia. *Blood.* 2006 107:859-861

[6] Hamblin TJ, Davis Z , Gardiner A, et al. Unmutated Ig V(H) genes are associated with a more aggressive form of chronic lymphocytic leukaemia. *Blood.* 1999 94 :1848-54.

[7] Dohner H, Stilgenbaur S, Benner A, et al. Genomic aberrations and survival in chronic lymphocytic leukemia. *N. Engl. J. Med.* 2000 343 :1910-6.

[8] Dohner H, Stilgenbaur S, James MR et al. 11q deletions identify a new subset of B-cell chronic lymphocytic leukemia characterized by extensive nodal involvement and inferior prognosis. *Blood.* 1997 89:2516-2522

[9] de Capoa A, Grapelli C, Febbo FR , et al. Methylation levels of normal and chronic lymphocytic leukemia B lymphocytes:Computer assisted quantitative analysis of anti-5-methyl cytosine antibody binding to individual nuclei. *Cytometry.* 1999 36:157-159

[10] Calin GA, Croce CM . Genomics of chronic lymphocytic leukemia microRNAs as new players with clinical significance. *Semin. Oncol.* 2006 33:167-173

[11] Kelsoe G. B cell diversification and differentiation in the periphery. *J. Exp. Med.* 1994 180: 5-6

[12] Damle RN, Ghiotto F, Valetto A et al. Chronic lymphocytic leukaemia : A proliferation of B cells at 2 distinct stages of differentiation. *Curr. Tops Microbiol. Immunol.* 2000 252:285-92

[13] Muramatsu M, Sankaranand VS, Anant S et al. Specific expression of activation induced cytidine deaminase (AID), a novel member of the RNA editing deaminase family in germinal centre B cells. *J. Biol. Chem.* 1999 274 : 18470-18476

[14] Heintel D,Kroemer E, Kienle D et al . High expression of activation-induced cytidine deaminase (AID) mRNA is associated with unmutated IGV_H gene status and unfavourable cytogenetic aberrations in patients with chronic lymphocytic leukaemia. *Leukemia.* 2004:18:756-762

[15] Oppezo P, Dighiero G. What do somatic hypermutation and class switch recombination teach us about chronic lymphocytic leukemia pathogenesis? *Curr. Top Microbiol. Immunol.* 2005 294 : 71-89

[16] Damle RN, Ghiotto F, Valetto A et al B-cell chronic lymphocytic leukemia cells express a surface membrane phenotype of activated, antigen experienced B lymphocytes. *Blood.* 2002 99:4087-4093

[17] Martin F, Kearny JF Marginal Zone B-cells. Positive selection from newly formed to marginal zone B cells depends on the rate of clonal production , CD19, and btk. *Nat. Rev. Immunol.* 2002 2:323-335

[18] Chen X, Martin F, Forbush KA, et al Evidence for selection of a population of multi-reactive B cells into the splenic marginal zone. *Int. Immunol.* 1997 9: 27-41

[19] Weller S ,Faili A, Garcia C et al. CD40-CD40L independent Ig gene hypermutation suggests a second B-cell diversication pathway in humans. *Proc. Natl. Acad. Sci. USA.* 2001 98: 1166-1170

[20] William J, Euler C, Christensen S , Shlomchik MJ. Evolution of autoantibody responses via somatic hypermutation responses outside of germinal centers. *Science.* 2002 297: 2066-2070

[21] Klein U, Dalla-Favera R. New insights into the phenotype and cell derivation of B cell chronic lymphocytic leukemia. *Curr. Top. Microbiol. Immunol.* 2005 294 ; 31-41

[22] Stevenson FK, Caligaria-Cappio F Chronic lymphocytic leukemia : revelations from the B-cell receptor. *Blood.* 2004 103:4389-4395

[23] Payelle-Brogard B, Magnac C, Mauro FR et al Analysis of B cell receptor B29(CD79B) gene in familial chronic lymphocytic leukaemia. *Blood.* 1999 94:3516-3522

[24] Alfarano A, Indraccolo S, Circosta P et al An alternatively spliced form of CD79b gene may account for altered BCR expression in B-chronic lymphocytic leukaemia. *Blood.* 1999 93; 2327-2335

[25] Vuillier F, Dumas G, magnac C et al Lower levels of surface B-cell-receptor expression in chronic lymphocytic leukemia are associated with glycosylation and folding defects of mu and CD79a chains. *Blood.* 2005 105:2933-2940

[26] Rosenwald A, Alizadeh AA, Widhopf G et al Relation of gene expression phenotype to immunoglobulin mutation genotype in B-cell chronic lymphocytic leukemia. *J. Exp. Med.* 2001 194 :1639-1647

[27] Chen L, Apgar J, Huynh L et al. ZAP- 70 directly enhances IgM signalling in chronic lymphocytic leukaemia. *Blood.* 2005 105:2036-2041

[28] Potter KN,Orchard J, Critchley E, et al Features of overexpressed V1-69 genes in unmutated subset of chronic lymphocytic leukaemia are distinct from those in healthy elderly repertoire. *Blood.* 2003:101 : 3082-3084

[29] Tobin G, Thunberg U, Johnson A et al. Somatically mutated Ig V(H)3-21 genes characterize a new subset of chronic lymphocytic leukemia. *Blood.* 2002 99:2262-2264

[30] Tobin G,Thunberg U, Johnson A, et al. Chronic lymphocytic leukemias utilizing the VH3 -21 gene display highly restricted V lambda 2-14 and use homologous CDR3s: implicating recognition of a common antigen epitope. *Blood.* 2003 101: 4952-4957

[31] Capello D, Zuchetto A, Degan M et al .Immunophenotypic characterisation of IgVH3-72 Chronic lymphocytic leukaemia (B-CLL). *Leuk. Res.* 2005

[32] Hamblin T . Is chronic lymphocytic leukaemia a response to a infectious agent? *Leuk. Res.* 2006

[33] Tobin G, Rosenquist R Prognostic usage of VH mutation status and its surrogate markers and the role of antigen selection in chronic lymphocytic leukemia. *Medical. Oncology.* 2005 22 217-228

[34] Vrhovac R, Delhemer A, Tang R . Prognostic significance of the cell cycle inhibitor in chronic B cell lymphocytic leukaemia leukemia. *Blood.* 1998 91: 4694 4700

[35] Allsop RC, Chang E, Kashefi-Aazam M et al Telomere shortening is associated with cell division in vitro and in vivo. *Exp. Cell Res.* 1995 220: 194-200

[36] Hultdin M, Rosenquist R, Thunberg U et al Association between telomere length and V(H) mutation status in chronic lymphocytic leukaemia : Clinical and biological implications. *Br. J. Cancer.* 88: 2003 593-598

[37] Damle RN, Batliwalla FM, Ghiotto F, et al . Telomere length and telomerase activity delineate distinctive replicative features of the B-CLL subgroups defined by immunoglobulin V gene mutations. *Blood.* 2004 103 : 375- 382

[38] Rassenti LZ, Haynh L, Toy TL et al. ZAP70 compared to immunoglobulin heavy-chain gene mutation status as a predictor of disease progression in chronic lymphocytic leukaemia. *N. Engl. J. Med.* 2004 351:856-857

[39] Falt S, Merup M, Gahrton G et al. Identification of progression markers in B-CLL by gene expression profiling. *Exp. Hematol.* 2005 883- 893

[40] Van Hoof C,Goris J. Phosphatases in apoptosis: to be or not to be is the heart of the question. *Biochem. Biophys. Acta.* 2003 1640 : 97-104

[41] Vuocolo S, Purev E, Zhang et al. Protein phosphatase 2A associates with Rb2/p130 and mediates retinoic acid - induced growth suppression of ovarian carcinoma cells. i2003 278 41881-41889

[42] Kienle D , Benner A, Krober A et al. Distinct gene expression patterns in chronic lymphocytic leukaemia defined by usage of specific VH genes. *Neoplasia.* 2006 107: 2090 – 2093.

[43] Falt S , Merup M , Tobin G et al Distinctive gene expression pattern in V_H 3-21 utilizing B-cell chronic lymphocytic leukemia. *Blood.* 2005 106: 681 – 689

[44] Chen-WY, Baylin SB. Inactivation of tumour suppressor genes: Choice between genetic and epigenetic routes. *Cell Cycle.* 2005 4: 10-12.

[45] Brito-Babapulle V, Rudenko H, Dearden C et al Bacterial artificial chromosome array CGH study of B-CLL with 17p13 deletion (ASH Annual meeting abstracts). *Blood*

[46] 2005 106:2110

[47] Rudenco HC, Brito-Babapulle-V,Gonzalez-D et al. combined aCGH and mutation analysis of CLLpatients with 17p deletion (ASH Annual meeting abstracts) *Blood* 2006 108: 2091

[48] Shiloh Y. ATM and related protein kinases:safe guarding genome intergrity. *Nat. Rev. Cancer.* 2003 3:155-168

[49] Stancovic T, Hubank M, Cronin et al.Microarray analysis reveals that TP53- and ATM -mutant B-CLLs share a defect in activating proapoptotic responses after DNA damage but are distinguished by major differences in activating prosurvival responses. *Blood.* 2004 103 :291-300

[50] Austen B, Powell J, Alvi-A et al . Mutations of the ATM gene lead to impaired overall and treatment- free survival that is independent of IGVH mutation status in patients with B-CLL. *Blood.* 106:3175-3182

[51] Pettit AR,Sherrington PD, Stewart G, Cawley JC et al. p53 dysfunction in B-cell chronic lymphocytic leukemia : inactivation of ATM as an alternative to TP53 mutation. *Blood.* 2001 98:814-822

[52] Calin GA, Sevigani C, Dumitru CD et al Human micro RNA genes are frequently located at fragile sites and genomic regions involved in cancers. *Proc. Natl. Acad. Sci. USA.* 2004 101: 2999- 3004

[53] Fink SR, Smoley SA, Stockero KJ et al Loss of p53 is due to rearrangements involving chromosome 17p10 to 17p12 in chronic lymphocytic leukemia. *Cancer Genet. Cytogenet.* 2006 167: 177-181

[54] Stankiewicz P, Shaw CJ, Dapper JD et al. Genome architecture catalyzes nonrecurrent chromosomal rearrangements. *Am. J. Hum. Genet.* 2003 72:1101-1116

[55] Thornton PD,Gruszka-Westwood AM, Hamoudi RA et al.Characterisation of TP53 abnormalities in chronic lymphocytic leukaemia. *Haematol. J.* 2004 5:47- 54

[56] Sankar M,Tanaka K, Kumaravel TS et al Identification of a commonly deleted region at 17p 13.3 in leukaemia and lymphoma associated with 17p abnormality. *Leukemia.* 1998 12:510-516

[57] Hoff C, Seranski P, Mollenhauer J et al .Physical and transcriptional mapping of the 17p13.3 region that is frequently deleted in human cancer. *Genomics.* 2000 70:26-33.

[58] Ferretti E, De Smaele E, Di marcotullio L et al . Hedgehog checkpoints in medulloblastoma:the chromosome 17p deletion paradigm. *Trends in molecular medicine.* 2005 11: 537-545

[59] Migliazza A, Bosch F, Komatsu H et al: Nucleotide sequence ,transcription map , and mutation analysis of the 13q14.3 chromosomal region deleted in B-cell chronic lymphocytic leukaemia. *Blood.* 2001 97:2198-2104

[60] Bullrich F, Fujii H,Calin GA et al.Characterisation of 13q14.3 tumour suppressor locus in CLL: identification of ALT1, and alternative splice variant of the LEU 2 gene. *Cancer Res.* 2001 61:6640-6648

[61] Calin GA, Dumitru CD, Shimizu M et al.Frequent deletions and down-regulation of micro-RNA genes miR15 and miR16 at 13q14.3 in chronic lymphocytic leukemia. *Proc. Natl. Acad. Sci. USA.* 2002 99:15524-15529

[62] Cimmino A, Calin GA, Fabbri M et al miR-15 and miR-16 induce apoptosis by targeting BCL2. *Proc. Natl. Acad. Sci.* 2005 102:13944-13949

[63] Calin, Ferracin M, Cimmino A et al A microRNA signature associated with prognosis and progression in chronic lymphocytic leukemia. *N. Engl. J. Med.* 2005 353:1793-1801

[64] Valganon M, Giraldo P, Agirre X et al. p53 abberrations do not predict individual response to fludarabine in patients with B-cell chronic lymphocytic leukaemia in advanced stages Rai III/IV. *British J. of Haematology.* 2005 129:53-59

[65] Venkatachalam S, Shi YP, Jones SN et al .Retention of wild type p53 in tumours from p53 heterozygous mice: reduction in p53 dosage can promote cancer formation. *J. EMBO.* 1998 17:4657-4667

[66] Pogribny IP, James JS. Reduction of p53 expression in human hepatocellular carcinoma is associated with promoter region methylation without coding mutation. *Cancer letters.* 2002 176:169-174

[67] Raval A, Lucas DM, Matkovic JJ et al TWIST2 demonstrates differential methylation in immunoglobulin variable heavy chain mutated and unmutated chronic lymphocytic leukaemia. *J. Clin. Oncol.* 2005 23:3877-3885

[68] Maestro R, Dei Tos AP, Hammori Y et al. Twist is a potential oncogene that inhibits apoptosis. *Genes Dev.* 1999 13:2207-2217

[69] Rush LJ, Raval A, Funchain P, et al Epigenetic profiling in chronic lymphocytic leukemia. *Cancer Res.* 64:2424-2433

[70] Raval A, Byrd JC, Plass C. Epigenetics in chronic lymphocytic leukemia. *Semin. Oncol.* 2006 33:157-166

[71] Hanada M, Delia D, Aiello A et al .BCL2 gene hypomethylation and high-level expression in B- cell chronic lymphocytic leukemia. *Blood.* 1993 82: 1820 – 1828

[72] Yuille MR, Condie A, Stone EM et al : TCL1 is activated by chromosomal rearrangement or by hypomethylation. *Genes Chromosomes and Cancer.* 2001 30: 336 -341

[73] Marti GE, Carter C, Abbasi F et al B-cell monoclonal lymphocytosis and B-cell abnormalities in the setting of familial B-cell chronic lymphocytic leukemia. Cytometry. Part B, *Clinical cytometry.* 2003 52:1-12

[74] Jonsson V, Houlston RS, Catovsky D et al CLL family 'Pedigree 14 revisited : 1947-2004. *Leukemia.* 2005 19:1025-1028

[75] Goldin LR, Pfeiffer RM, Li X, Hemminki K . Familial risk of lymphoproliferative tumours in families of patients with chronic lymphocytic leukaemia : results from the Swedish family-cancer database. *Blood.* 2004 104 :1850-1854

[76] Wiernik PH, Ashwin M, Hu XP et al. Anticipation in familial chronic lymphocytic leukaemia. *Br. J. Haemtology.* 2001 113:407-414

[77] Yuille MR, Houlston RS, Catovsky D: Anticipation in chronic lymphocytic leukemia families. *Leukemia.* 1998 12 1696-8

[78] Ishibe N. Sgambati MT, Fontaine L et al. Clinical Characteristics of familial B-CLL in the National cancer institute familial registry. *Leuk. Lymphoma.* 2001 42:99-108

[79] Sellick GS, Webb EL, Allinson R. et al. A high- density SNP genome wide linkage scan for chronic lymphocytic leukemia –susceptibility loci. *Am. J. Human Genetics.* 2005 77 420-429

[80] Rawstron AC, Yuille MR, Fuller J et al.Inherited predisposition to CLL is detectable as sublinical monoclonal B-lymphocyte expansion. *Blood.* 2002 100:2289-2290

[81] Rawstron AC, Green MJ, Kuzmicki A et al monoclonal B lymphocytes with 'indolent' chronic lymphocytic leukemia are present in 3.5% of adults with normal blood counts. *Blood.* 2002 100:635-639.

[82] Rawstron A, Hillmen P, Houlston R et al Clonal lymphocytes in persons without known chronic lymphocytic leukemia (CLL): Implications of recent findings in family members of CLL patients. *Semin. Hematol.* 2004 41 : 192 -200

In: Chronic Lymphocytic Leukemia Research Focus
Editor: Chadi Nabhan, pp. 143-163
ISBN: 978-1-60021-526-1

Chapter VII

The "In Vitro Apoptosis" Profile in CLL Correlates with Some Prognostic Markers, Treatment History and In Vivo Response to Treatment

A. Janssens, F. Offner, H. Thierens and J. Philippé
Department of Haematology, University Hospital Ghent, Department of Radiopharmacy and Biomedical Physics, Ghent University, Department of Clinical Chemistry, Microbiology and Immunology, University Hospital Ghent, Belgium

Abstract

The objective of the study was to evaluate if prognostic factors as stage, lymphocyte morphology, lymphocyte doubling time, CD38 expression, genomic aberrations, Ig V_H mutational status and ZAP-70 expression correlate with the in vitro apoptosis of CLL lymphocytes. We also wondered if lymphocytes from patients who were treatment naive or were exposed in the past to different treatment regimens would show a different apoptosis profile. We were interested too if this in vitro assay could predict treatment outcome.

In vitro apoptosis of the mononuclear fraction was measured after 24h culturing using the flow cytometric method with Annexin V binding combined with 7-amino actinomycin D (7-AAD). Cells were cultured in cell medium alone, were exposed to pharmacological doses of chlorambucil (Chl), fludarabine (FdA) or methylprednisolone (MP) and were irradiated with 2 Gy.

A total of 187 samples of 110 patients with CLL were successfully analysed

Mean spontaneous apoptosis of lymphocytes was 27.42 %, with a broad range. 2 Gy, Chl, FdA and MP increased apoptosis respectively with an average of 17.9%, 13.25%, 14.99% and 22.04%.

A higher spontaneous and lower Chl and FdA induced apoptosis was found in high risk stage patients, who were almost all treated in the past. A rapid lymphocyte doubling time was associated in untreated and treated patients with a higher MP sensitivity. Less irradiation and more MP induced apoptosis was seen in samples with unmutated V_H

genes. The higher MP sensitivity applied to the treated subgroup only. 11q deletion sensitized lymphocytes to MP apoptosis while 17p deletion made lymphocytes more resistant to FdA without suspected differences between treatment naïve and treated patients. More MP apoptosis was also observed in ZAP-70 positive lymphocytes.

53 patients were treated after assessing in vitro apoptosis, adding up to a total of 73 treatment outcomes, which prospectively could be correlated with the in vitro results. The in vitro assay forecasted the clinical response correctly in 37/44 patients treated with Chl, in 14/17 patients treated with FdA and in 12/12 patients treated with splenic irradiation resulting in an overall predictive accuracy of 86%.

All these results indicate that the Annexin V-7-AAD apoptosis assay is a valuable tool to test drug sensitivity in CLL. Old and new prognostic factors as well as treatment history can modify spontaneous, irradiation or drug induced apoptosis. A good correlation between in vitro results and in vivo responses is translated in a high overall predictive accuracy .

Introduction

Chronic B-lymphocytic leukemia (CLL) is the most common adult leukemia in the western world. CLL is caused by a progressive accumulation of small, morphologically mature B lymphocytes in the blood, bone marrow and lymphoid tissues. These B-lymphocytes have a common immunophenotype that confirms clonality. Typically they are positive for CD5, CD23, CD19 and negative for surface CD22 , CD79b and FMC7. Surface immunoglobulin (usually IgM and IgD) is sparse. [1] Although CLL lymphocytes look mature they are immunologically incompetent. Since most cells are arrested in early G1 of the cell cycle accumulation is not only explained by an excess in proliferation but mainly by a reduction in programmed cell death or apoptosis. Apoptosis is a physiological process whereby most cells, including B-lymphocytes, are eliminated leading to homeostasis. [2] Morphological and biochemical changes characterise apoptosis. Morphologically cell shrinkage, membrane blebbing, chromatin condensation and the formation of apoptotic bodies are seen. This is caused by biochemical events where endonuclease activation leads to execution of the cell by DNA-fragmentation. [3] Although leukemic B-cells show a prolonged life span in vivo they display an enhanced degree of apoptosis in liquid cell cultures. [4-5] The loss of survival factors (cytokines, cell-cell contact, cell-matrix contact) has been held responsible for this change in behaviour. [6-10]

Today there is enough evidence that cytotoxic drugs, corticosteroids and γ-irradiation kill tumour cells through the mechanisms of apoptosis. These agents interact with a specific target such as DNA, RNA or microtubules. This interaction causes injury or dysfunction which is signalled trough the cell and induces apoptosis. It seems that injured cancer cells choose more often the apoptosis pathway in contrast with normal cells which show an arrest of the cell cycle to enable the cell to repair the damage. [11] Since the introduction of cytotoxic chemotherapy for the treatment of cancer it has been clear that patients with identical disease given the same drugs respond differently. De novo or acquired chemo- and radioresistance are major obstacles encountered in the treatment of malignancies, including CLL. The biologic basis for the different therapeutic responses of CLL patients remains greatly unknown. An in vitro assay predicting the clinical response to a particular agent would be of value in the selection of drugs with which to treat CLL. Correlation between in vitro cytotoxicity and in vivo response in CLL has been suggested. [12-15]

Flow cytometry could be the methodology of choice to study apoptosis in CLL because a homogeneous monocellular population of malignant B lymphocytes is easily obtained. [16] In the early stages of apoptosis phosphatidylserine (PS) is translocated from the inner side of the plasma membrane to the outer layer. PS can be bound by Annexin V. PS translocation occurs before morphological changes are visible and before DNA fragmentation occurs. Pepper et al. has shown that the flow cytometric assessment of the Annexin V-PS-binding is more sensitive than morphological quantification, the golden standard till then, for measurement of apoptosis. [17]

CLL patients are staged according to Rai or Binet. [18,19] These stagings are useful in assessing prognosis and starting treatment although these classifications are not accurate enough to identify subgroups with an indolent or a rapidly progressive behaviour. In addition to clinical stage classical parameters with proven prognostic value are age, gender, performance status, number of lymphocytes in the peripheral blood, bone marrow infiltration pattern, atypical morphology and lymphocyte doubling time (LDT). Other variables also have recently been shown to have prognostic value. These include serum markers like LDH, β2-microglobulin, thymidine kinase, serum CD23 and molecular markers like CD38, cytogenetic abnormalities, mutational status of the immunoglobulin heavy chain genes (IGV_H) as well as ZAP-70. [20] Damle initially demonstrated that CD38 expression correlates with prognosis. The relationship with mutational status and which cut-off for CD38 to apply remain a matter of debate. [21-24] Damle and Hamblin have shown in their pioneer studies a significant correlation of the V_H gene mutational status with disease course and survival. This correlation also applied to Binet A patients only. [21,25] Recent studies confirmed the value of Ig V_H mutational status as a prognostic factor independent of conventional clinical markers and genomic aberrations. [21,22] Genomic aberrations are detected with fluorescence in situ hybridisation (FISH) in over 80% of CLL cases today. [28] Genetic subgroups with distinct clinical features have been identified. 11q- is associated with marked lymphadenopathy and rapid disease progression while 17p- predicts for treatment failure with alkylating agents as well as fludarabine and short survival times. The incidence of these high risk genetic aberrations is significantly higher in the patients with unmutated V_H genes. [29] 17p- and/or mutant p53 is also identified as an independent prognostic factor by multivariate analysis. [23,24,26] Because gene sequencing is not available in each diagnostic laboratory, is time consuming and expensive, the search for surrogate markers was necessary. Although the gene expression patterns of mutated and unmutated V_H genes are generally similar DNA-microarray studies have shown that the two subtypes can be distinguished by the differential expression of a small number of genes one of which encodes ZAP-70. [30] ZAP-70 is an intracellular tyrosine kinase with a critical role in T-cell-receptor signalling and is normally not expressed in mature B-cells. However it seems anomalously expressed in CLL cells with unmutated V_H genes. [31] Several studies report on flow cytometric assays in which the intracellular content of ZAP-70 correlates with the IGV_H mutational status. [32,33] The proportion of discordant cases ranged from 5 to 25%. [32-35] It has been shown recently that the number of discordant cases was particularly high in the subgroups with V_H 3-21 usage and 17p or 11q deletion. [35] Therefore it seems too soon to abandon the sequencing of immunoglobulin genes and replace it by ZAP-70 expression.

The objective of this study was as follows: first we wanted to evaluate if prognostic factors as stage, lymphocyte morphology, LDT, CD38 expression, genomic aberrations, Ig V_H mutational status, ZAP-70 expression correlate with the in vitro apoptosis of CLL lymphocytes; secondly we investigated if treatment history influenced these results and thirdly we wondered if there would be a correlation between the in vitro results and the in vivo responses.

Patients, Materials and Methods

CLL Patients

Since 1992, we collected data from all CLL patients seen at the Department of Haematology (Ghent University Hospital). Ninety percent of these patients can be considered as unselected patients while 10% is referred due to progressive disease or treatment problems. Eighty percent of patients are staged as Binet A at diagnosis. The diagnostic criteria for CLL includes a persistent lymphocytosis of more than 5000/µl and a typical immunophenotypic picture defined by Moreau et al. [1] Patients are not treated unless they present with symptomatic advanced disease (Binet stage B or C) or progressive disease. Progressive disease is defined as a lymphocyte doubling time of less than 12 months, development of anemia and/or thrombocytopenia, upgrading of stage, development of systemic symptoms or evidence of transformation to a more aggressive disease. Treatment consists predominantly of chlorambucil in first and second line. Fludarabine is chosen for resistant or chronically relapsing disease and alemtuzumab for resistant or rapidly relapsing disease after fludarabine in accordance with the Belgian state health insurance reimbursement policies.

Morphology

May-Grünwald-Giemsa stained peripheral blood smears were observed by 2 examiners. A total of 200 lymphoid cells were counted per blood sample. Patients showing more than 55% prolymphocytes were excluded as PLL. CLL with typical morphology was characterised by small mature appearing lymphocytes with < 11% of atypical lymphocytes. Atypical CLL is subdivided in CLL/PL when more than 10% prolymphocytes are seen and CLL/mixed cell when 15% atypical lymphocytes are circulating. Prolymphocytes have clear and abundant cytoplasm and a single evident vesicular nucleolus. Large lymphocytes, lymphoplasmocytoid cells , small cleaved cells and prolymphocytes are all counted as atypical lymphocytes. Large lymphocytes are similar to small lymphocytes but the size is larger than 2 RBC. Lymphoplasmocytoid cells show features of large cells with an eccentric and frequently shaped nucleus with abundant basophilic cytoplasm. Small cleaved lymphocytes present with a shallow or a deep narrow nuclear cleft. [36,37]

Cell Cultures

Peripheral blood lymphocytes (PBL) were isolated from EDTA anticoagulated venous blood, using density-gradient centrifugation on Ficoll-Hypaque (Lymphoprep, Nyegaard, Oslo, Norway). PBL were kept in RPMI 1640 culture medium (Gibco Europe, Ghent, Belgium), supplemented with L-glutamine (0.05%), sodium pyruvate (1 mL/100 mL), penicillin (50 IU/mL), streptomycin (50 μg/mL)), and 10% heat inactivated fetal calf serum (all from Gibco BRL, Paisley, Scotland), at a concentration of 1.10^6 cells/mL. One fraction was cultured in medium alone. A second fraction was irradiated in a waterbath at 37°C with gamma rays of a ^{60}Co source at a dose rate of 1.2 Gy minute^{-1} with an absorbed dose of 2Gy (2Gy). Three other fractions of cells were incubated respectively with chlorambucil (Chl) (2.5 μM), fludarabine (FdA) (5 μM) (Schering AG,Berlin, Germany) or methylprednisolon (MP) (10 μM) (SoluMedrol, Upjohn). In order to induce cytotoxicity levels in vitro comparable with those induced in vivo, dosages were used that were within the range of the plasma concentrations of these drugs given at standard doses. The cells were kept in culture for 24h at 37°C in a 95% air plus 5% CO_2 fully humidified incubator (Nuair, Brussels, Belgium). Subsequently the cells were washed and analysed for apoptosis.

Analysis of Apoptotic Cells

The measurement of Annexin V binding to translocated PS, was performed simultaneously with a dye exclusion test. In this way early apoptotic cells could be distinguished from the late stages of apoptosis, called secondary necrosis. 7-Amino-actinomycin D (7-AAD) was used to distinguish cells that had lost membrane integrity (Annexin V positive, 7-AAD positive) from early apoptotic (Annexin V positive, 7-AAD negative) and live cells (Annexin V negative, 7-AAD negative). After incubation with the different ex vivo treatment modalities the cells were washed and stained with CD19-PE (Becton Dickinson, San Jose), followed by staining with AnnexinV (FITC-labeled recombinant human Annexin V, Bender MedSystems, Vienna, Austria) and 7-AAD by resuspending 3.10^5 cells (300 μL cell medium) in 700 μL Hepes buffer (pH 7.4). After centrifugation the pellet was incubated with 1 μL Annexin V-FITC for 15 minutes at room temperature in the dark. Hepes buffer (400 μL) and 7-AAD (1 μg) were added immediately before cells were analyzed on the flow cytometer. A minimum of 1.10^4 cells were analyzed per sample. Using Cellquest Software, the percentage AnnexinV positive cells in the CD19 gated cells was measured. The apoptosis score is the percentage of apoptosis in treated samples minus the percentage of spontaneous apoptosis in cells incubated for 24 hours in control medium only.

Ig V_H Mutational Analysis

Preparation of cDNA and DNA

The preferred source material was RNA, since this reduces the likelihood of amplifying an aberrantly rearranged V_H gene; cDNA was synthesized by reverse transcription using random hexamers (Applied Biosystems, Foster City, USA). If genomic DNA was extracted, we used the QIAamp blood kit (Qiagen, Hilden, Germany).

Amplification of V_H Genes

cDNA (if starting material was RNA) or genomic DNA was amplified by PCR using a mixture of oligonucleotide 5'primers specific for each leader sequence of the V_H *1* to V_H *6* families together with mixed 3'primers complementary to the germ line J_H regions. In our hands, the V_H1 leader primer also amplifies sequences from the closely related V_H7 family. In all cases, PCR was performed in a final volume of 25 µL with 6 pmol of each primer, 24 µmol dNTPs, and 1 U Taq DNA polymerase with reaction buffer (Applied Biosystems). Amplification consisted of an initial denaturation step of 10 minutes at 95°C followed by 40 cycles of 95°C, 55°C, and 72°C for 1 minute each, with a final extension step of 7 minutes at 72°C. For each PCR, a control with no added template was used to check for contamination. Since for leader primers V_H5 and V_H6 we did not get optimal results, we used primers complementary to FR1 regions for these families and the PCR programme was slightly adapted (annealing temperature was increased to 60°C).

Polyacrylamide Gel Electrophoresis

After heteroduplex analysis of the PCR products (10 µL) (10 minutes at 95°C, 60 minutes at 4°C) a polyacrylamide gel analysis was performed during 40 minutes over 200 V and visualised with ethidium bromide under UV light. In this way the monoclonality could be identified and the clonal PCR product was purified using a QIAquick column.

Sequencing of PCR Products

V_H gene DNA sequences were determined by reamplifying 5 µl of the original cDNA using the appropriate V_H leader and C_H primers. These reactions were carried out as follows: denaturation at 96°C for 10 seconds; annealing at 50°C for 5 seconds; and extension at 60°C for 4 minutes. After 25 cycles PCR products were sequenced after purification with Centri-SepTM spin columns (Princeton Separations, Adelphia, USA) or after ethanol precipitation, using an automated sequencer ABI 310 (Applied Biosystems, Foster City, CA). Immediately before sequencing samples were denatured by heating the samples up to 95°C during 2 minutes and keeping them on ice afterwards.

Analyses of V_H, D, and J_H Sequences

Nucleotide sequences were aligned to EMBL/GenBank and current databases (BLAST or V-BASE sequence directories). Mutation status was determined by the following formula:*"100 - [(number of mutations/number of nucleotides compared to the germ line sequence)x 100]"*. In some instances the algorithm of Chang and Casali was used to determine whether "antigen selection" of the replacement ® mutations had occurred, taking into account the inherent susceptibility of CDR to R mutations. Thus, the expected number of R mutations in CDR and FR was calculated using the formula R = $n \times$ *CDR Rf* (or *FR Rf*) $\times$ *CDRrel* (or *FRrel*). Where n is the total number of observed mutations, *Rf* is the replacement frequency inherent to the CDR or FR, and *CDRrel* and *FRrel* are the relative sizes of the segments. A binomial probability model was used to evaluate whether the excess of R mutations in CDR or the scarcity in FR was due to chance.

Detection of CD38 Expression

For measurement of CD38 expression on CD5+ B-cells, we used the CD38 monoclonal antibody (clone T16) from Beckman Coulter (Miami, USA) labelled with FITC. Measurements were performed on a FACSort instrument (BD Biosciences). A negative internal control was used by setting the cut-off on the T-cell population. If a score of higher than 30% was obtained, the sample was considered positive.

Flow Cytometric Analysis of ZAP-70 Expression

Flow cytometric analysis of intracellular ZAP-70 expression was performed on whole blood samples using the method described by Crespo et al. [9]. Briefly, cells were fixed and permeabilized using the Fix and Perm kit (ImTec Diagnostics, Antwerp, Belgium) according to the manufacturer's instructions. Cells were incubated with ZAP-70 antibody (clone 2F3.2, Upstate Biotechnology, Waltham, MA), washed and incubated with PE-labelled goat-anti-mouse antibody (Caltag, Burlingame, California). After washing, cells were incubated with normal mouse serum, washed again and stained with monoclonal antibodies (CD3 PerCP-Cy 5.5, CD5/56 APC, CD19 FITC, all from BD Biosciences, New Jersey, USA). Ten thousand events, unless otherwise specified, were acquired for four-color flow cytometry performed by means of a FACSort flow cytometer and analyzed with Cell Quest Pro Software (BD Biosciences). First, lymphocytes were gated based on forward and side scatter, excluding debris, monocytes or doublets. CD3+ T-cells and CD19+/CD5+/CD3- B-cells were subsequently gated. The approach was based on the percentage ZAP-70 positive CLL B-cells. Bi-parametric dotplot graphs were obtained for CD3 and ZAP-70 or CD19 and ZAP-70. Cut-off for ZAP-70 expression on B-CLL cells was set for 95% ZAP-70 positivity of T-cells.

Fluorescence in Situ Hybridisation (FISH)

Each CLL sample was tested for trisomy 12, deletion of 17p and deletion of 11q by FISH using the following locus specific probes: CEP12 (Abbott Laboratories, North Chicago, Illinois, USA), LSI p53 (17p13.1) (Abbott Laboratories, North Chicago, Illinois, USA) and BAC clone RP11-241D13 labeled in Spectrum Green and Spectrum Red encompassing the

ATM locus at 11q22.3. Hybridization and detection were performed as previously described. [38]

Statistical Analysis

Apoptosis scores of patient groups were compared using the Student's t-test or the Wilcoxon test when the distribution of the variable was not normal. A p-value of less than 0.05 was considered significant.

Results

A total of 187 samples of 110 patients with CLL seen at the Ghent University Hospital were successfully analysed. Blood samples were obtained after informed consent in accordance to institutional guidelines. 66 patients were studied once, 28 patients twice and 16 patients got 3 or more analyses at different time points. Patient characteristics are shown in table 1.

Spontaneous, Irradiation and Drug-Induced Apoptosis

CLL lymphocytes showed a variable spontaneous apoptosis rate (mean 27,42%) with a broad range (standard deviation (SD) 15.24%; 95% confidence interval (CI) 24.51 to 30.32%) at 24 hours. Exposure of CLL cells to γ-irradiation followed by 24h of incubation increased apoptosis with an average of 17.9% (SD 11.56%; 95% CI 15.67 to 20.12%) while exposure of the cells to 2.5µM Chl, 5µM FdA and 10 mM MP followed by 24h of incubation resulted in an increase of apoptosis with an average of respectively 13.25% (SD 11.72%; 95% CI 10.67 to 15.82%), 14.99% (SD 12.81%; 95% CI 12.51 to 17.47%) and 22.04% (SD 12.84%; 95% CI 19.47 to 24.61%). No effect of addition of Chl, FdA, MP or irradiation (2Gy) (defined as less than 5% apoptosis increase compared to spontaneous apoptosis) was seen in 20/82, 24/107, 6/100 and 15/108 patients respectively.

Higher Spontaneous and Lower Chl and FdA Induced Apoptosis in High Risk Stage Patients (figure 1)

Staging according to Binet could be achieved in all patients at the time of the in vitro apoptosis test. 69 patients were staged as Binet A, 16 as Binet B and 25 as Binet C. Spontaneous apoptosis was higher in Binet B and C patients compared to Binet A patients *($p=0.019$)*. Apoptosis increase after Chl and FdA incubation was significantly lower in high risk stage patients than in patients with low tumour burden (Binet C versus A) *($p=0.015$ and $p=0.049$)*. Apoptosis sensitivity to irradiation or MP seems not to be influenced by stage.

Table 1. Characteristics of the study population

Gender	**Patient Characteristics**
male	66/110
female	44/110
Binet staging	
Stage A	69/110
Stage B	16/110
Stage C	25/110
Lymphocyte morphology	
typical	68/110
atypical	42/110
CLL/PL	14/42
CLL/mixed	28/42
Immunophenotype	
CLL score 5/5	76/110
CLL score 4/5	27/110
CLL score 3/5	4/110
kappa light chain	51/98
lambda light chain	47/98
Lymphocyte doubling time	
>12 months	61/110
≤12 months	49/110
V*H* gene rearrangement	
V_H genes < 98%	58/99
V_H genes ≥ 98%	41/99
CD38 expression	
<30%	71/89
≥30%	18/89
ZAP-70 expression	
< 20%	38/61
≥20%	23/61
Genetic aberrations by FISH	
trisomy 12	11/102
11q deletion	25/100
17p deletion	10/102
no t(12), 11q-, 17p-	59/99
Treatment	
naïve	53/110
treated	57/110
alkylators once	21/57
alkylators >1	21/57
different treatments	15/57
Outcome	
alive	64/110
death	46/110

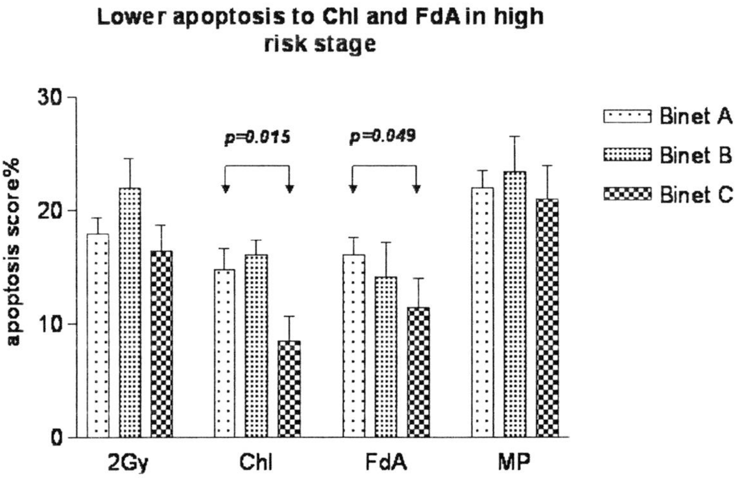

Figure 1. The apoptosis score after Chl and FdA incubation is significantly lower in high risk stage patients than in patients with lower tumour burden. (The apoptosis score is the percentage of apoptosis in treated samples minus the percentage of spontaneous apoptosis in cells incubated for 24 hours in control medium only).

Lymphocytes from Patients with a Rapid LDT Are More Sensitive to MP Apoptosis (figure 2)

LDT could be assessed in all patients at the time of testing. 61 patients showed a LDT of more than 12 months while 49 patients doubled there lymphocytes in less than 12 months. The two groups showed a similar degree of spontaneous apoptosis and apoptosis increase after culture with Chl, FdA or irradiation. However lymphocytes from patients with a rapid doubling time were clearly more sensitive to MP than slowly proliferating ones *(p=0.005)*.

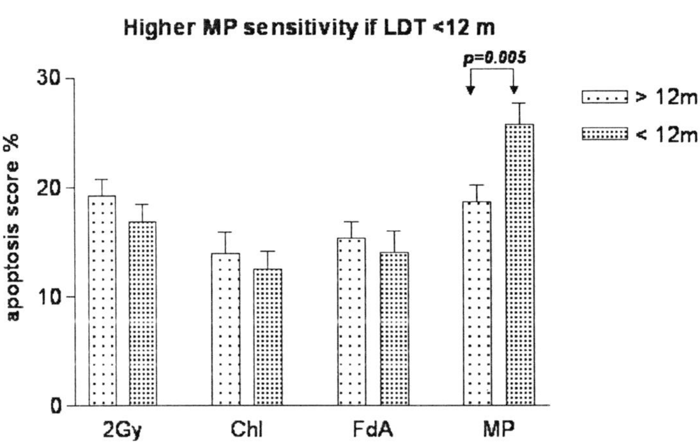

Figure 2. Lymphocytes from patients with a rapid doubling time are more sensitive to MP than slowly proliferating ones.

Lymphocyte Morphology Does Not Influence the Apoptosis Profile

Cytology smears of all patients could be reassessed . No statistical differences were seen between patients with typical and atypical morphology, nor for spontaneous apoptosis nor for irradiation or drug-induced cell death.

The Apoptosis Profile Is Not Altered by CD38 Expression

CD38 expression was measured in 89 patients. 18/89 patients showed ≥30% CD38 expression. No statistical differences were seen between patients with less and patients with more than 30% CD38 expression nor for spontaneous apoptosis nor for irradiation or drug-induced apoptosis.

Less Irradiation and More MP Induced Apoptosis in Samples with Unmutated V_H Genes (figure 3)

Gene sequencing was performed in 99/110 patients. 41 cases showed no rearrangement while 58 patients showed mutated V_H genes. Relative radioresistance was obvious in the unmutated group *(p=0.041)*. The latter seemed also more sensitive to MP compared to the mutated group *(p=0.025)*. Radioresistance persisted in unmutated Binet A stage patients. No difference was seen between the mutated and unmutated subgroups for spontaneous apoptosis or apoptosis to Chl or FdA.

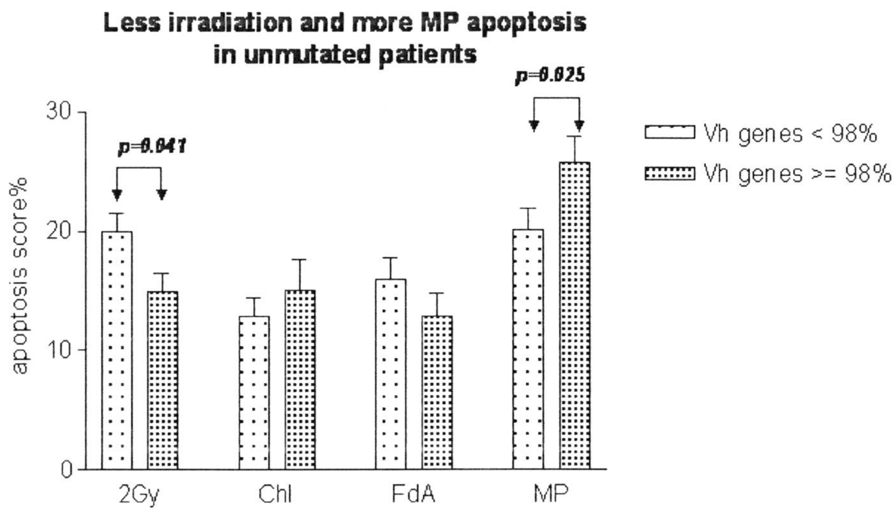

Figure 3. Lymphocytes with unmutated V_H genes show less irradiation and more MP induced apoptosis.

11q Deletion Sensitizes Lymphocytes to MP Apoptosis, 17p Deletion Seems to Make Lymphocytes More Resistant to Fludarabine (figure 4)

25/100 patients analysed by FISH showed the 11q deletion. Patients with the 11q deletion were more sensitive to MP *(p=0.044)* while no differences were seen for spontaneous, irradiation, Chl and FdA induced apoptosis. 17p deletions were found by FISH in 10 /102 patients. In the ten patients with the 17p deletion there was a trend to fludarabine resistance *(p=0.05),* and to Chl and 2Gy to a minor extent. Trisomy 12 was found in 11 of 102 analyzed samples. No differences in apoptosis were seen between patients with or without this trisomy.

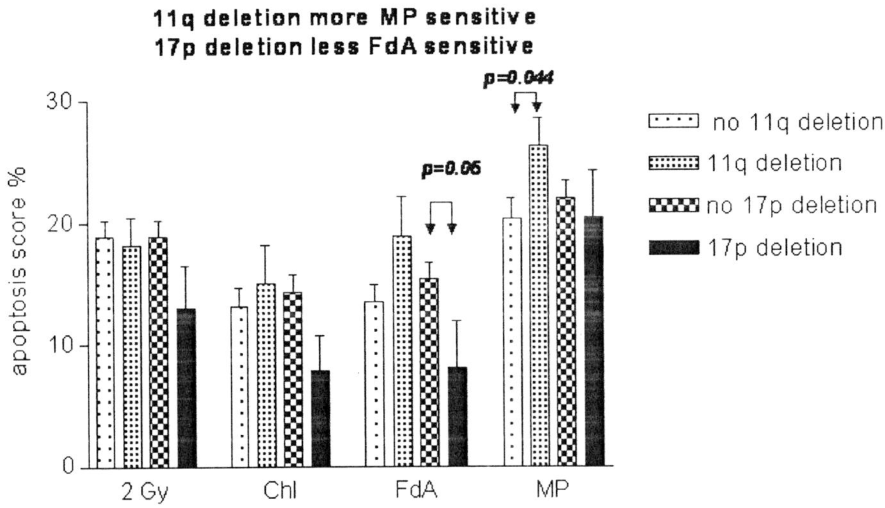

Figure 4. 11q deletion sensitizes lymphocytes to MP apoptosis, 17p deletion seems to make lymphocytes more resistant to fludarabine.

ZAP-70 Expression Renders Lymphocytes More Prone to MP Apoptosis (figure 5)

ZAP-70 results became available for 61 patients. The 23 patients showing ZAP-70 positivity were more sensitive to MP apoptosis than the 38 patients who were ZAP-70 negative *(p= 0.009).*

In Vitro Apoptosis According to Treatment History

53 of the 110 patients studied were treatment naive whereas 57 patients had been treated in the past (>3 months before the apoptosis test). 21 patients got only one treatment course with chlorambucil, 21 patients got multiple courses with alkylating agents only and 15 patients were treated with different cytotoxic agents (fludarabine (13/15), irradiation (splenic 8/15, abdominal 1/15, cervical 1/15), combination chemotherapy) because of alkylator

resistance. Spontaneous apoptosis was comparable for the treatment naive group and the three treatment subgroups. The apoptosis profile for irradiation, Chl, FdA or MP showed no significant difference between the treatment naive and the alkylating agents only treated groups. However in the multi-therapy subgroup apoptosis induction due to Chl *($p<0.05$)* and FdA *($p<0.05$)* was less pronounced. (figure 6) We saw that lymphocytes with a rapid doubling time were clearly more sensitive to MP *($p=0.005$)*. This sensitivity was seen in the untreated as well as in the treated subgroup. The radioresistance observed in the unmutated group *($p=0.023$)* was seen in the treatment naive and the treated patients. The higher MP sensitivity seen in patients with no IGV_H rearrangement applied to the treated subgroup *($p=0.048$)*. (figure 7) The 24 patients with the 11q deletion were more sensitive to MP apoptosis *($p=0.03$)* than patients without deletion. No difference in apoptosis was seen between the four untreated patients and the 20 treated patients (median 25.1%, SD 17.88% vs median 26% SD 10.3%). The trend to fludarabine resistance in the 10 patients with a 17p deletion *($p=0.059$)* was obvious in the six treatment naive and the four treated patients (median 5% SD 15.23% vs median 5% SD 4.44%).

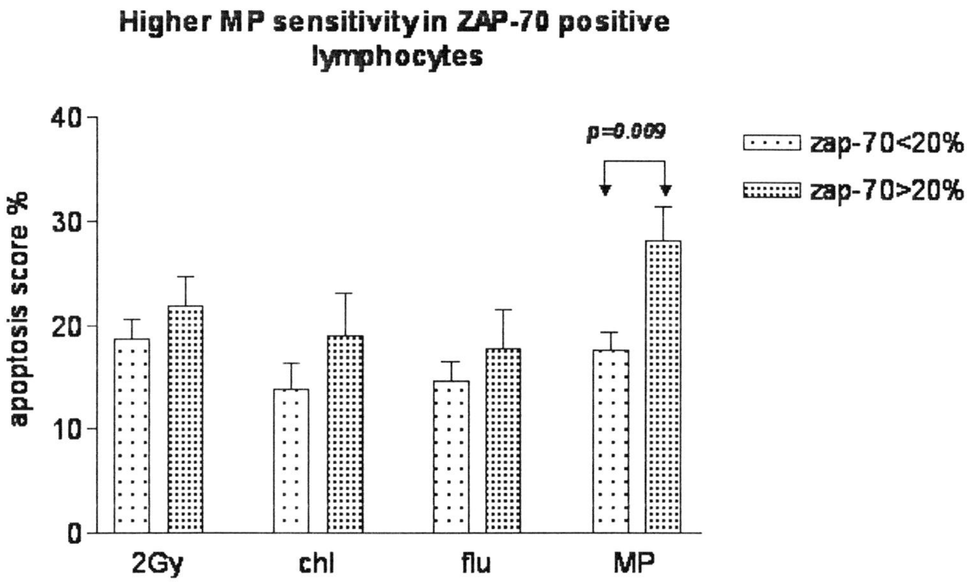

Figure 5. ZAP-70 expression renders lymphocytes more prone to MP apoptosis.

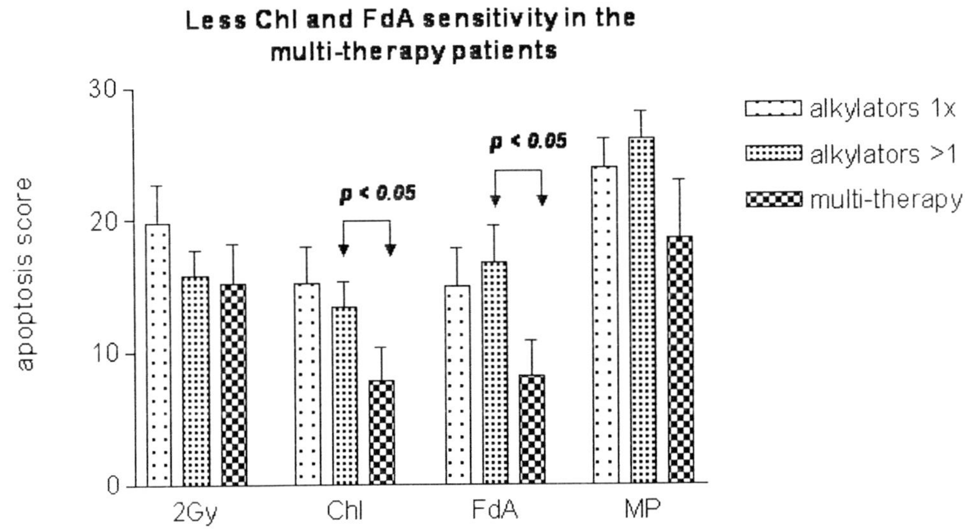

Figure 6. Less pronounced apoptosis induction due to Chl and FdA in the multi-therapy subgroup.

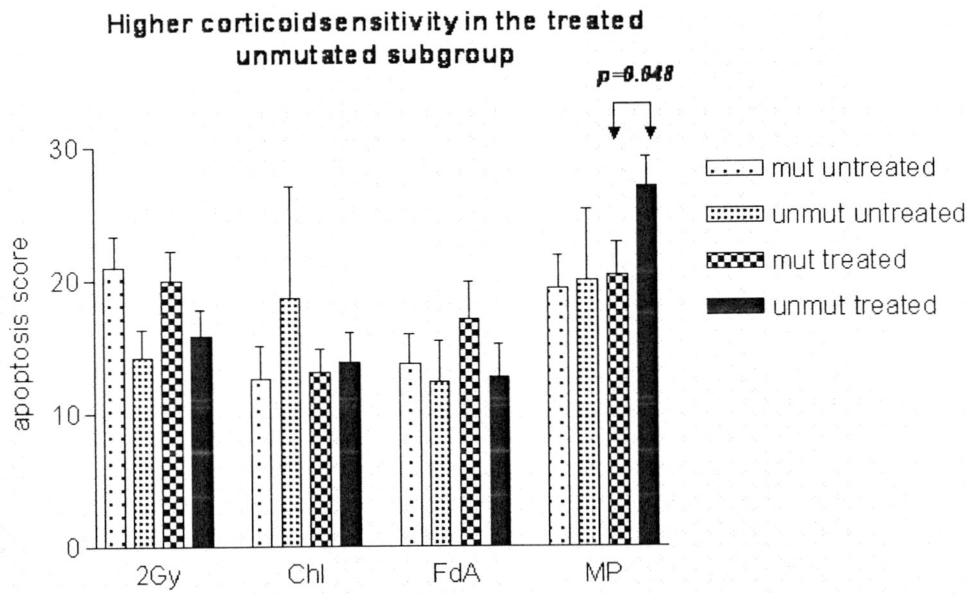

Figure 7. The higher MP sensitivity seen in patients with no IGV_H rearrangement apply to the treated subgroup.

Correlation between In Vitro Apoptosis and In Vivo Response to Chemo- and Radiotherapy

53 patients were treated after assessing in vitro apoptosis, adding up to a total of 73 treatment outcomes. Each treatment outcome was prospectively correlated with the most recently available in vitro test, taken in account that no other cytoreductive treatment had been given in the meanwhile. This was done for 44 treatment episodes with Chl, 17 episodes with FdA and 12 episodes with splenic irradiation. The following figures were calculated for the total group: a sensitivity of 88.5%, a specificity of 70%, a positive predictive value of 95%, a negative predictive value of 56% giving an overall predictive accurary of 86% and an overall response rate of 83.5%. The degree of correlation was extremely high for the irradiation group where no false negative or false positive outcomes were seen. (table 2)

Table 2. Correlation between in vitro apoptosis and in vivo response to chemo- and radiotherapy

	In vitro and in vivo correlation	
	in vivo	
in vitro	sensitive	resistant
sensitive to Chl	33	2
resistant to Chl	5	4
sensitive to FdA	11	1
resistant to FdA	2	3
sensitive to 2Gy	10	0
resistant to 2 Gy	0	2

Conclusion

In vitro chemosensitivity assays have been developed since the nineteen-seventies. Although it is impossible that these tests mimic the complexities of drug delivery and metabolism or mimic the interaction between cancer cells and the microenvironment, a high level of agreement between laboratory and clinical results has been reported. We choose to measure in vitro apoptosis of CLL cells by the flowcytometric assessment of the Annexin V-PS-binding. Like many others we could confirm in our cohort of 110 patients that in vitro apoptosis showed a wide intra- and inter-individual variation. [5,39]

The first and second goal of our study was to look if some old and some more recent prognostic factors together with treatment history could influence spontaneous, irradiation or drug induced in vitro apoptosis. In this study spontaneous apoptosis at 24 hours was not altered by lymphocyte morphology, LDT, CD38 and ZAP-70 expression, cytogenetic aberrations and mutation status. The level of spontaneous cell death was similar for treatment naive patients and patients treated in different ways in the past. Only staging according to Binet revealed a higher spontaneous apoptosis for Binet B and C patients. Rojas et al. observed a higher number of apoptotic cells in his 8/41 high risk Rai patients, compared to the intermediate and good risk stages. [40] Robertson et al. however described greater levels of DNA fragmentation in 8/17 Rai 0 and 1 stage and Kitada et al. did not observe correlations between spontaneous apoptosis and stage. [5, 41] The relationship between the cytogenetic status of the CLL cells and their tendency to undergo apoptosis in vitro was explored by Jahrsdörfer et al. in 22 patients. They found an association between poor prognosis cytogenetics (17p deletion, trisomy 12 and 11q deletion, n=11) and a higher rate of spontaneous in vitro apoptosis. On the contrary CLL cells with normal karyotype and 13q deletion (n=11) showed less spontaneous apoptosis. These authors found no correlation between spontaneous survival in vitro and white blood cell count, Rai stage, CD38 expression, β2-microglobulin level and platelet count. [42] Ringhausen et al. saw no difference in spontaneous apoptosis for ZAP-70 positive (n=13) and ZAP-70 negative (n=10) cases. [43]

Indolent non-Hodgkin's lymphoma and CLL are highly sensitive to radiotherapy. However the indications for irradiation as a treatment modality in CLL become more and more restricted with the availability of effective antineoplastic drugs. In some clinical situations however splenic irradiation or irradiation of localised lymph nodes can be considered as an effective palliative treatment with little toxicity and a low cost. Masdehors et al. showed that CLL-lymphocytes could be hypersensitive to irradiation or completely resistant to any applied dose. In some rare cases irradiation even inhibited spontaneous apoptosis. [44] In our hands, in vitro apoptosis induced by irradiation seems not to be influenced by stage, lymphocyte morphology, LDT, CD38 and ZAP-70 expression, cytogenetic aberrations and treatment history. However, samples from patients with unmutated V_H genes were less radiosensitive than the mutated counterpart. The lack in radiosensitivity was the same for untreated and treated patients. Kivekas et al. explored radiosensitivity in 16 unmutated and 8 mutated patients previously untreated. No significant difference was seen for irradiation. [45]

Chlorambucil is the oldest and best known treatment for CLL. It is available as a cheap and well-tolerated oral preparation. Treatment with chlorambucil can induce partial remissions in approximately 70% of previously untreated patients with only a few complete responders. After a response of variable duration the drug can be reintroduced, however with a lower response rate and a shorter duration of response. Finally resistance may develop. Combinations with alkylating agents induce similar or better responses compared to Chl but no improvement in survival has been seen. Lower Chl sensitivity in vitro was seen in our patients with high risk stages. Most patients (23/25) were already treated before the in vitro testing. Chl induced apoptosis seemed not to be dependent of lymphocyte morphology, LDT, CD38 and ZAP-70 expression, cytogenetic aberrations and mutation status. No difference in apoptosis was seen between untreated and treated patients and patients treated with alkylating agents only once or several times. Patients treated with different cytotoxic agents were more resistant to Chl and FdA than patients treated with alkylating agents only.

Fludarabine has been the most widely tested and used purine analogue in CLL. In many countries it has become the standard treatment for relapsing or refractory CLL. In treatment naive patients fludarabine induced even more remissions and complete responders than alkylating agents. Despite the higher efficacity and the longer median duration of remission no gain in overall survival has been noticed till today. The results gathered for FdA in vitro apoptosis were very similar to those of Chl except that the lower FdA sensitivity in high risk stage patients was not statistically different. Consoli et al. saw also a higher toxicity of fludarabine in Rai 0-1 compared to Rai 2-3-4 while Castejon et al. could not demonstrate a difference in apoptosis induced by 2-chlorodeoxyadenosine between Rai 0-1 (n=11) and more advanced stages (n=11). (46,47) Furthermore there was also a trend to fludarabine resistance in the 10 patients with the unfavourable 17p deletion. Koski et al. saw that the patients with a 17p deletion (n=3) were the most resistant to 2-chloroseoxyadenosine and fludarabine, although none of them was previously treated with purine analogs. Two of these 3 patients were also very resistant to Chl. They also observed drug resistance to Chl and FdA in the subgroup with complex genomic changes (n=11). [48]

It has been known for a long time that corticosteroids are lymphotoxic. Therefore steroids are still an essential part of the treatment of acute lymphoblastic leukaemia and mature lymphomas. However the efficacy of steroids at conventional dose given as monotherapy or in combination with alkylating agents or purine analogues in CLL was not impressive. Samples from CLL patients with unfavourable prognostic factors like a rapid lymphocyte doubling time, unmutated V_H genes, 11q deletion and ZAP-70 expression were more sensitive to MP induced apoptosis in our in vitro assay. Kivekas et al. confirmed that unmutated CLL cases were more sensitive to prednisolone than the mutated cases. *(p=0.04)* Treatment with alkylating agents only increased also the MP sensitivity. Bosanquet et al. already published in 1995 that cells from previously treated patients were significantly more sensitive to MP than those from untreated patients. [14] Koski et al. showed that the prednisolone sensitivity was retained in patients with complex chromosomal abnormalities (7/35). [48]

The third goal of our study was to look if there was a correlation between the in vitro apoptosis results and the in vivo responses. This assay forecasts the clinical response correctly in 37/44 patients treated with Chl, in 14/17 patients treated with FdA and in 12/12 patients treated with splenic irradiation. An overall predictive accuracy of 86% was calculated. This compared well to the differential staining cytotoxicity (DiSC) results. The DiSC assay, a 4-day in vitro chemosensitivity test designed to identify the sensitivity or resistance of fresh tumour cells isolated from patients, compared over 500 assay results with subsequent CLL patient responses and found a predictive accuracy of 83%. These authors concluded that the choice of treatment could be optimized and the response rates improved by using the assay results. [49] They also showed that in vitro drug sensitivity is an important independent prognostic variable and that health resources could be used more cost-effective. [50] Silber et al. showed concordance between in vitro response to Chl, using the 3-(4,5 dimethylthiazol-2-yl)-2,5-diphenyltetrazolium bromide (MTT) assay and the effect of therapy on the patient's clinical and absolute lymphocyte count. [12] Chemosensitivity to fludarabine was assessed with the terminal deoxynucleotidyl transferase (TdT) assay by Bromidge et al. The assay predicted the clinical response correctly in 10/12 patients. However discordance was seen in patients with p53 mutations. [15] Bosanquet et al. showed that treating fludarabine-test-resistant patients with fludarabine resulted in poor response compared with fludarabine-test-sensitive patients (7% *v* 69%) and short survival (median 7.9 *v* 41.7 months). [51] Bosanquet et al. reported also on the outcome of six patients who underwent splenic irradiation and from which CLL samples where irradiated in vitro. The five patients who showed extreme radioresistance did not respond to splenic irradiation while the sixth patient being radiosensitive in vitro obtained a complete response. [52] A high correlation between tumour response and in vitro response to irradiation was reported in 14 patients by Decaudin et al. They suggested that the sensitivity of leukemic cells to irradiation should be first evaluated in an in vitro assay to spare refractory patients from the useless toxicity of radiotherapy. [53] The DiSC assay learned that some end stage patients were highly sensitive to steroids. When these patients were treated with high dose methylprednisolone (1g/m^2 x 5 days) partial responders were seen. [14]

All these results indicate that the Annexin V-7-AAD apoptosis assay is a valuable tool to test drug sensitivity in CLL. Old and new prognostic factors as well as treatment history can modify spontaneous, irradiation or drug induced apoptosis. A good correlation between in vitro results and in vivo responses is translated in a high overall predictive accuracy .

References

[1] Moreau, E. and Matutes, E. (1997). Improvement of the chronic lymphocytic leukemia scoring system with the monoclonal antibody SN8 (CD79b). *Am. J. Pathol,* 108, 378-382

[2] Krammer, P. and Behrmann, I. (1994). Regulation of apoptosis in the immune system. *Curr. Opin. Immunol*, 6, 279-289

[3] Wyllie, A. and Kerr, J. (1980). Cell death: the significance of apoptosis. *Int. Rev. of Cytol*, 68, 251-306

[4] Collins, R.and Verschuer, L. (1989). Spontaneous programmed cell death (apoptosis) of B-chronic lymphocytic leukemia cells following their culture in vitro. *Br. J. Haematol*, 71, 343-350

[5] Robertson, L. and Chubb, S. (1993). Induction of apoptotic cell death in chronic lymphocytic leukaemia by 2-chloro-2'deoxyadenosine and 9- □-D-arabinosyl-2-fluoroadenine. *Blood*, 81, 143-150

[6] Panayiotidis, P. and Jones, D. (1996). Human bone marrow stromal cells prevent apoptosis and support the survival of chronic lymphocytic leukemia cells in vitro. *Br. J. Haematol*, 92, 97-103

[7] Jewell, A. and Yong, K. (1997). Regulation and function of adhesion molecules in B-cell chronic lymphocytic leukemia. *Acta Haematol*, 97, 67-72

[8] Lagneaux, L. and Delforge, A. (1998). Chronic lymphocytic leukemic B cells but not normal B cells are rescued from apoptosis by contact with normal bone marrow stromal cells. *Blood*, 91, 2387-239

[9] Burger, J. and Tsukada, N. (2000). Blood-derived nurse cells protect chronic lymphocytic leukemia B cells from spontaneous apoptosis through stromal cell-derived factor-1. *Blood*, 96, 2655-2663

[10] Munk Petersen, I. and Reed, J. (2004). Microenvironmental interactions and survival of CLL B-cells. *Leuk. Lymphoma*, 45, 2365-2372

[11] Hannun, Y. (1997). Apoptosis and the dilemma of cancer chemotherapy. *Blood*, 89, 1845-1853

[12] Silber, R.and Degar, B. (1994). Chemosensitivity of lymphocytes from patients with B-cell chronic lymphocytic leukemia to chlorambucil, fludarabine, and camptothecin analogs. *Blood*, 84, 3440-3446

[13] Begleiter, A. and Lee, K. (1994). Chlorambucil induced apoptosis in chronic lymphocytic leukemia (CLL) and its relationship to clinical efficacy. *Leukemia*, 8, S103-106

[14] Bosanquet, A. and McCann, S. (1995). Methylprednisolone in advanced chronic lymphocytic leukemia: rationale for, and effectiveness of treatment suggested by DiSC assay. *Acta Haematol*,1995, 93, 73-79

[15] Bromidge, T. and Turner, D. (1998). In vitro chemosensitivity of chronic lymphocytic leukemia to purine analogues-correlation with clinical course. *Leukemia*, 12, 1230-1235

[16] Vermes, I. and Haanen, C. (2000). Flow cytometry of apoptotic cell death. *J. Immunol. Methods*, 243, 167-190

[17] Pepper, C. and Thomas, A. (1998). Flow cytometric assessment of three different methods for the measurement of in vitro apoptosis. *Leuk. Res*, 22, 439-444

[18] Rai, K. and Sawitsky, A. (1975). Clinical staging of chronic lymphocytic leukemia. *Blood*, 46, 219-234

[19] Binet, JL. and Auquier, A. (1981). A new prognostic classification of CLL derived from a multivariate survival analysis. *Cancer*, 48, 198-206

[20] Montserrat, E. (2002).Classical and new prognostic factors in chronic lymphocytic leukemia: where to now? *Hematology J*, 3, 7-9

[21] Damle, R. and Wasil, T. (1999). Ig V gene mutation status and CD38 expression as novel prognostic indicators in chronic lymphocytic leukemia. *Blood*, 94, 1840-1847

[22] Ibrahim, S. and Keating, M. (2001). CD38 expression as an important prognostic factor in B-cell chronic lymphocytic leukemia. *Blood*, 98, 181-186

[23] Lin, K. and Sherrington, P. (2002). Relationship between p53 dysfunction, CD38 expression, and IgV_H mutation in chronic lymphocytic leukemia. *Blood*, 100, 1404-1409

[24] Chevallier, P. and Penther, D. (2002). CD38 expression and secondary 17p deletion are important prognostic factors in chronic lymphocytic leukemia. *Br. J. Haematol*, 116, 142-150

[25] Hamblin, T. and Davies, Z. (1999). Unmutated Ig V_H genes are associated with a more aggressive form of chronic lymphocytic leukemia. *Blood*, 94, 1848-1854

[26] Oscier, D. and Gardiner, A. (2002). Multivariate analysis of prognostic factors in CLL: clinical stage, IgV_H gene mutational status, and loss or mutation of the *p53* gene are independent prognostic factors. *Blood*, 100, 1177-1184

[27] Kröber, A. and Seiler, T. (2002). V_H mutation status, CD38 expression level, genomic aberrations and survival in chronic lymphocytic leukemia. *Blood*, 100, 1410-1416

[28] Döhner, H. and Stilgenbauer, S. (2000). Genomic aberrations and survival in chronic lymphocytic leukemia. *N. Engl. J. Med*, 343, 1910-1916

[29] Stilgenbauer, S. and Bullinger, L. (2002). Genetics of chronic lymphocytic leukemia: genomic aberrations and V_H gene mutation status in pathogenesis and clinical course. *Leukemia*, 16, 993-1007

[30] Rosenwald, A. and Alizadeh, A. (2001). Relation of gene expression phenotype to immunoglobulin mutation genotype in B cell chronic lymphocytic leukaemia. *J. Exp. Med*, 194, 1639-1647

[31] Wiestner, A. and Rosenwald, A. (2003). ZAP-70 expression identifies a chronic lymphocytic leukemia subtype with unmutated immunoglobulin genes, inferior clinical outcome and distinct gene expression profile. *Blood*, 101, 1944-4951

[32] Crespo, M. and Bosch, F. (2003). ZAP-70 expression as a surrogate for immunoglobulin-variable-region mutations in chronic lymphocytic leukemia. *N. Engl. J. Med*, 348, 1764-1775

[33] Orchard, J. and Ibbotson, R. (2004). ZAP-70 expression and prognosis in chronic lymphocytic leukemia. *Lancet*, 363, 105-111

[34] Rassenti, L. and Huynh, L. (2004). ZAP-70 compared with immunoglobulin heavy-chain gene mutation status as a predictor of disease progression in chronic lymphocytic leukemia. *N. Engl. J. Med*, 351, 893-901

[35] Kröber, J. and Bloehdorn, J. (2006). Additional genetic high-risk features such as 11q deletion, 17p deletion, and V3-21 usage characterize discordance of ZAP-70 and V_H mutation status in chronic lymphocytic leukemia. *J. Clin. Oncol*, 24, 969-975

[36] Matutes, E. and Polliack, A. (2000). Morphological and immunophenothypic features of chronic lymphocytic leukemia. *Rev. Clin. Exp. Hematol*, 4.1, 22-47

[37] Criel, A. and Verhoef, G. (1997). Further characterization of morphologically defined typical and atypical CLL: a clinical, immunophenotypic, cytogenetic and prognostic study on 390 cases. *Br. J. Haematol*, 97, 383-391

[38] Van Roy, N. and Laureys, G. (1994). 1;17 translocations and other chromosome 17 rearrangements in human primary neuroblastoma tumors and cell lines. *Genes Chromosomes Cancer*, 10, 103-114.

[39] Kivekäs, I. and Vilpo, L. (2002). Relationships of in vitro sensitivities tested with nine drugs and two types of irradiation in chronic lymphocytic leukemia. *Leuk. Res*, 26, 1035-1041

[40] Rojas, R. and Roman, J. (1996). Inhibition of apoptotic cell death in B-CLL by interferon gamma correlates with clinical stage. *Leukemia,* 10, 1782-1788

[41] Kitada, S. and Anderson, J. (1998). Expression of apoptosis-regulating proteins in chronic lymphocytic leukemia: correlations with in vitro and in vivo chemoresponses. *Blood*, 91, 3379-3389

[42] Jahrsdörfer, B. and Wooldridge, J. (2005). Good prognosis cytogenetics in B-cell chronic lymphocytic leukemia is associated in vitro with low susceptibility to apoptosis and enhanced immunogenicity. *Leukemia*, 19, 759-766

[43] Ringhausen, I. and Oelsner, M. (2006). Mechanisms of apoptosis-induction by rottlerin: therapeutic implications for B-CLL. *Leukemia*, 20, 514-520

[44] Masdehors, P. and Omura, S. (1999). Increased sensitivity of CLL-derived lymphocytes to apoptotic death activation by the proteasome-specific inhibitor lactacystin. *Br. J. Haematol*, 105, 752-757

[45] Kivekäs, I. and Tobin, G. (2003). Ex vivo drug and radiation sensitivities in hypermutated and unmutated forms of chronic lymphocytic leukemia cells. *Leuk. Res,* 27, 337-341

[46] Consoli, U. and El-Tounsi, I. (1998) Differential induction of apoptosis by fludarabine monophosphate in leukemic B and normal T cells in chronic lymphocytic leukemia. *Blood*, 91, 1742-1748

[47] Castejon, R. and Vargas, J. (1997). Induction of apoptosis by 2-chlorodeoxyadenosine in B-cell chronic lymphocytic leukemia. *Leukemia*, 11, 1253-1257

[48] Koski, T. and Karhu, R. (2000). Complex chromosomal aberrations in chronic lymphocytic leukemia are associated with cellular drug and irradiation resistance. *Eur. J. Hematol*, 65, 32-39

[49] Bosanquet, A. and Bell, P. (1996). Enhanced *ex vivo* drug sensitivity testing of chronic lymphocytic leukemia using refined DiSC assay methodology. *Leuk. Res*, 20, 143-153

[50] Mason, J. and Drummond, M. (1999). The DiSC assay: a cost-effective guide to treatment for chronic lymphocytic leukemia? *Int. J. Technol. assess in health care*, 15:1, 173-184

[51] Bosanquet, A. and Johnson, S. (1999). Prognosis for fludarabine therapy of chronic lymphocytic leukaemia based on ex vivo drug response by DiSC assay. *Br. J. Haematol*, 106, 71-77

[52] Bosanquet, A. (1991). Correlations between therapeutic response of leukaemias and in-vitro drug sensitivity assay. *Lancet*, 337, 711-714

[53] Decaudin, D. and Delic, J. (2002). Clinical efficacity of irradiation in CLL patients: predictive value of in vitro radio-induced apoptosis. *Leuk. Lymphoma*, 43, 827-829

In: Chronic Lymphocytic Leukemia Research Focus
Editor: Chadi Nabhan, pp. 165-181
ISBN: 978-1-60021-526-1

Chapter VIII

CD38 Expression in B-Cell Chronic Lymphocytic Leukemia: What Have We Learned?

Giovanni D'Arena [1], *Tiziana Vaisitti* [2], *Semra Aydin* [2] *and Silvia Deaglio* [2]
[1] Hematology Oncology Unit, National Cancer Institute, IRCCS Fondazione "G. Pascale", Naples, Italy
[2] Lymphocyte Signaling Unit, Laboratory of Immunogenetics, Department of Genetics, Biology and Biochemistry and CeRMS, University of Torino, Medical School, Torino, Italy.

Abstract

The considerably variable clinical course (indolent to aggressive) of B-cell chronic lymphocytic leukemia (B-CLL) addressed clinical scientists to the search for biological prognostic indicators. The tumor burden-based staging systems developed almost three decades ago was followed by a quest involving a vast array of clinical and biological parameters. More recently, the analysis of IgVH mutation status provided the most powerful prognostic indicator in B-CLL. Beside Ig, the expression of CD38 was reported as having an independent prognostic value and to correlate – at least partially - with the absence of mutations in the IgVH genes. The consequence is that CD38 was proposed as an easy test, and an inexpensive surrogate of the Ig status. However, the validity of CD38 expression as a prognostic marker was not confirmed by other investigators and gave origin to debate still open. This chapter represents a contribute to the analysis of issues and problems linked to the use of CD38 as a prognostic marker in B-CLL.

Introduction

Clinical presentation and course in B-cell chronic lymphocytic leukemia (B-CLL) are far from uniform [1]. One third of B-CLL patients survives for a long time without treatment, while another third has an initial indolent phase followed by a progressive disease. The final third presents with an aggressive disease from the start and requires prompt treatment [2,3]. These observations prompted the search for prognostic parameters allowing a dependable identification of the high-risk patients, *i.e.* those who will benefit from aggressive therapies immediately after diagnosis [4].

The staging systems for B-CLL, proposed by Rai and Binet were based on tumor burden and allowed a rough discrimination among patients with good, intermediate and poor prognosis [5,6]. Thus, attention was turned to laboratory parameters, including lymphocyte doubling time, β_2-microglobulin levels, cell morphology and pattern of bone marrow involvement, among the others [7,8].

More recently, two groups independently reported that the mutational status of the IgVH genes is highly predictive of clinical outcome [9,10]. Absence of mutations in the IgVH genes correlates with a poor prognosis, while their presence is associated with an indolent course of the disease [11]. CD38 was initially proposed by the Chiorazzi group as a surrogate marker of the absence of IgVH mutations [9]. Although subsequent studies failed to show a perfect correlation between absence of mutations and CD38 expression, it is now clear that CD38 is an independent negative prognostic marker in B-CLL, irrespective of the genetic background of the patients. Open questions still remain concerning methodological issues as well more conceptual ones on the biological role of CD38 in B-CLL.

CD38: A Molecule with a Widespread Expression and Multifunctional Roles

CD38, a type II single chain transmembrane glycoprotein of ~ 45 kDa encoded by a gene localized on the short arm of chromosome 4 (4p15), was firstly identified by Reinherz et al in 1980 [12,13]. It is a nonlineage-restricted molecule, with a distinct pattern of expression within the B cell compartment, where B cell precursors in the bone marrow and plasma cells display the highest levels of expression (table 1) [14-23]. In recent years, CD38 has emerged as a multi-functional protein simultaneously endowed with ecto-enzymatic and receptorial functions [17]. A catalytic site was mapped in the extracellular region of the molecule, close to the COOH terminus [24]. The substrates so far identified are NAD and NAAPD, which are metabolized to ADPribose with the generation of cyclic ADP ribose as a potent calcium-mobilizing intermediate compound. Apparently independently of the catalytic activities, CD38 also works as a receptor through the interaction with CD31, a cell surface bound and non substrate ligand [25]. The binding by agonistic monoclonal antibodies (mAbs) - which can be mimicked in vitro - starts a signaling cascade which differs in the cell lineages where CD38 is expressed. The signaling outcome in the B cell lineage include opposite events, such as inhibition of differentiation of B cell precursors in the bone marrow and rescue from apoptosis in the germinal center [20-23].

Table 1. Cellular Distribution and Putative Functions of CD38 Molecule

Tissue	Cellular distribution	Function
Bone marrow	Hematologic precursors plasma cells	Homing and apoptosis; marker of precursor cell commitment
Thymus	Throughout thymic development	Unknown
Spleen/lymph nodes	Germinal center B cells	Rescue from apoptosis
Blood	T, B, NK and monocyte subsets; platelets and erythrocytes	Interaction with endothelium
Gut	Intra epithelial and *lamina propria* lymphocytes	Mucosal immunity
Brain	Purkinje cells; neurofibrillary tangles	Memory process
Prostate	Epithelial cells	Unknown
Pancreas	β-cells	Insulin secretion
Bone	Osteoclasts	Bone resorption
Eye	Retinal cells	Vision process
Muscle	Sarcolemma of smooth and striated muscle	Muscle contraction (?)

CD38 and B-CLL

According to the mutational status of immunoglobulin variable (IgV) genes, B-CLL can be classified in two major subgroups. The first one carrying somatic mutations and possibly stemming from "memory" B-cells challenged by antigens in the germinal center of secondary follicles; the second one originating from the so-called "naïve" antigen inexperienced B-cells in which no Ig V gene mutations can be detected (figure 1) [9,10,26-29]. Somatic mutations were found in about half of the cases and displayed a more benign course of the disease with respect to B-CLL without mutations. In addition, Damle et al firstly showed a strong inverse correlation between CD38 expression on neoplastic B-cells and IgV mutations suggesting the former as a surrogate and a similarly reliable prognostic marker for B-CLL [19,30]. However, the correlation was not confirmed later by some other groups, despite demonstrating the independent prognostic significance of both these markers (table 2). Other investigators showed the impact of the expression of CD38 on disease-free and overall survival along with response to therapy [31-54].

Table 3 shows that the median number of B-CLL CD38-positive cases, selected according to different cu-off of positivity (7%, 20% and 30%) ranged from 14.8% to 52%. With the exclusion of one group [39], there is a consensus on the independent prognostic impact of CD38 expression on survival. A cumulative picture of the observations made on field are collected in table 3. A crucial issue is the ability of CD38 expression to identify a subgroup of patients with Rai stage I-II (indolent-intermediate disease) but characterized by an aggressive course of the disease, as well as a subset of patients with a less aggressive disease characterized by the lack of CD38 expression in Rai stage III-IV. Consistently, CD38 expression identified a subgroup of patients with favourable prognosis among those with high β_2-microglobulin levels [32]. Thornton et al, were able to demonstrate a correlation between

CD38 expression and a shorter treatment-free interval in stage A patients, using a similar cut-off value of ≥ 7% [53].

Table 2. CD38 Expression and IgVH Mutational Status

Author	Correlation
Damle et al, 1999	Yes
Jelinek et al, 2001	No
Thunberg et al, 2001	No
Matrai et al, 2001	Yes
Krober et al, 2002	Yes (*)
Lin et al, 2002	Yes
Hamblin et al, 2002	Yes (*)
Ghia et al, 2003	Yes
Degan et al, 2004	Yes

(*) but in one third of cases gave discordant results.

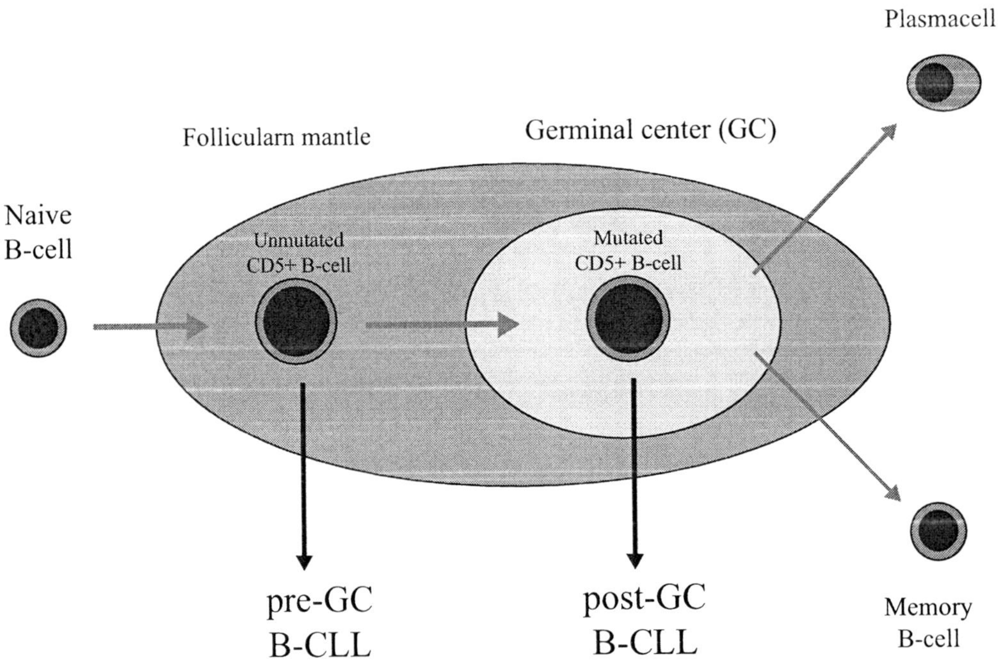

Figure 1. *B-cell differentiation and B-CLL.* Virgin B-cells enter the follicular mantle, then the germinal center progressing from centroblasts to centrocytes. Finally, they differentiate into plasma cells or memory B-cells. Transiting through the germinal center, B-cells acquire somatic hypermutation of variable regions of immunoglobulin (IgVH) genes. Based on the presence or absence of such mutations B-CLL may be distinguished into two categories: i) B-CLL that shows IgVH mutations and originating from cells that have passed through the germinal center; ii) B-CLL with unmutated IgVH sequences and originating from naïve B-cells.

Table 3. Incidence of CD38-Positivity in B-CLL, Impact on Overall Survival (OS) and Correlations with other Variables

Author	CD38+/total patients (% of cases)	Cut-off of CD38 positivity	Correlation with poorer OS	Other correlations
Damle et al, 1999*	17/37 (46%)	30%	Yes	None
Ibrahim et al, 2001	96/218 (44%)	20%	Yes	Higher incidence of lymph node involvement, lower hemoglobin level, hepatomegaly, higher β_2-microglobulin level.
Del Poeta et al, 2001	50/168 (29%)	30%	Yes	Intermediate/high risk modified Rai stages, higher soluble CD23 and β_2-microglobulin levels, higher tumor burden, lymphocyte doubling time < 12 months shorter progression-free survival, increased risk of partial or no response to fludarabine
Morabito et al, 2001	61/161 (32.4%)	30% (and score >3)	No Yes	Older age (> 60 years)
D'Arena et al, 2001	22/61 (38%)	30%	Yes	Atypical morphology, higher peripheral blood lymphocytosis, diffuse bone marrow infiltration, trisomy 12
Heintel et al, 2001	19/81 (23%)	30%	Ne because of a short median observation time	Advanced clinical stages, sCD23 and β_2-microglobulin levels
Jelinek et al, 2001	Nr/131 (%Nr)	20 and 30%	Yes	Progression-free survival
Thunberg et al, 2001	13/48 (27%)	30%	No	None
Matrai et al, 2001	13/40 (32%)	30%	Yes	None
Chevallier et al, 2002	33/123 (27%)	30%	Yes	Binet B/C stages, atypical morphology, 11q deletion, shorter free-treatment interval
Durig et al, 2002	56/133 (42%)	20%	Yes	Unfavourable clinical course with a more advanced disease stage, poor responsiveness to chemotherapy, short time to initiation of first treatment
Domingo-Domenech et al, 2002	30 /155 (19%)	20 and 30%	Yes	Atypical morphology, diffuse bone marrow pattern, Rai stage ≥ 2, high lactate dehydrogenase, high β_2-microglobulin, higher lymphocyte count, treatment requirement and CLL-related mortality **
Krober et al, 2002	82/157 (52%)	7 %	Yes	None

Table 3. (Continued).

Author	CD38+/total patients (% of cases)	Cut-off of CD38 positivity	Correlation with poorer OS	Other correlations
Mainou-Fowler et al, 2002	17/81 (21%)	30% (and ABC ≥ 250)	Yes Yes	Time to treatment and disease progression, quantitative evaluation of CD38 antigen
Lin et al, 2002	19/71 (26%)	20%	Yes	Lack of association with p53 dysfunction
Hamblin et al, 2002	61/145 (42%)	30%	Yes	Independent prognostic factor (multivariate analysis) along with Binet stage and IgVH mutations
Ghia et al, 2003	16/108 (14.8%) ***	Pattern of expression	Yes	Risk status, disease status, treatment requirement
Hsi et al, 2003	67/131 (51%)	30% (and ABC ≥ 100)	Yes Yes	None
Chang et al, 2003	11/24 (45%)	30%	Yes	Treatment requirement
Thornton et al, 2004	150/289 (52%)	30%	Not evaluated	Using cut-off ≥ 7% correlation with advanced stages, treatment-free interval and trisomy 12
Degan et al, 2004	51/140 (36%)	30%	Yes	None

Legend: Nr: not reported; Ne: not evaluated; * confirmed by the same Authors that updated their data on 61 patients [30]; ** CD38 expression in the multivariate analysis lost its statistical significance; *** only omogenously positive cases were included (see the text).

Open Questions

I) Defining a Cut-Off Value of CD38 Expression

An emerging important issue is the definition of a cut-off value for CD38 expression. Most reports adopted a cut-off of $CD38^+$ neoplastic B-cells which ranged from 20 to 30%; 7% was also found as the cut-off score providing the best separation for prognostic subgroups (see Krober et al and recently confirmed by Thornton et al) [44-52]. Morabito et al designed an original scoring system, which integrated the percentage and the mean fluorescence intensity values of CD38 (CD38 score was calculated on the basis of the 25th, 50th and 75th percentile of mean fluorescence intensity and percentage of CD38 positive cells). This scoring system was adopted in the evaluation of CD38 expression in a cohort of untreated B-CLL patients [34]. The results obtained did not highlight significant differences in overall survival between patients with $CD38^+$ cells < 30% *vs* > 30%, while a significantly longer survival was revealed in patients with CD38 score ≤ 3. Lately, the same group demonstrated the additional ability of CD38 expression to predict the progression-free survival in a cohort of B-CLL patients after first-line therapy with high-dose chlorambucil [35]. Mainou-Fowler et al reported that CD38 score, in that instance evaluated as antibody binding capacity (ABC) by means of flow cytometry, improves the prognostic value of the CD38% positivity in a cohort of 81 B-CLL patients [45,46]. The adoption of a CD38 positivity expressed as ABC over 250 consented a good prediction of disease progression and survival, mostly for patients <60 years or with stage A disease [47]. Ghia et al adopted as score the pattern of CD38 expression, rather than a numerical cut-off value based on the percentage of B-cells co-expressing CD38 molecule [49]. This made it possible the identification of a group of B-CLL patients homogeneously $CD38^-$, a group homogeneously $CD38^+$ and a group with a bimodal profile, due to the concomitant presence of two discrete populations ($CD38^+$ and $CD38^-$) at variable proportions. A second finding was the presence of a statistically significant correlation between the expression of CD38 and the mutational status of IgVH genes along with platelet count, Hb and β_2-microglobulin levels, risk (Rai modified stage) and disease status and treatment requirement. A third observation was that $CD38^-$ patients had a significantly longer survival than those with CD38 bimodal expression and the $CD38^+$ subsets.

On the other hand, Hsi et al, did not observe a statistically significant impact on overall survival when CD38 expression was evaluated as a continuous variable in a cohort of B-CLL patients requiring treatment. Using quantitative flow cytometry, B-CLL patients with >100 CD38 molecules per cell ABC were arbitrarily defined as $CD38^+$ cases and were characterized by a significantly worse overall survival with respect to the $CD38^-$ group [50].

Lately, Boonstra et al compared the use of three flow cytometric methods in evaluating CD38 expression in B-CLL: i) the percentage of $CD38^+$ cells; ii) the number of CD38 ABC; iii) the evaluation of CD38 histogram shape [55]. In fact, authors described five different CD38 histogram types called as follows: Type I) unimodal negative; Type II) bimodal with < 50% cells with strong expression; Type III) bimodal with > 50% cells with strong expression; Type IV) unimodal dim expression; Type V) unimodal strong expression. Bimodal expression was defined as a condition in which a second cell population contained more than 5%, but less than 95% of selected events. Interestingly, a negative prognostic impact on overall survival was identified by both > 30% $CD38^+$ cells and a median CD38 ABC > 100. With respect to the histogram type, only type I histogram identified a relatively good overall survival. All other groups, including those with 5-30% of $CD38^+$ cells, showed a decreased survival.

In conclusion, different techniques have been used to detect CD38 expression by B-CLL cells. While the negative prognostic value of the expression of the molecule is indisputable, further work is required to determine the most predictive experimental method.

Figure 2 shows the pattern of CD38 expression by lymphocytes from four representative patients with B-CLL.

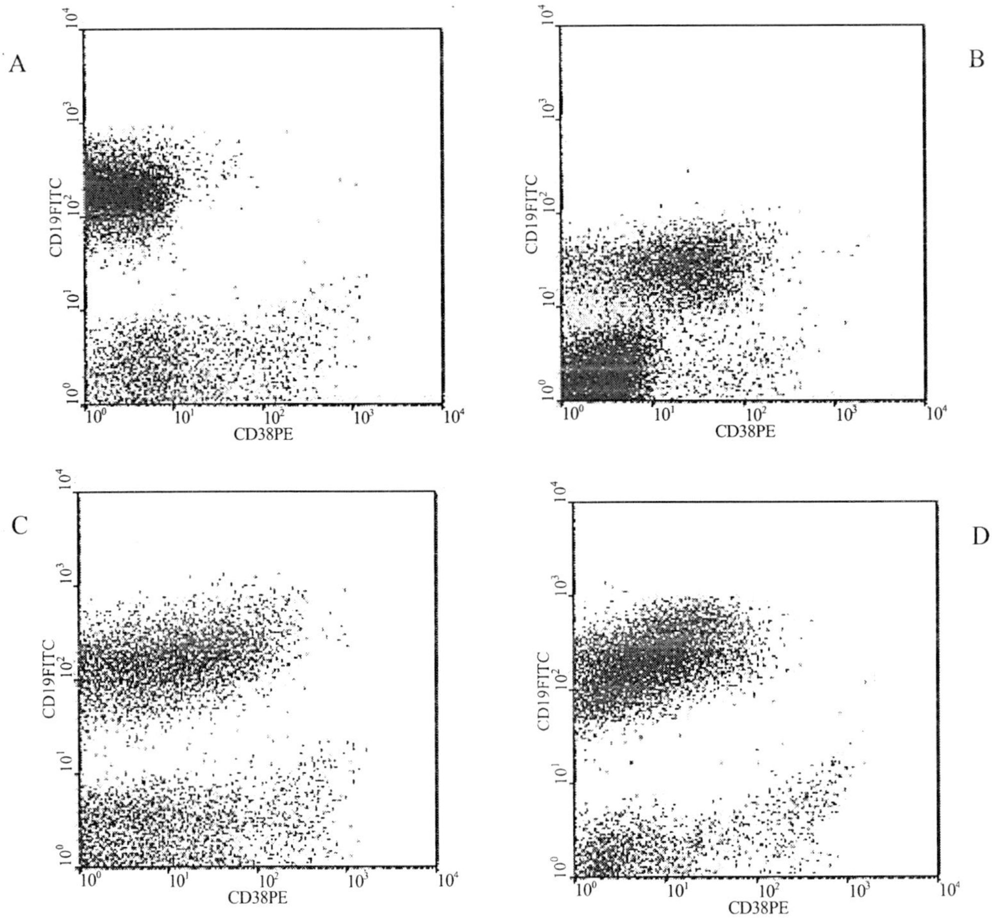

Figure 2. Flow cytometric profiles of CD38 expression on four paradigmatic B-CLL cases. In dot plots are shown: a case omogenously negative for CD38 expression (A); a case omogenously positive for CD38 expression (B); two cases with non omogeneous CD38 expression profile (C and D).

II) Does CD38 Expression Change During the Course of the B-CLL Disease?

The issue of the stability of the expression of CD38 in B-CLL during the course of the disease is still matter of a recent debate. Observations on few isolated patients provided only anedoctal evidence: studies on larger samples provided discordant results (table 4) [9,32,41,49,52,56-59]. The analysis of sequential cryopreserved 41 B-CLL samples showed an increased expression by 8 patients CD38^{+}, a decrease was observed in 2 patients, while 31 featured a striking stability on time [56]. Data showed that CD38 expression increases in 5 patients during treatment with chlorambucil and fludarabine, suggesting that a selective killing of CD38^{-} cells may occour. Moreover, Ghia et al reported that CD38 expression is more up-modulated in treated patients than in untreated patients. This observation may reflect secondary effects induced by drugs or, specularly, hints to a selective role of drugs in killing leukemic B-cells [59]. Probably linked to this observation are the reports of the *in vitro* ability of some cytokines (*e.g.*, interferon-α, -β and -γ or interleukin-2 and tumor necrosis factor-α) to up-regulate the expression of CD38 [60,61].

Table 4. CD38: A Changeable Molecule

Author	No. of patients evaluated	Change:Yes/No (no. of patients)	Observations
Damle et al, 1999	16	No	None
Ibrahim et al, 2001	7	Yes (1/7)	Association with worsening of disease. Frozen samples.
Hamblin et al, 2002	41	Yes (10/41)	In 8 pts increased, in 2 decreased, 31 were stable. Frozen samples.
Chevallier et al, 2002	45	Yes (7/32)	7 out of 32 (22%) CD38- cases became CD38+, whereas none of the 13 CD38+ patients exhibited a loss of CD38 expression. Frozen samples.
D'Arena et al, 2002	29	No	Fresh samples.
Chang et al, 2002	2	Yes (2/2)	Plus 1 in which CD13 levels also changed. Fresh samples.
Krober et al, 2002	12	Yes (3/12)	All but one patients received chemotherapy. Frozen samples.
Ghia et al, 2003	63	Yes (4/63)	Modifications after chemotherapy. Fresh and frozen samples.
Thornton et al, 2004	17	Yes (1/17)	Change associated with disease progression. Fresh samples.
Ghia et al, 2004	111	Yes (12/111)	A higher proportion of treated patients (8/31) with respect to untreated patients (4/80) showed a change
Schroers et al, 2005	60	Yes (6/60)	Only in one patient (1,7%) the utilized CD38$^+$ 30% positivity cut-off was crossed

III) CD38 and B-CLL: A Simple Marker or a Pathogenetic Element?

The conceptual open question deals with the presence of CD38 in B-CLL: is the molecule a simple epiphenomen, representative of a more activated status of the neoplastic clone, or does it plays a direct role in determining a more aggressive behavior of the disease? Both hypotheses are worth considering, even if we would like to considering the second and more daring one [62]. The starting evidence was based on the consideration that CD38 is a receptor on B-CLL [63]. Indeed the use of agonistic mAbs were both able to elicit CD38-mediated signals. Using this approach, CD38 was found to induce proliferation and increase survival of the neoplastic clone. More importantly, a significant proportion of neoplastic cells was found to undergo immunoblast transformation, a situation reminiscent of Richter's syndrome. These effects were all reproduced using instead of surrogate moAb murine fibroblasts transfected with human CD31, mimicking the events that likely take place in microenvironments [64].

Two hypotheses may be provided to link negative prognostic impact and expression of CD38 in B-CLL. The first one relies on the receptorial function of the molecule, the second one stems from the involvement in a network of adhesion molecules. Contrasting with these working hypotheses are the observations of Zupo et al, who found that stimulated $CD38^+$ B-CLL cells had a marked propensity to undergo apoptosis compared to $CD38^-$ cells [65]. However, the same group demonstrated that CD38 ligation by an agonistic moAb was able to prevent apoptosis of human germinal center B-lymphocytes [23]. Looking at the side of the ligand, it was shown that CD31, an adhesion molecule present on platelets and endothelial cells (better known as platelet endothelial adhesion molecule-1, PECAM-1) undertakes with CD38 a selectin-type adhesion. CD31 is a molecule endowed with multiple functions and is reported as exerting pivotal roles in adhesion and trafficking of hematopoietic cells [17]. Another observation against the suggested working hypothesis is that CD38 ligation is able to suppress normal human lymphopoiesis [22]. Lastly, it was recently shown that B-CLL patients with low expression levels of both CD38 and CD31 had a good prognosis [66].

Deaglio et al showed that patients with low CD38 were inept in signaling, but this ligation might be recovered by addition of interleukin-2 (IL-2) which induces CD38 up-modulation. Once a threshold expression was reached, CD38 becomes a receptor able in inducing Ca^{2+} fluxes and marked increased proliferation. Further CD38 signaling appears to prolong survival of IL-2 treated B-CLL and induce the appearance of a subset of plasmablasts [63,67].

Zeta-associated protein 70 (ZAP-70) gene was found up-regulated in the cohort of $CD38^+$ patients. ZAP-70 is tyrosine phosphorylated in response to antibody-mediated CD38 stimulation in natural killer cells and may play a role in CD38 signaling in B-CLL [68-71]. High level of CD38 expression was found to correlate with ZAP-70 expression by Chen et al [72].

Furthermore, Crespo et al showed that ZAP-70 expression strongly correlated with IgVH mutational status, disease progression, and survival in B-CLL [73]. Further, the results of a multiple regression analysis indicated that only ZAP-70, and not CD38 expression, maintained a correlation with IgVH mutational status. The conclusion was that ZAP-70 evaluation is the most dependable surrogate for IgVH mutational status. Again, the combined analysis of CD38 and ZAP-70 has recently been demonstrated to predict disease progression in B-CLL patients [74-77]. In particular, Del Principe et al showed that the combined analysis of CD38 and ZAP-70 expression allowed to separate a large cohort of patients with B-CLL into three prognostic subgroups: $ZAP\text{-}70^-CD38^-$ with good clinical features, $ZAP\text{-}70^+CD38^+$ with a poor prognosis and the so-called "discordant" $ZAP\text{-}70^/CD38$ expression ($ZAP\text{-}70^-CD38^+$ or $ZAP70^+CD38^-$) with an intermediate outcome [77]. Almost all discordant cases were found to be $ZAP\text{-}70^+$ and $CD38^-$ thus suggesting the superior clinical impact of ZAP-70 over CD38. Moreover, a high proportion of patients with mutated IgV_H status were reported in this B-CLL subgroup.

Finally, Dewald et al found that CD38 expression was significantly associated with high-risk FISH categories (+12, 11q-, and 17p-, according to Döhner risk model of FISH abnormalities in B-CLL) [78]. in B-CLL patients [79]. In a similar fashion, performing comparative genomic hybridization on purified peripheral blood B-cells of patients with B-

CLL, Ottaggio et al found a clear correlation between the presence of chromosomal imbalances (most of them correlated with a poor prognosis) and CD38 expression [80].

In an attempt to identify the molecular differences potentially responsible for the different clinical behaviors of CD38$^+$ and CD38$^-$ B-CLL groups, gene profiles of 76 patients were recently taken in consideration by Dürig et al, using an oligonucleotide-based DNA chip microarrays representative of approximately 5,600 genes [81]. The result of an unsupervised clustering showed that the gene expression profile is largely independent of CD38 expression, but using a more detailed statistical analysis a significant differential expression of 14 genes was identified.

Conclusion

Taken together, these data suggest that CD38 expression is directly linked to the implementation of a signaling pathway ultimately leading to B-CLL proliferation and increased survival and potentially acting in synergy with other negative prognostic markers in B-CLL. Further analyses will be required in order to show a direct functional interaction between CD38 and ZAP-70.

Acknowledgements

The work was supported by grants from the Chronic Lymphocytic Leukemia Global Research Foundation (CLL-GRF), the Associazione Italiana Ricerca Cancro (AIRC), PRIN (SD), and the University of Turin.

References

[1] Rozman C and Montserrat E. (1995). Chronic Lymphocytic Leukemia. *New Engl. J. Med,* 333, 1052-1057.

[2] Rai KR and Chiorazzi N. (2003). Determining the clinical course and outcome in chronic lymphocytic leukemia. *New Engl. J. Med,* 348, 1797-1799.

[3] D'Arena G, Di Renzo N, Brugiatelli M, Vigliotti ML, Keating MJ. Biological and clinical heterogeneity of B-cell chronic lymphocytic leukemia. (2003). *Leuk. Lymphoma,* 44, 223-228.

[4] Molica S. (2001). Prognostic factors in Chronic Lymphocytic Leukemia. In B. Cheson (Ed), *Chronic Lymphoid Leukemias.* (pp. 231-260). N.Y.: Marcel Dekker, Inc.

[5] Rai KR,. Swats A, Cronkite AP, Chanana AD, Levy RN, Pasternack BS. (1975). Clinical staging of chronic lymphocytic leukemia. *Blood,* 46, 219-234.

[6] Binet JL, Auquier A, Dighiero G, Chastang C, Piguet H, Goasguen J, Vaugier G, Potron G, Colona P, Oberling F, Thomas M, Tchernia G, Jacquillat C, Boivin P, Lesty C, Duault MT, Monconduit M, Belabbes S, Gremy F. (1981). A new prognostic

classification of chronic lymphocytic leukemia derived from a multivariate survival analysis. *Cancer,* 48, 198-206.

[7] Shanafelt T, Geyer S, Kay N. (2004). Prognosis at diagnosis: integrating molecular biologic insights into clinical practice for patients with CLL. *Blood,* 103, 1202-1210.

[8] Binet JL, Caligaris-Cappio F, Catovsky D, Cheson B, Davis T, Dighiero G, Döhner H, Hallek M, Hillmen P, Keating M, Montserrat E, Kipps TJ, Rai K. (2006). Perspective on the use of new diagnostic tools in the treatment of chronic lymphocytic leukemia. *Blood,* 107, 859-861.

[9] Damle RN, Wasil T, Fais F, Ghiotto F, Valetto A, Allen SL, Buchbinder A, Budman D, Dittmar K, Kolitz J, Lichtman SM, Schulman P, Vinciguerra VP, Rai KR, Ferrarini M, Chiorazzi N. (1999). Ig V gene mutation status and CD38 expression as novel prognostic indicators in chronic lymphocytic leukemia. *Blood,* 94, 1840-1847.

[10] Hamblin TJ, Davis Z, Gardiner A, Oscier DG, Stevenson FK. (1999). Unmutated Ig VH genes are associated with a more aggressive form of chronic lymphocytic leukemia. *Blood,* 94, 1848-1854.

[11] Ferrarini M and Chiorazzi N. (2004). Recent advances in the molecular biology and immunobiology of chronic lymphocytic leukemia. *Semin. Hematol,* 41, 207-223.

[12] Reinherz EL, Kung PC, Goldstein G, Levey RH, Schlossman SF. (1980). Discrete stages of human intrathymic differentiation: analysis of normal thymocytes and leukemic lymphoblasts of T-cell lineage. *Proc. Natl. Acad. Sci. USA,* 77, 1588-92.

[13] Katz F, Povey S, Parkar M, Schneider C, Sutherland R, Stanley K, Solomon E, Greaves M. (1983). Chromosome assignment of monoclonal antibody-defined determinants on human leukemic cells. *Eur. J. Immunol,* 13, 1008-13.

[14] Malavasi F, Funaro A, Roggero S, Horenstein, Calosso L, Mehta K. (1994). Human CD38: a glycoprotein in search of a function. *Immunology Today,* 96, 95-97.

[15] Mehta K, Shahid U, Malavasi F. (1996). Human CD38, a cell-surface protein with multiple functions. *FASEB J,* 10, 1408-1417.

[16] Deaglio S, Mehta K, Malavasi F. (2001). Human CD38: a (r)evolutionary story of enzymes and receptors. *Leuk. Research,* 25, 1-12.

[17] Deaglio S and Malavasi F. (2002). Human CD38: a receptor, an (ecto)enzyme, a disease marker and lots more. *Mod. Asp. Immunobiol,* 2, 121-125.

[18] Sun L, Adebanjo OA, Moonga BS, Corisdeo S, Anandatheerthavarada HK, Biswas G, Arakawa T, Akeda Y, Koval A, Sodam B, Bevis PJ, Moser AJ, Lai FA, Epstein ST, Troen BR, Kumegawa M, Zaidi M. (1999). CD38/ADP-ribosyl cyclase: a new role in the regulation of osteoclastic bone resorption. *J. Cell Biol,* 146, 1161-1172.

[19] Khoo KM and Chang CF. (1999). Characterization and localization of CD38 in the vertebrate eye. *Brain Res,* 821, 17-25.

[20] Lund FE, Yu N, Kim KM, Reth M, Howard MC. (1996). Signaling through CD38 augments B cell antigen receptor (BCR) responses and is dependent on BCR expression. *J. Immunol,* 157, 1455-1467.

[21] Funaro A, Morra M, Calosso L, Zini MG, Ausiello CM, Malavasi F. (1997). Role of the human CD38 molecule in B cell activation and proliferation. *Tissue Antigens,* 49, 7-15.

[22] Kumagai M, Coustan-Smith E, Murray DJ, Silvennoinen O, Murti KG, Evans WE, Malavasi F, Campana D. (1995). Ligation of CD38 suppresses human B lymphopoiesis. *J. Exp. Med,* 181, 1101-1110.

[23] Zupo S, Rugari E, Dono M, Taborelli G, Malavasi F, Ferrarini M. (1994). CD38 signaling by agonistic monoclonal antibody prevents apoptosis of human germinal center B cells. *Eur. J. Immunol,* 24, 1218-1222.

[24] Hoshino S, Kukimoto I, Kontani K, Inoue S, Kanda Y, Malavasi F, Katada T. (1997). Mapping of the catalytic and epitopic sites of human CD38/NAD+ glycohydrolase to a functional domain in the carboxyl terminus. *J. Immunol,* 158, 741-747.

[25] Deaglio S, Morra M, Mallone R, Ausiello CM, Prager E, Garbarino G, Dianzani U, Stockinger H, Malavasi F. (1998). Human CD38 (ADP-ribosyl cyclase) is a counter-receptor of CD31, an Ig superfamily member. *J. Immunol,* 160, 395-402.

[26] Chiorazzi N, Rai KR, Ferrarini M. (2005). Chronic lymphocytic leukemia. *N. Engl. J. Med,* 352, 804- 815.

[27] Naylor M and Capra JD. (1999). Mutational status of Ig VH genes provides clinically valuable information in B-cell chronic lymphocytic leukemia. *Blood,* 94, 1837-1839.

[28] Maloum K, Davi F, Merle-Beral H, Pritsch O, Magnac C, Vuillier F, Dighiero G, Troussard X, Mauro FF, Benichou J. (2000). Expression of unmutated VH genes is a detrimental prognostic factor in chronic lymphocytic leukemia (letter). *Blood,* 96, 377-379.

[29] Dyer MJS and Oscier DG. (2002). The configuration of the immunoglobulin genes in B cell chronic lymphocytic leukemia. *Leukemia,* 16, 973-984.

[30] Damle R, Wasil T, Allen S, Schulman, Rai KR, Chiorazzi N. (2000). Updated data on V gene mutation status and CD38 expression in B-CLL. *Blood,* 95, 2456-2457.

[31] Hamblin TJ, Orchard JA, Gardiner A, Oscier DG, Davis Z, Stevenson FK. (2000). Immunoglobulin V genes and CD38 expression in CLL (letter). *Blood,* 95, 2455-256.

[32] Ibrahim S, Keating M, Do K-A, O'Brien S, Huh YO, Jilani I, Lerner S, Kantarjian H, Albitar M. (2001). CD38 expression as an important prognostic factor in B-cell chronic lymphocytic leukemia. *Blood,* 98, 181-186.

[33] Del Poeta G, Maurillo L, Venditti A, Buccisano F, Epiceno AM, Capelli G, Tamburini A, Suppo G, Battaglia A, Del Principe MI, Del Moro B, Masi M, Amadori S. (2001). Clinical significance of CD38 expression in chronic lymphocytic leukemia. *Blood,* 98, 2633-2639.

[34] Morabito F, Mangiola M, Oliva B, Stelitano C, Callea V, Deaglio S, Iacopino P, Brugiatelli M, Malavasi F. (2001). Peripheral blood CD38 expression predicts survival in B-cell chronic lymphocytic leukemia. *Leuk. Research,* 25, 927-932.

[35] Morabito F, Mangiola M, Stelitano C, Deaglio S, Callea V, Malavasi F. (2002). Peripheral blood CD38 expression predicts time to progression in B-cell chronic lymphocytic leukemia after first-line therapy with high-dose chlorambucil. *Haematologica,* 87, 217-218.

[36] D'Arena G, Musto P, Cascavilla N, Dell'Olio M, Di Renzo N, Perla G, Savino L, Carotenuto M. (2001). CD38 expression correlates with adverse biological features and predicts poor clinical outcome in B-cell chronic lymphocytic leukemia. *Leuk. Lymphoma,* 42, 109-114.

[37] Heintel D, Schwarzinger I, Chizzali-Bonfadin C, Thalhammer R, Schwarzmeier J, Fritzer-Szekeres M, Weltermann A, Simonitsch I, Lechner K, Jaeger U. (2001). Association of CD38 antigen expression with other prognostic parameters in early stages of chronic lymphocytic leukemia. iLeuk. Lymphoma, 42, 1315-1321.

[38] Jelinek DF, Tschumper RC, Geyer SM, Bone ND, Dewald GW, Hanson CA, Stenson MJ, Witzig TE, Tefferi A, Kay NE. (2001). Analysis of clonal B-cell CD38 and immunoglobulin variable region sequence status in relation to clinical outcome for B-chronic lymphocytic leukemia. *British J. Haematol,* 115, 854-861

[39] Thunberg U, Johnson A, Roos G, Thorn I, Tobin G, Sallstrom J, Sundstrom C, Rosenquist R. (2001). CD38 expression is a poor predictor for VH gene mutational status and prognosis in chronic lymphocytic leukemia (letter). *Blood,* 97, 1892-1893.

[40] Matrai Z, Lin K, Dennis M, Sherrington P, Zuzel M, Pettitt AR, Cawley JC. (2001). CD38 expression and Ig V_H gene mutation in B-cell chronic lymphocytic leukemia (letter). *Blood,* 97, 1902-1903.

[41] Chevallier P, Penther D, Loiseau-Avet H, Robillard N, Ifrah N, Mahe B, Hamidou M, Maisonneuve H, Moreau P, Jardel H, Harousseau JL, Bataille R, Garand R. (2002). CD38 expression and secondary 17p deletion are important prognostic factors in chronic lymphocytic leukemia. *Br. J. Haematol,* 166, 142-150.

[42] Dürig J, Naschar M, Schmucker U, Kohler-Renzing K, Holter T, Huttmann A, Duhrsen U. (2002). CD38 expression is an important prognostic marker in chronic lymphocytic leukemia. *Leukemia,* 16, 30-35.

[43] Domingo-Domenech E, Domingo-Claros A, Gonzalez-Barca E, Beneitez D, Alonso E, Romagosa V, De Sanjos C, Petit J, Granena A, Fernandez de Sevillo A. (2002). CD38 expression in B-cell chronic lymphocytic leukemia: association with clinical presentation and outcome in 155 patients. *Haematologica,* 87, 1021-1027.

[44] Krober A, Seiler T, Benner A, Bullinger L, Brucale E, Lichter P, Dohner H, Stilgenbauer S. (2002). VH mutation status, CD38 expression level, genomic aberrations, and survival in chronic lymphocytic leukemia. *Blood,* 100, 1410-1416.

[45] Mainou-Fowler T, Dignum H, Taylor PRA, Dickinson AM, Saunders PWG, Proctor SJ, Summerfield GP. (2002). Quantification improves the prognostic value of CD38 expression in B-cell chronic lymphocytic leukaemia. *Br. J. Haematol,* 118, 755-761.

[46] Mainou-Fowler T, Dignum HM, Proctor SJ, Summerfield GP. (2004). The prognostic value of CD38 expression and its quantification in B-cell chronic lymphocytic leukemia (B-CLL). *Leuk. Lymphoma,* 45, 455-462.

[47] Dignum HM, Summerfield GP, Proctor SJ, Mainou-Fowler T. (2004). Quantification of CD38 expression in B-cell chronic lymphocytic leukemia (B-CLL): a comparison between antibody binding capacity (ABC) and relative median fluorescence (RMF). *Leuk Lymphoma,* 45, 1167-1173.

[48] Lin K, Sherrington PD, Dennis M, Matrai Z, Cawley JC, Pettitt AR. (2002). Relationship between p53 dysfunction, CD38 expression, and IgVH mutation in chronic lymphocytic leukemia. *Blood,* 100, 1404-1409.

[49] Ghia P, Guida G, Stella S, Gottardi D, Geuna M, Strola G, Scielzo C, Caligaris-Cappio F. (2003). The pattern of CD38 expression defines a distinct subset of chronic

lymphocytic leukemia (CLL) patients at risk of disease progression. *Blood,* 101, 1262-1269.

[50] Hsi ED, Kopecky KJ, Appelbaum FR, Boldt D, Frey T, Loftus M, Hussein MA. (2003). Prognostic significance of CD38 and CD20 expression as assessed by quantitative flow cytometry in chronic lymphocytic leukaemia. *Br. J. Haematol,* 120, 1017-1025.

[51] Chang C-C, Liu CZ, Cleveland RP. (2003). Relative importance of CD38 expression over myeloid-associated markers expression in predicting the clinical course of B-CLL patients. *Leuk. Lymphoma,* 44, 977-982.

[52] Thornton PD, Fernandez C, Giustolisi GM, Morilla R, Shayne A, A'Hern RP, Matutes E, Catovsky D. (2004). CD38 expression as a prognostic indicator in chronic lymphocytic leukaemia. *Hematol. J,* 5, 145-151.

[53] Degan M, Bomben R, Dal Bo M, Zucchetto A, Nanni P, Rupolo M, Steffan A, Attadia V, Ballerini PF, Damiani D, Pupillo C, Poeta GD, Colombatti A, Gattei V. (2004). Analysis of IgV_H gene mutations in B cell chronic lymphocytic leukaemia according to antigen-driven selection identifies subgroups with different prognosis and usage of the canonical somatic hypermutation machinery. *Br. J. Haematol,* 126, 29-42.

[54] Chang CC and Cleveland RP. (2002). Conversion of CD38 and/or myeloid-associated marker expression status during the course of B-CLL: association with a change to an aggressive clinical course (letter). *Blood,* 100, 1106.

[55] Boonstra JG, van Lom K, Langerak AW, Graveland WJ, Valk PJM, van't Veer M, Gratama JW. (2006). CD38 as a prognostic factor in B cell chronic lymphocytic leukaemia (B-CLL): comparison of three approaches to analyze its expression. *Cytometry B Clin. Cytom,* 70B, 136-141.

[56] Hamblin TJ, Orchard JA, Ibbotson RE, Davis Z, Thomas PW, Stevenson FK, Oscier DG. (2002). CD38 expression and immunoglobulin variable region mutations are independent prognostic variables in chronic lymphocytic leukemia, but CD38 expression may vary during the course of the disease. *Blood,* 99, 1023-1029.

[57] D'Arena G, Nunziata G, Coppola G, Vigliotti ML, Tartarone A, Carpinelli N, Matera R, Colella Bisogno R, Pistolese G, Di Renzo N. (2002). CD38 expression does not change in B-cell chronic lymphocytic leukemia (letter). *Blood,* 100, 3053-3053.

[58] Schroers R, Griesinger F, Trümper L, Haase D, Kulle B, Klein-Hitpass L, Sellmann L, Duhrsen U, Durig J. (2005). Combined analysis of ZAP-70 and CD38 expression as a predictor of disease progression in B-cell chronic lymphocytic leukemia. *Leukemia,* 19, 750-758.

[59] Ghia P, Guida G, Scielzo C, Geuna M, Caligaris-Cappio F. (2004). CD38 modifications in chronic lymphocytic leukemia: are they relevant ? *Leukemia,* 18, 1733-1735.

[60] Bauvois B, Durant L, Laboureau J, Barthelemy E, Rouillard D, Borilla G, Delterre P. (1999). Upregulation of CD38 gene expression in leukemic B cells by interferon types I and II. *J. Interferon Cytokines Res,* 19, 1059-1066.

[61] Vilpo J, Hulkkonen J, Hurme M, Vilpo L. (2002). Surface membrane antigen expression changes induced in vitro by exogenous growth factors in chronic lymphocytic leukemia cells. *Leukemia,* 16, 1691-1698.

[62] Deaglio S, Vaisitti T, Aydin S, Ferrero E, Malavasi F (2006). In-tandem insight from basic science combined with clinical research: CD38 as both marker and key component of the pathogenetic network underlying chronic lymphocytic leukemia. *Blood*, 108, 1135-1144.

[63] Deaglio S, Capobianco A, Bergui L, Dürig J, Morabito F, Duhrsen U, Malavasi F. (2003). CD38 is a signaling molecole in B-cell chronic lymphocytic leukemia cells. *Blood*, 102, 2146-2155.

[64] Deaglio S, Vaisitti T, Bergui L, Bonello L, Horenstein AL, Tamagnone L, Boumsell L, Malavasi F. (2005). CD38 and CD100 lead a network of surface receptors relaying positive signals for B-CLL growth and survival. *Blood*, 105, 3042-3050.

[65] Zupo S, Isnardi L, Megna M, Massara R, Malavasi F, Dono M, Cosulich E, Ferrarini M. (1996). CD38 expression distinguishes two groups of B-cell chronic lymphocytic leukemias with different responses to anti-IgM antibodies and propensity to apoptosis. *Blood*, 88, 1365-1374.

[66] Ibrahim S, Jilani I, O'Brien S, Rogers A, Manshouri T, Giles F, Faderl S, Thomas D, Kantarjian H, Keating M, Albitar M. (2003). Clinical relevance of the espression of the CD31 ligand for CD38 in patients with B-cell chronic lymphocytic leukemia. *Cancer*, 97, 1914-1919.

[67] Hamblin T. (2003). CD38: what is it there for? *Blood*, 102, 1939-1940.

[68] Dürig J, Nuckel H, Huttmann A, Kruse E, Holter T, Halfmeyer K, Fuhrer A, Rudolph R, Kalhori N, Nusch A, Deaglio S, Malavasi F, Moroy T, Klein-Mitpass L, Duhrsen U. (2003). Expression of ribosomal and translation-associated genes is correlated with a favourable clinical course in chronic lymphocytic leukemia. *Blood*, 101, 2748-2755.

[69] Rosenwald A, Alizadeh A, Widhopf G, Simon R, Davis RE, Yu X, Yang L, Pickeral OK, Rassenti LZ, Powell J, Botstein D, Byrd JC, Grever MR, Cheson B, Chiorazzi N, Wilson WH, Kipps TJ, Brown PO, Staudt LM. (2001). Relation of gene expression phenotype to immunoglobulin mutation genotype in B cell chronic lymphocytic leukemia. *J. Exp. Med*, 194, 1639-1647.

[70] Rosenwald A and Staudt LM. (2002). Clinical translation of gene expression profiling in lymphomas and leukemias. *Semin. Oncol*, 29, 258-263.

[71] Deaglio S, Zubiaur M, Gregorini A, Bottarel F, Ausiello CM, Dianzani U, Rancho J, Malavasi F. (2002). Human CD38 and CD16 are functionally dependent and physically associated in natural killer cells. *Blood*, 99, 2490-2498.

[72] Chen L, Widhopf G, Huynh L, Rassenti L, Rai KR, Weiss A, Kipps TJ. (2002). Expression of ZAP-70 is associated with B cell receptor signalling in chronic lymphocytic leukemia. *Blood*, 100, 4609-4614.

[73] Crespo M, Bosch F, Villamor N, Bellosillo B, Colomer D, Rozman M, Marce S, Lopez-Guillermo A, Campo E, Montserrat E. (2002). ZAP-70 expression as a surrogate for immunoglobulin-variable-region mutations in chronic lymphocytic leukemia. *New Engl. J. Med*, 348, 1764-1775.

[74] Del Giudice I, Morilla A, Osuji N, Matutes E, Morilla R, Burford A, Maravelaki S, Owusu-Ankomah K, Swansbury J, A'Hern R, Brito-Babapulle V, Catovsky D. (2005) ζ-chain associated protein 70 and CD38 combined predict the time to first treatment in patients with chronic lymphocytic leukemia. *Cancer*, 104, 2124-2132.

[75] Schroers R, Griesinger F, Trümber L, Haase D, Kulle B, Klein-Hitpass L, Sellmann L, Dührsen U, Dürig J. (2005). Combined analysis of ZAP-70 and CD38 expression as a predictor of disease progression in B-cell chronic lymphocytic leukemia. *Leukemia,* 19, 750-758.

[76] Hus I, Pdhorecka M, Bojarska-Junak A, Rolinski J, Schmitt M, Sieklucka M, Wasik-Szczepanek E, Dmoszynska A. (2006). The clinical significance of ZAP-70 and CD38 expression in B-cell chronic lymphocytic leukemia. *Ann. Oncol,* 17, 683-690.

[77] Del Principe I, Del Poeta G, Buccisano F, Maurillo L, Venditti A, Zucchetto A, Marini R, Niscola P, Irno Consalvo MA, Mazzone C, Ottaviani L, Panetta P, Bruno A, Bomben R, Suppo G, Degan M, Gattei V, De Fabritiis P, Cantonetti M, Lo Coco F, Del Principe D, Amadori S. (2006). Clinical significance of ZAP-70 protein expression in B-cell chronic lymphocytic leucemia. *Blood.* (Epub ahead of print, April 6).

[78] Döhner H, Stilgenbauer S, Benner A, Leupolt E, Krober A, Bullinger L, Dohner K, Bentz M, Lichter P. (2000). Genomic aberrations and survival in chronic lymphocytic leukemia. *New Engl. J. Med,* 343, 1910-1916.

[79] Dewald GW, Brockman SR, Paternoster SF, Bone ND, O'Fallon JR, Allmer C, James CD, Jelinek DF, Tschumper RC, Hanson CA, Pruthi RK, Witzig TE, Call TG, Kay NE. (2003). Chromosome anomalies detected by interphase fluorescence in situ hybridisation: correlation with significant biological features of B-cell chronic lymphocytic leukaemia. *Br. J. Haematol,* 121, 287-295.

[80] Ottaggio L, Viaggi S, Zunino A, Zupo S, Rossi E, Spriano M, Abbondandolo A, Ferrarini M. (2003). Chromosome aberrations evaluated by comparative genomic hybridization in B-cell chronic lymphocytic leukemia: correlation with CD38 expression. *Haematologica,* 88, 769-777.

[81] Dürig J, Nuckel H, Huttmann A, Kruse E, Holter T, Halfmeyer K, Fuhrer A, Rudolph R, Kalhori N, Nusch A, Deaglio S, Malavasi F, Moroy T, Klein-Hitpass C, Duhrsen U. (2003). Expression of ribosomal and translation-associated genes is correlated with a favourable clinical course in chronic lymphocytic leukemia. *Blood,* 101, 2748-2755.

In: Chronic Lymphocytic Leukemia Research Focus
Editor: Chadi Nabhan, pp. 183-213

ISBN: 978-1-60021-526-1

Chapter IX

Novel and Uncommon Chromosome Aberrations in Chronic Lymphocytic Leukemia: Cytogenetic, FISH and Clinical Evaluation

Roxana Cerretini*[*1,2], Christian Chena[2] and Irma Slavutsky[2]

[1] Departamento de Genética Experimental, Centro Nacional de Genética Médica, Buenos Aires, Argentina.

[2] Departamento de Genética, Instituto de Investigaciones Hematológicas "Mariano R. Castex", Academia Nacional de Medicina, Buenos Aires, Argentina.

Abstract

The identification of the non-random genetic alterations associated with cancer has great importance not only to basic research but also to disease management as diagnostic and prognostic markers. In contrast to other types of lymphomas, translocations are rare events in chronic lymphocytic leukemia/small lymphocytic lymphoma (CLL/SLL) and not any CLL/SLL-specific translocation has been found yet. Moreover, translocations have been reported to be a prognostic marker for an unfavorable clinical outcome. The purpose of our study was to contribute to the understanding of the chromosome alterations associated with this group of lymphoid malignancies. The combination of conventional cytogenetics with FISH (fluorescence *in situ* hybridization) analysis allowed us a more accurate definition of the genomic aberrations involved in CLL/SLL. Novel and uncommon chromosome rearrangements were found in our series. In the present work, we describe the cytogenetic and FISH analyses of 12 CLL/SLL patients out of 150 cases with successful cytogenetic study, that interestingly showed in their

* Correspondence to: Roxana Cerretini; Departamento de Genética, Instituto de Investigaciones Hematológicas "Mariano R. Castex", Academia Nacional de Medicina; Pacheco de Melo 3081, Ciudad Autónoma de Buenos Aires, Argentina, CP: 1425. FAX: (5411) 4803-9475 ; e-mail: rcerretini@argentina.com; e-mail: rcerretini@hotmail.com

karyotypes new or infrequent clonal chromosome rearrangements, all of them associated to adverse clinical course with partial or null response to different chemotherapeutic agents. We identified 16 different chromosome rearrangements and translocations were the most frequent aberration (7/16; 43,8%). Some of these findings were recently reported by us as a new alterations in SLL/CLL patients: t(2;17)(p21;q23); der(17)t(12;17)(q13;q25); der(14)(t(11;14)(q13;q24), t(5;12;19)(q15;p11;q13); psu dic (17;2)(p11.2;p21), psu dic (12;21)(q24;q10), t(10;14)(q22;q32), del(3)(p11) and del(5)(q15). Translocation t(11;13)(p15;q22) and del(18)(p11) were not previously described, meanwhile, dic(17;18)(p11.2;p11.2), dup(12)(q13q24), del(7)(q34), t(2;14)(p11;q32) and del(7)(q11) were infrequently reported in the literature. Translocation t(2;17)(p21;q23) is recurrent in our series and t(2;14)(p11;q32) became recurrent from our data. Chromosomes 17, 2, 12 and 14 were the most frequently involved. Twenty one different breakpoints were found and 38.1% (8/21) of them, such as 2p21, 5q15, 17p11.2, 17q23, 12q13, 12q24, 14q32 and 18p11 were recurrent in our series. All but one patient of our series showed an unfavorable clinical outcome. A detailed analysis and a review of the cytogenetic and molecular cytogenetics findings in CLL/SLL were performed. Our findings may facilitate the understanding of genetic events associated with development and progression of CLL/SLL and may have implications in the clinical management of patients with this disease.

1. Introduction

Chronic lymphocytic leukemia (CLL) is a monoclonal disease of mature-appearing lymphocytes that are functionally incompetent and accumulate in bone marrow, blood, lymph nodes, spleen, liver and other organs, primarily of the lymphatic system. A diagnosis of small lymphocytic lymphoma (SLL) can be made histologically in the absence of bone marrow or blood involvement. Most cases (>95%) are characterized by monoclonal lymphocytes expressing normal B-cell surface proteins including immunoglobulin (Ig), CD19 and CD20 and aberrantly expressing CD5, a protein normally found on T-cells [1].

CLL is the most frequent type of leukemia among adults in the western world, accounting for about 30% of all leukemias [2-5]. However, its incidence is very low in Asia [6, 7]. Particularly in Japan, it accounts for just 3-4% of all leukemias and interestingly, CLL is quite rare not only among Japanese living in Japan, but also among Japanese immigrants and their descendants living in USA [8-10]. This suggests that the rareness of CLL in Japanese is caused by genetic rather than environmental factors. The basis for this ethnic variation is unknown.

This disease affects mainly people of advanced age, with a median age of 65 years, but about 20% of patients are younger than 55 years. The gender ratio varies from approximately 1.5 to 2.1 males to 1 female [11]. The clinical diagnosis of CLL requires an absolute lymphocytosis (more than 5000 mature-appearing lymphocytes/μL). Nevertheless, it is common to find small percentages of larger or atypical cells, cleaved cells, or prolymphocytes [12].

This pathology follows an extremely variable clinical course with overall survival times ranging from months to decades. Some patients have no or minimal signs and symptoms during their entire disease course and have a survival time similar to age-matched controls.

However, other patients progress rapidly to symptomatic disease, needing immediate treatment [13].

A number of clinical and biologic markers of prognostic relevance have been identified. These include clinical characteristic, such as age, stage, gender, and performance status, as well as laboratory parameters reflecting the tumor burden or disease activity (i.e. lymphocyte count, higher lactate dehydrogenase [LDH] levels, bone marrow infiltration pattern, or lymphocyte doubling time) [14-16].

More recently, prognostic markers related to the biology of the disease have been identified: serum parameters such as soluble CD23 (sCD23), β_2-microglobuline (β_2-MG) or thymidine kinase (TK) [17, 18] and genetic markers of the CLL cells, such as genomic aberrations, the mutation status of the variable segment of immunoglobulin heavy chain genes (IgV_H), or surrogate markers for these factors [19, 16, 20-22].

Over the recent years the focus of research has moved towards the molecular genetic and epigenetic levels, which may not only provide insight into the biology and transforming events of CLL but may also allow the definition of mechanisms directly responsible for the course of the disease with regard to clinical progression, response to treatment, and overall survival.

Among epigenetic modifications, DNA methylation and post-translational modifications of histones tail proteins, in the form of acetylation, methylation, ubiquitination, and phosphorylation are the best documented epigenetic changes. These modifications are heritable and influence the chromatin structure on a genome wide scale. Many novel genes that are epigenetically silenced in CLL have been identified. At present, the field of epigenetics is at an evolving stage but there is no doubt that further unraveling of its cause and effects in transformed cells will bring a new revolution in cancer therapeutics [23].

One of the most important molecular genetic parameters defining pathogenic and prognostic subgroups of CLL is the mutation status of the IgV_H genes. The presence or absence of somatic mutations in IgV_H genes can be used to stratify CLL patients into two groups [24, 25]. In CLL with mutated IgV_H genes, the proliferating B cell may have transited through germinal centers, the physiologic site of hypermutation, whereas in CLL with unmutated IgV_H genes the malignant B cell may derive from a pre-germinal center naïve B cell [12]. Mutated and unmutated CLL patients clearly differ in terms of prognosis and may also differ with respect to oncogenic mechanisms. Patients whose CLL cells express unmutated IgV_H have a shorter median survival than those expressing mutated IgV_H [24].

In despite of these clinical and molecular differences, recent studies on gene expression profiling of CLL cells showed that this pathology is characterized by a common gene expression signature which is independent of IgV_H mutational status. This profile of gene expression differs from other lymphoid cancers and normal lymphoid subpopulations, suggesting that CLL cells share a common mechanism of transformation and/or cell of origin [26, 27].

However, in despite of sharing a common signature, CLL expressing mutated and unmutated IgV_H genes differentially express more than 100 genes. Among these differentially expressed genes, overexpression of *ZAP-70*, lipoprotein lipase (*LPL*), *BCL-7a*, dystrophin and gravin are observed in the aggressive unmutated cases, while stable mutated cases overexpress *Wnt3*, *CTLA-4*, *NRIPI* nuclear receptor gene, *ADAM* and the transcription factor *TCF7* [26-29].

ZAP-70, a receptor associated protein tyrosine kinase expressed by T lymphocytes, natural killer (NK) cells, activated tonsillar and splenic B cells and cases of CLL with unmutated IgV_H is the best surrogate considering IgV_H mutational profiles. The mode of action of *ZAP-70* may be to enhance B-cell receptor signaling, promoting the survival and/or proliferation of CLL [30]. Recently, Catherwood et al [31] confirmed that IgV_H unmutated CLL cells have a high expression of *ZAP-70* in comparison with IgV_H mutated-CLL B cells. *ZAP-70* is differentially methylated in CLL and a close association exits between *ZAP-70* methylation and expression and IgV_H mutational status [32]. Expressions of *ZAP-70* and CD38 have been identified by several groups as predictors of disease progression and overall survival in CLL [33]. The prognosis information given by *ZAP-70* and CD38 expression is complementary and these two markers have allowed for the separation of three CLL patient subgroups with good, intermediate and poor prognosis [34].

2. Genomic Aberrations

Genomic aberrations are the other genetic parameters shown to be of pathogenic and clinical relevance in this disease. The study of the genomic aberrations provides insights into the pathogenesis of the disease since they point to loci of candidate genes and identify subgroups of patients with distinct clinical pictures.

Cytogenetic analysis is a routine part of the diagnosis and management of a significant number of lymphoid malignancies. Particularly, in CLL, this study is problematic as malignant B cells tend to have a low spontaneous mitotic index, usually a poor response to most common mitogens, and in addition, the growth advantages of normal T cells in the cell culture [35]. Early studies reported normal karyotypes for most cases of CLL, almost certainly as the metaphase spreads obtained in these cases originated from normal cells present in the blood or bone marrow culture.

The addition of B-cell mitogens such as 12-0-tetradecanoylphorbol-13-acetate (TPA) and Pokeweed have been shown to be effective in stimulating malignant B cells to divide [36] and enabled the detection of cytogenetic abnormalities in approximately 40-50% of CLL cases using chromosome banding [37-39]. Peripheral blood is generally the most appropriate sample for cytogenetic analysis of CLL.

From patients with abnormal karyotype by standard cytogenetic, as many as 65% have one chromosome abnormality, 25% have two abnormalities and the reminder ones have more complex alterations [40].

The development of molecular cytogenetic techniques, such as fluorescence *in situ* hybridization (FISH), has enabled sensitive detection of specific changes in interphase and metaphase cells and has shown a higher incidence of abnormalities than cytogenetics. The incidence of FISH detection of genome aberrations is estimated to be 77-80% [19,22]. In CLL cases with normal karyotype by banding analysis but clonal aberrations by interphase FISH, the metaphase cells are derived either from non-clonal T-cells or clonal cells with cryptic interstitial deletions, and therefore do not reflect the karyotype of the malignant B-CLL cells [41, 42].

The most common cytogenetic and/or FISH anomalies in B-CLL involve chromosome 6, 11, 12, 13, 14 and 17 [19, 22, 36, 41, 43-47]. Meanwhile by interphase FISH, the most frequent abnormality is 13q14 deletion, followed by 11q22-23 deletion, trisomy 12, 17p13 deletion, 6q deletion and t(14q32) [19]. These anomalies have been associated with different prognoses.

13q Deletion: del(13)(q14.3)

The del(13)(q14.3), is the most common isolated chromosome abnormality in CLL, and is found in about 10-20% of the cases by cytogenetics or 55-64% of the cases by FISH [19, 22, 48]. CLL metaphases with 13q- are not always detected because these deletions are often cryptic and not apparent by conventional G-banding [41].

The 13q deletion in CLL uniformly involves 13q14. Commonly used FISH probes to detect the deletion in the lymphoid disorders are chromosome 13q14 specific, including *RB1*, D13S319 and D13S25. The most commonly deleted marker in CLL initially reported was D13S319, between the retinoblastoma and D13S25 loci. The deletion of an allele of *RB1* is detected by molecular cytogenetics in one-fourth of all CLL cases [49, 50]. The 13q14 deletions usually do not lead to inactivation of the *RB1* gene, and del(13q) is associated with the presence of one intact *RB1* gene on the homologous chromosome, implying the role of an adjacent locus that may harbor important tumor suppressor gene(s) [19, 49, 51].

A critically deleted region (CDR) of approximately 350 kb was identified at 13q14 with the proximal border around 120 kb centromeric from D13S319, encompassing an area rich in expressed sequence tagged sites and containing *DLEU1*, *DLEU2*, for which a pathogenic role has not yet been established, and *BCMS* (B-cell neoplasia associated gene with multiple splicing). Sequence analysis of the RNA variants suggest that *BCMS* transcripts belong to the group of non-coding RNAs (ncRNAs). The alignment of the gene with all critical subregions provides a strong argument for *BCMS* being a very good candidate for the tumor suppressor gene in 13q14 involved in the leukemogenesis of CLL. Due to the limited understanding of functional RNAs, however, the pathogenic role of *BCMS* has not yet been proven [52, 53]. Moreover, during the past few years, molecular studies have discovered hundred of genes that encode small RNA molecules, named microRNAs (miRNAs). More recently it has been shown the presence of two microRNA genes, *miR15* and *miR16*, located at 13q14 within the minimal deleted region (MDR) for CLL patients. Both genes are deleted or down-regulated in the majority of CLL cases [54]. In addition, a unique miRNA signature was associated with prognostic factors and disease progression [55], suggesting an important role for these genes in this pathology.

11q Deletion: del(11)(q22-q23)

Loss of genomic material on the long arm of chromosome 11 is frequently observed in various types of lymphoproliferative disorders, most frequently involving 11q23 [56, 57]. The recurring loss of material from chromosome bands 11q22-q24 suggests that this region harbors a tumor suppressor gene and many candidate genes. All 11q deletions affect a minimal consensus region of 2-3 Mb in size in bands 11q22.3-q23.1 [56, 58]. The ataxia telangiectasia mutated (*ATM*) gene, which is responsible for the autosomal recessive disorder ataxia telangiectasia, is located within this minimal consensus region in bands 11q22.3-q23.1 and is implicated in the pathogenesis of different lymphoid malignancies, mainly mantle cell lymphoma [59].

Patients with inherited inactivation of both alleles of the *ATM* gene by deletion and/or mutation are highly predisposed to the development of leukemia and lymphoma. However, mutation of the remaining *ATM* allele is found in about 15-23% of CLL cases, which indicates a possible involvement of additional genes in this region in CLL [19, 60]. By contrast, in mantle cell lymphoma, in which the 11q22-q23 deletion occurs in 50% of the cases, mutation of the remaining allele is observed in all cases [59, 61, 62].

ATM is a pleiotropic protein, which activities are induced by double-stranded chromosome breaks that occur spontaneously or after exposure to DNA-damaging agents. ATM protects the integrity of the genome by arresting the cell cycle and activating DNA-repair pathways [63]. If the DNA damage is more severe, ATM induces apoptosis. After DNA damage, the serine-threonine kinase activity of ATM increases, leading to the phosphorylation of downstream effectors involved in the regulation of cell-cycle checkpoints, DNA repair and apoptosis [64]. The most important of these effectors seems to be the *TP53* tumor suppressor gene [65]. Constitutional ATM deficiency in cells of ataxia telangiectasia patients and in atm-l- mice leads to the deregulation of apoptosis and cell-cycle checkpoint control and causes defects in DNA recombination [66].

Trisomy 12

Trisomy 12 is found in one third of cytogenetically abnormal CLLs by conventional karyotype [40, 67, 68] and in about 11-42% of cases when interphase FISH is used [19, 48, 69-71].

Using restriction fragment length polymorphism analysis, the extra chromosome 12 has been shown to derive from duplication of one chromosome 12, with retention of the other homolog rather than from triplication of one homolog. Trisomy 12 is thought to be a secondary change, associated with disease progression, since combined immunological and cytogenetic studies showed that it is present in only a part of the neoplastic B-cells [72-74].

The role of genes potentially involved in the pathogenesis of CLL with trisomy 12 is not established. While the critical region of most aberrations in CLL could be narrowed down to a set of candidate genes, this has been difficult in the case of trisomy 12. Cases with partial duplication 12q were analyzed by FISH, and a "minimal duplicated region" could be defined in segment 12q13-12q15 [75, 76]. Bands 12q13-q22 were the smallest duplicated segment in four CLL cases in the series of Dierlamm et al [75] and by DNA microarray technology it was identified the smallest replicated genomic regions in bands 12q13-q21 in this disorder [77, 78].

Recent RNA expression analyses showed an overexpression of genes located on chromosome 12 in CLL cases with trisomy 12 as compared to cases without this aberration [79, 80]. Genes of oncogenic potential, like *CDK4*, *GLI* and *MDM2* are localized in this genomic region, but they have not shown pathogenetic relevance for CLL to date. Recently, it has been identified a novel gene, CLL up-regulated gene 1 (*CLLU1*), in band 12q22, uniquely overexpressed in CLL but not in normal tissue or in other hematological disease [81].

6q Deletion: del (6)(q21)

Lymphoid malignancies are characterized by recurrent chromosome changes, involving deletion in bands 6q21-q23. The frequency of the deletions is difficult to estimate by conventional cytogenetic analysis because small interstitial deletions are beyond the sensitivity of the technique; furthermore, many studies have reported conflicting data on the putative region of overlap and the number of regions involved. The breaks occur predominantly in 6q21, but 6q15 is also often described; overall, del(6q) cases encompassed the 6q21 band [82, 83].

Comparison of the extent of the deletions among different cases by FISH allowed the identification of two distinct regions of minimal deletion (RMD) at 6q25 to 6q27 (RMD-1) and at 6q21 to 6q23 (RMD-2). These data support a role for 6q deletions in B-cell non-Hodgkin's lymphomas (NHL) pathogenesis and provide a basis for identifying the corresponding tumor suppressor genes [84-86], defined the deletion breakpoints exactly and narrowed this region further. They found 3-cM (4-5 Mb) commonly deleted region in 6q21, involved in the pathogenesis of both low and high grade NHL and acute lymphoblastic leukemias (ALL).

Deletion in band 6q21 is often observed in malignant transformation and development of both T- and precursor B-ALL and it is the band most frequently loss in CLL. This lack of specificity for a particular immunophenotype may imply that the gene or genes affected by 6q abnormalities are broadly active in the multistep process of lymphoid leukemogenesis [87].

17p Deletion: del(17)(p13)

TP53 gene is located near the telomere of the chromosome 17 p-arm at 17p13, its absence can be detected by FISH in interphase nuclei of 8-10% patients with CLL. Results of chromosome studies suggest that loss of *TP53* is often caused by unspecified structural anomalies of chromosome 17 rather than by monosomy 17 [40, 88]. However, this type of chromosome 17 anomalies and their mechanisms of origin in CLL have not been well characterized.

Isochromosome 17q, or "i(17q)," is one of the most common structural abnormalities observed in human neoplasms. A breakpoint cluster region for i(17q) in 17p11.2 has been identified and hypothesized that genome architectural features could be responsible for this clustering. A complex genomic architecture in the i(17q) breakpoint cluster region, characterized by large (approximately 38-49-kb), palindromic, low-copy repeats, strongly suggest that somatic rearrangements are not random events but rather reflect susceptibilities due to the genomic structure [89]. Fink et al. [42] showed that loss of *TP53* results from the absence of nearly the entire chromosome 17 p-arm rather than to monosomy 17 or deletions of *TP53*. Translocation or isochromosome formations at sites of low copy DNA repeats (LCR) in 17p10 to 17p12 appear to be the mechanism for the loss of *TP53* in CLL. These results show that 17p10-p12 is a hot spot for structural anomalies in this pathology.

Further studies will be useful to identify if regions on other chromosomes have LCR that are homologous to 17p10-p12 and mediate translocation with chromosome 17. This cytogenetic mechanism involving LCR is responsible for most, if not all, instances for loss of *TP53* in CLL.

In addition, the high rate of *TP53* mutations in the remaining allele with 17p13 deletions suggests that *TP53* is the tumor suppressor gene affected by the aberrations.

Rearrangement of the IgH locus in Band 14q32

Although 14q32 involvement was frequent in early CLL studies, currently t(11;14)(q13;q32) with *CCND1/IgH* rearrangement is considered to be the hallmark of mantle cell lymphoma (MCL). Thus, lymphoproliferative disorders exhibiting the translocation t(11;14) should therefore be carefully examined for morphological and immunophenotypical characteristics of MCL. Other rare 14q32 rearrangements, such as t(14;19)(q32;q13), is not restricted to CLL but it has also been described in low and high grade NHLs [90].

A pathogenetic role was also suggested for *BCL2* at 18q21. While the breakpoint of t(14;18) in follicular lymphoma occur in the major breakpoint region (MBR) or in the minor cluster region (mcr) within the 3′portion of *BCL2*, the breakpoints in CLL tumors have been localized at the 5′end with preferential juxtaposition of *BCL2* to the immunoglobulin light chains. However, subsequent studies show that *BCL2* rearrangements are rare events in CLL [19, 91], more often as a secondary event occurring late during disease progression [92].

Thus, translocations involving 14q32/*IgH* locus are either rare events or likely associated with other types of lymphoproliferative disorders. Disregarding t(11;14)(q13;q32) and t(14;18)(q32;q21), 14q32 rearrangements are found in 4% of cases studied by FISH [19].

3. Clinical Impact of Genomic Aberrations in CLL

In general, complex karyotypes and a high percentage of abnormal metaphases appear to confer a worse prognosis than single or dual chromosomal abnormalities. The median survival in cases with diploid normal karyotype is about 15 years and about 6-7.7 years for those with complex karyotypes [40]. However, multivariate analysis confirmed neither the presence of clonal aberrations nor the number of clonal aberrations to be independent prognostic variables.

13q Deletions

Patients that have monoallelic deletion 13q14 (13q14x1) as a single anomaly have good prognosis. They are characterized by slow disease progression, initial stages of the disease at diagnosis and survival times similar to those with a normal karyotype, with a median survival of 133 months [19, 22, 88]. This chromosome anomaly predominates among patients displaying mutated profile of IgV_H genes [12, 16, 93].

By FISH, recent studies have observed about 10% of cases carrying biallelic deletion 13q14 (13q14x2) into MDR region. This abnormality may result from chromosomal evolution and would represent a more aggressive FISH anomaly than 13q14x1 [22, 47].

Deletion 11q22-Q23

Presence of the 11q deletion in CLL patients is associated with the diagnosis at younger age, more advanced clinical stages, extensive lymphadenopathy, and more rapid disease progression with a shorter treatment-free interval (9 months versus 43 months). The prognostic effect of 11q deletion on survival depends on the age such as patients under 55 years, median survival is significantly shorter in the deletion group (64 months versus 209 months) while there is not difference in those aged 55 years or older (94 months versus 111 months) [16]. So, deletion of 11q is considered an independent prognostic factor and is an important predictor of survival in younger CLL patients [16, 19, 94]. This chromosome anomaly predominates among patients displaying unmutated IgV_H genes [12, 16, 93].

Trisomy 12

Some studies found that CLL patients with trisomy 12 by banding analysis had both shorter treatment-free interval and shorter overall survival, but such adverse prognostic impact has not been confirmed by other groups [43, 67, 68, 95, 96].

However, when the data of conventional chromosome banding analysis were included together with FISH studies, median survival in patients with trisomy 12 was significantly shorter than that observed in patients with a normal karyotype (7.8 versus 14.4 years). In other series, however, the prognosis in the group with trisomy 12 (median survival time 114 months) was intermediate to that 13q deletion, 11q22-23 or 17p13 [19, 97].

At the present time it is known that trisomy 12 is associated with an atypical morphology and immunophenotype, a high proliferative rate, advanced clinical stage and intermediate outcome but remain unresolved its pathogenic impact. This chromosome anomaly shows a similar distribution among patients displaying unmutated and mutated IgV_H genes, suggesting that CLL patients with trisomy 12 have considerable heterogeneity [16, 19, 74, 93]

Deletion 6q

By cytogenetic studies, deletion 6q was observed as a product of clonal evolution, mainly affecting region 6q21-q25 [44, 98]. Recently, an association was found between del(6)(q21) by FISH and a high white blood cell count, more extensive adenopathy at presentation, atypical morphology, classical inmunophenotype with CD38+, and a therapy-demanding disease. They show an intermediate incidence of IgV_H somatic hypermutation between the patients with 11q-/17q- and the patients with isolated 13q-, suggesting that the 6q- patients may be biologically heterogeneous as is the case with CLL with trisomy 12. Clinico-biological features and outcome show that this cytogenetic subset of CLL should be allocated in an intermediate risk group [99].

Deletion 17p

CLL patients with a *TP53* deletion have aggressive clinical course, advanced stages of the disease, shorter survival than those without deletion and poor response to treatment with purine nucleoside analogues and alkylating agents [19, 22].

Multivariate analysis identified *TP53* gene deletion as the strongest independent prognostic factor for survival [16, 83, 100]. The presence of a 17p deletion predominates among patients displaying unmutated IgV_H genes [12, 16, 93].

Conclusion

The cytogenetic status correlates with the biological behavior of CLL in vitro. Poor cytogenetic prognosis was associated with more rapid spontaneous apoptosis in vitro and lower inmunogenicity. Good cytogenetic prognosis have lower levels of spontaneous apoptosis, but higher basal expression of molecules involved in co-stimulation, cellular adhesion and antigen presentation, and induced significantly T-cell proliferation in mixed lymphocyte cultures [22, 101].

Different genomic aberration allow to identify distinct risk subgroups being 17p- and 11q- the most aggressive, +12 and 6q- as intermediate risk and normal karyotype and 13q14x1 as the less aggressive ones. Recent data from the literature [22] and from our laboratory (unpublished data) suggest that 13q14x2 could represent a new aggressive risk category. These data reveal the importance of FISH panel to monitor genomic changes with different clinical outcome.

Based on different studies, it appears that IgV_H mutation status and genomic aberrations are among the strongest currently available parameters and are of independent value to predict outcome in CLL. The IgV_H mutational profile has the advantage that it remains constant during clonal evolution, which contrasts with genomic aberrations [12, 93].

In contrast to other types of lymphoproliferative disorders, translocations are rare events in CLL and not any CLL-specific translocation has been found yet. Moreover, translocations have been reported to be a prognostic marker for an unfavorable clinical outcome [102]. In this context, the purpose of our study was to contribute to the understanding of the genomic aberrations associated with this group of lymphoid malignancies. The combination of conventional cytogenetics with FISH analysis allowed us a more accurate definition of the genomic aberrations involved in CLL/SLL. Novel and uncommon chromosome rearrangements were found in our series. Interestingly, among them, translocations were the most frequent aberration.

Material and Methods

Patients

Between 1993 and 2006, 191 patients with CLL/SLL were referred to our institution and cytogenetic analysis was successfully performed on 150 (78.5%) of them (103 males and 88 females; mean age 63.8 yr; range 39-89 yr). Twelve cases, (8%, 12/150) (8 males and 4 females; mean age 57.6 yr; range 46-72 yr) (table 1) were selected for inclusion in the present work based on the presence of non frequent clonal structural chromosome aberrations.

All patients were histologically reclassified according to World Health Organization classification of lymphoid neoplasias [103, 104]. Three patients (2, 5 and 9) were diagnosed as SLL by clinical, immunophenotypic, and histopathologic analysis. In all cases the peripheral white blood counts were normal. The remaining 9 patients were diagnosed as CLL, based on the National Cancer Institute Working Group guidelines for diagnosis of CLL [105]. According to the morphologic evaluation, six patients were classified as typical (cases 1, 3, 4, 6, 10 and 12) and three as atypical CLL (cases 7, 8 and 11). Clinical stages were established according to the Ann Arbor classification for SLL [106] and to the Rai classification for CLL [107]. Immunophenotypic analysis showed expression of pan-B antigen (CD19, CD20, CD22) with coexpression of CD5 and CD23. All of them were CD5+ and CD19+. Case 1 showed Hodgkin's disease (HD) of the mixed cellularity subtype after the diagnosis of CLL [108] and Case 7 showed a transformation to a diffuse large B-cell lymphoma (Ritcher'syndrome) after the second cycle of treatment [92].

Clinical features, treatment and survival information are shown in table 1. This series included patients whose karyotypes were performed at diagnosis or at variable points during disease evolution. Six patients had sequential cytogenetic studies. Cytogenetic and FISH analysis are detailed in table 2. All but one (case 6) patients had an adverse clinical course, with partial or null response to different chemotherapeutic regimens. Seven patients had died at the time of this study (table 1).

Table 1. Clinical characteristics of CLL/SLL patients

Case	Sex/ Age	Lymph (x10^9/l)	Clinical stage[1]	Treatment	Response	Survival (month) treatment-free	total
1	F/50	22.4	I[2]/HL[3]	CLB/P[2]	PR	7	
				ABVD[3]	Null		
				PrVCLBP[3]	PR		45[4]
2	M/70	2.3	IV[2]	CHOP[2]	PR	1	
				F/D[3]	PR		
				Anti CD20[3]	CR		56
3	M/56	9.1	0[2]	F/C[3]		2	
				Anti CD20[3]	CR		
				AT			106[4]
4	M/61	94.6	II[2]	CAPVE[2]/			
				CLB/P[2]/F	Null		24[4]
5	M/58	5	IV[2]	CHOP-MINE[2]	Null	5	9[4]
6	F/72		0[2]	No Treatment			28
7	M/54	82	III[2]	CHOP[2]	Null	2	
				F/C/D[3]	Null		
				ESHAP[3]	PR		
				Hyper-CVAD[3]	Null		9[4]
8	F/51	104	II[2]	CLB[3]	PR	48	
				F[3]	null		
				F/C/ Anti CD20[3]			155
9	M/46	1.5	II[2]	F/C[3]		4	73
10	M/50	140	I[2]	CLB/P[3]	PR	2	
				F/C[3]	CR		
				CHOP/Anti CD20[3]			51
11	F/66	68	II[2]	CLB/P[2]	PR	20	
				CLB/P[2]	Null		62[4]
12	M/57	21	I[2]	CLB[2]	PR	28	
				CLB[3]	Null		
				F/C[3]	Null		
				CHOP[3]			72

[1] CLL: Rai., SLL: Ann Harbor, [2] initial stage, [3] progression stage, [4] deceased.

PR, partial response; Null, without response; CR, complete response; HL, Hodgkin´s lymphoma; CLB, chlorambucil; P, prednisone; CAVPE, cyclophosphamide, adryamicine, vincristine, prednisone, etoposide; F, fludarabine; C, cyclophosphamide; D, dexamethasone; ABVD, doxorrubicin, bleomycin, vinblastine, dacarbazine; Pr: procarbazine. CHOP: cyclophosphamide, adryamicine, vincristine, prednisone. MINE: uromitexan, ifosfamide, navelbine, etoposide; ESHAP, etoposide, cisplatin, cytosine, arabinoside; CVAD, cyclophosphamide, vincristina, adryamicine, dexamethasone; AT, autologous transplantation.

Cytogenetic Analysis

Conventional cytogenetic studies were performed on peripheral blood, bone marrow and/or lymph nodes samples. Cells were cultured in F-10 medium supplemented with 15% fetal calf serum by direct method (24hs), synchronized cultures (48hs) and with stimulation of different mitogens such as lipopolysaccharide, pokeweed and phytohemagglutinin A for 72-96hs. G and C-banding techniques were used. Clonal chromosome abnormalities were described according to the International System for Human Cytogenetic Nomenclature (ISCN) [109].

FISH Analysis

For FISH analysis #2, #5, #12, #17, #18 and #21 whole chromosome painting probes (CAMBIO, Cambridge, UK), chromosomes #2, # 12 and #17 specific ∝-satellite DNA probes (VYSIS, Downer Grove, IL) and the locus specific DNA probes *Cyclin D1* at 11q13 (LSI Cyclin D1 Spectrum Orange/Cep 11 Spectrum Green, dual color), D13S319 at 13q14 band, *TP53* at 17p13.1 (Spectrum Orange), IgH Spectrum Green/BCL2 Spectrum Orange (VYSIS, Downer Grove, IL) and *ATM* (ataxia telangiectasia mutated) locus at 11q23 Spectrum Green (Q BIO gene) were used. FISH was performed according to manufacturer´s protocols. Briefly, probe mix was denatured for 2 minutes at 70°C and maintains at 45-50°C till use. Simultaneously, fresh spread slides were denatured for 2 minutes in 70% formamide 2 x SSC at 70°C and dehydrated in graded ethanol series. The probe mix was then applied to the slides and hybridized overnight at 37°C in a moist chamber. After hybridization, the slides were washed in 0.4 x SSC/ 0.3% NP40 for 2 minutes at 73°C, rinsed in 2 x SSC/ 0.1% NP40 for 1 minute at room temperature. The slides were counterstained with *4',6-diamidino-2-fenilindol diclorhidrato*_(DAPI). Four hundred interphase nuclei from the patients and 10 controls were scored for each probe. The cut-off for positive values (mean of normal controls+ 3SD) was 1.0, 3.0, 10.2, 5.1 and 7.7% for *IgH/BCL-2*, trisomy 12 and monosomies of D13S319, *TP53* and *ATM*, respectively.

Results

Table 2 shows the cytogenetic and FISH studies of 12 CLL/SLL patients of our series that presented non-frequent clonal structural chromosome aberrations in their karyotypes. In all cases, near diploid karyotypes (39-48 chromosomes) were observed.

Table 2. Cytogenetic and FISH results in the CLL/SLL patients

			Interphase (FISH (%)			
Case	Sample	Combined karyotype (GTG and FISH)	+12	13q-	17p-	11q-
1	BM LNB	46,XX[20] 45,XX,t(2;17)(p21;p11.2)[11].ish psu dic(17;2)(p11.2;p21)(wcp2+,D2Z1+,wcp17+,D17Z1+)/46,XX[2].	0.8 0.4	ND 7.7	7.0 85.0	ND 7.1
2	BM	47,XY,+12[4]/47,XY,t(2;17)(p21;q23),+12[9]/ 47,XY, t(2;17)(p21;q23),-10,+12,+17[2]/ 46,XX[16].	60.0	2.0	0.9	1.8
3	BM BM	47,XY,+12[3],/ 46,XY[19] 46,XY,t(2;17)(p21;q23),del(11)(q23)[cp4]/46,XY[8].	11.0 2.8	7.0 8.0	3.5 3.0	9.0 13.0
4	BM	39-45,XY,del(3)(p11),-10,der(14)t(11;14)(q13;q24), der(17)t(12;17)(q13;q25),-13,+r[cp23].	0.5	ND	ND	ND
5	BM	48,XY,+3,t(5;12)(q15;p11),+del (5)(q15),del(7)(q11), der(19)t(19;?)(q13;?)[13].ish.t(5;12;19)(q15;p11;q13) (wcp5+,wcp12++),del(5)(wcp5+)/46,XX[20].	ND	ND	ND	ND
6	BM	46-48,XX,t(2;14)(p11;q32),+10,del (11)(q23),+14[cp11]/ 46,XY[14]	1.7	22.4	4.7	2.5
7	BM BM	43,XY,psu-dic(12;21)(q24;q10),dic(17;18)(p11.2;p11.2)[6]/46;XY[16]. 43,XY,psu-dic(12;21)(q24;q10),dic(17;18)(p11.2;p11.2)[6]/46;XY[12]	1.0 0.7	18.0 11.7	43.8 77.3	ND ND
8	BM PB	46,XX[11] 47,XX,t(10;14)(q22;q32),+12[8] / 46,XX[5]	39 36.3	ND 4.8	ND 2.9	ND ND
9	BM	46,XX[10] 47,XX,del(7)(q34),+12[3]/ 46,XX [30]	3.3 ND	ND 16.0	ND ND	ND ND
10	BM	ND 46,XY,t(11;13)(p15;q22)[5]/ 46,XY[6]	0.7 0.2	50.0 27.3	1.7 1.9	ND 0.9
11	PB	46,XX,dup(12)(q13q24)[9]/ 46,XX [9]	2.4	20.6	4.4	ND
12	PB	46 ,XY,del(18)(p11)[4]/ 46,XY [11]	ND	ND	ND	ND

Bold chromosome rearragements correspond to unpublished or infrequent anomalies ; bold numbers correspond to abnormal clones; GTG: G-bands by Trypsin using Giemsa; ND: not determined; BM: bone marrow; LNB: lymph node biopsy; PB: peripheral blood.

As a whole, we identified 16 different novel or infrequent chromosome rearrangements by cytogenetic analysis: 7 translocations altogether, 4 of which were balanced, 2 unbalanced and 1 complex translocation; and 3 dicentric or psu dicentric chromosomes, 1 partial chromosome duplication and 5 deletions. Some of these findings were recently reported by us as a new alterations in SLL/CLL patients: t(2;17)(p21;q23); der(17)t(12;17)(q13;q25); der(14)(t(11;14)(q13;q24), t(5;12;19) (q15;p11;q13); psu dic (17;2)(p11.2;p21), del(3)(p11), del(5)(q15) [108], psu dic (12;21) (q24;q10) [92] and t(10;14)(q22;q32) [110]. Translocation t(11;13)(p15;q22) and del(18)(p11) were not previously described, meanwhile, dic(17;18)(p11.2;p11.2), dup(12)(q13q24), del(7)(q34), t(2;14)(p11;q32) and del(7)(q11) were infrequently reported in the literature [111]. In addition, the novel translocation t(2;17)(p21;q23)) was recurrent in our series. Chromosomes 17 (five cases), 2 and 12 (four cases each) and 14 (three cases each) were the most frequently involved in this series.

Twenty one different breakpoints were found in these infrequent structural rearrangements, and all of them were previously reported though with different partner chromosomes or in a different type of chromosome rearrangements. However, 38.1% (8/21) breakpoints, such as 2p21, 5q15, 17p11.2, 17q23, 12q13, 12q24, 14q32 and 18p11, were recurrent in our series.

Among these aberrations, psu dic(17;2) (p11.2;p21) (case 1), t(11;13)(p15;q22) (case 10), dup(12)(q13q24) (case 11) and del(18)(p11) (case 12), were observed as the sole cytogenetic abnormality. Associated clonal alterations were identified in the remaining patients. Several aberrations previously described in this neoplasm were noted, namely a 11q23 deletion in cases 3 and 6 and trisomies 3, 10, 12, 14 and 17, and monosomies 10 y 13.

In Case 1 [108] cytogenetic and FISH studies at the lymph node biopsy showed the presence of a psu dic (17;2)(p11.2;p21) (figure 1). Interphase FISH showed monoallelic loss of TP53 in 85% of cells. This chromosome rearrangement was seen at the moment in which the patient develop a mixed cellularity HD coexisting with the SLL in the same lymph node. A previous study on BM cells, at CLL diagnosis, had a low number of metaphases with normal karyotype but interphase FISH also showed a small clone (7% of cells) with *TP53* gene deletion, indicating the presence of the marker chromosome in pre-HD diagnostic tissue. In addition, EBV (Epstein Barr Virus) was detected in the Reed Sternberg (RS) cells, while the CLL/SLL tumour cells were EBV negative.

Cases 2 and 3, previously reported [108] showed the same translocation t(2;17)(p21;q23) (figure 2). In Case 2, a clear clonal evolution that included trisomies 12 and 17 and monosomy 10 was observed. Case 3 had two cytogenetic studies, the first one before treatment, showed a clone with trisomy 12 (13% of cells) and only one cell with t(2;17). A second study post-treatment (37 months later), showed a composite karyotype with t(2;17) and del(11)(q23), confirming the initial finding. Interphase FISH using ATM probe showed monoallelic loss of this locus with frequencies of 9% and 13% for the first and second study, respectively (table 2).

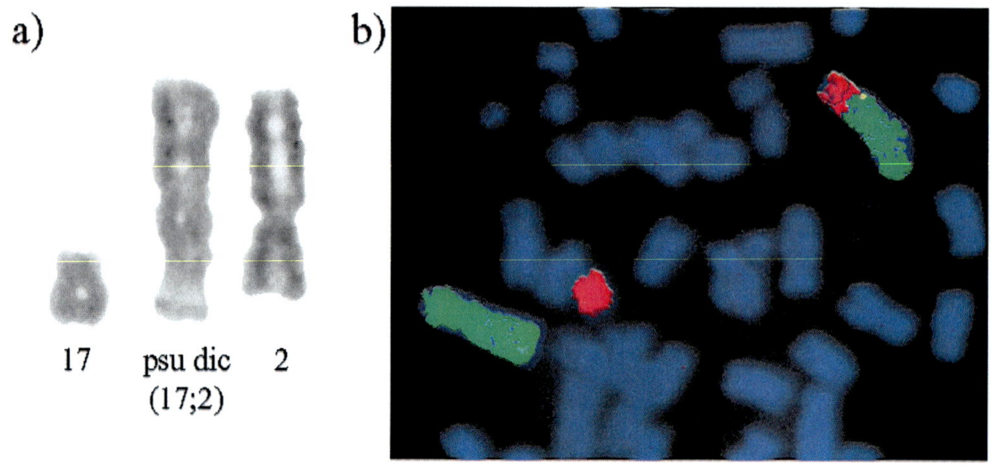

Figure 1. a) Partial karyotype showing psu dic(17;2)(p11.2;p21) from Case 1; b) Partial metaphase hybridized with chromosomes 2 and 17 painting probes showing normal chromosomes 2 (green) and 17 (red) and psu dic(17;2)(p11.2;p21) (arrow).

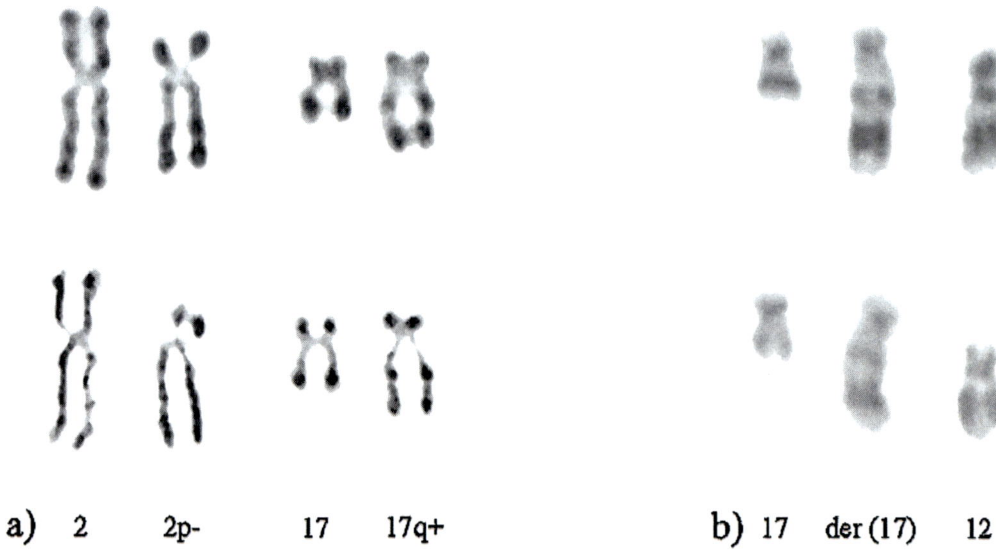

Figure 2. a) Partial karyotypes showing t(2;17)(p21;q23) from Cases 2 and 3; b) Partial karyotypes showing der(17)t(12;17)(q13;q25) from Case 4.

Cases 4, 5 and 6 had complex karyotypes including: del(3)(p11), der(14)t(11;14)(13;q24) and der(17)t(12;17)(q13;q25) (figure 2); t(5;12;19) (q15;p11;q13), del(5)(q15) and del(7)(q11); and t(2;14)(p11;q32), respectively. In addition of t(2;14), the karyotype of case 6 included trisomies of chromosomes 10 and 14, and del(11)(q23). Interestingly, FISH analysis did not showed ATM deletion (table 2), indicating that the breakpoint was telomeric to this gene. After 24 months from diagnosis, progression to Rai´s stage I without requirement of treatment was observed.

Case 7 [92], was studied in two opportunities, at diagnosis and during Richter's transformation showing the same karyotype: 44,XY,der(12)t(12;21)(q24;q11), der(17;18)(q10;q10). Both rearrangements were confirmed by FISH using painting probes. C-banding technique revealed the presence of the inactivated centromere of chromosome 21, indicating the pseudo dicentric nature of der(12): psu dic(12;21)(q24;q10). FISH using ∝-satellite DNA probes of chromosomes 17 and 18 showed a signal of each probe on the der(17;18), confirming the dicentric condition of this chromosome: ish dic (17;18)(p11.2;p11.2).

Cases 8 and 9 were first studied at diagnosis showing normal karyotypes but trisomy 12 by FISH (39% and 3.3%, respectively). Second cytogenetic study, during disease progression, showed t(10;14)(q22;q32) and del(7)(q34), respectively, and trisomy 12, suggesting clonal evolution. Moreover, case 9 showed a small clone with 13q14 deletion (16% of cells).

Case 10 was first studied at diagnosis. At this moment, the cytogenetic study was not succeeded, but FISH analysis showed monoallelic deletion of 13q14 (50% of interphase nuclei). The second study two years later, after treatment (Chlorambucil/Prednisone) and during accelerated lymphocytosis, showed a karyotype with t(11;13)(p15;q22) translocation. Simultaneously, a reduction in the del(13)(q14) clone was observed (27.3% of cells).

Case 11 [110] showed a clone with a dup(12)(q13q22) (50% of cells), confirmed by FISH. The remaining cells were normal. The analysis with D13S319 probe showed 20% of cells with monoallelic deletion.

Finally, in case 12 was detected a clone with del(18)(q11) as a sole anomaly. This study was done two years after diagnosis, without previous treatment, when patient was in disease progression. FISH analysis cannot be performed.

Discussion

The identification of the non-random genetic alterations associated with cancer has great importance not only to basic research but also to disease management as diagnostic and prognostic markers. We present the cytogenetic and FISH analyses of 12 CLL/SLL patients with new or infrequent clonal chromosome rearrangements in their karyotypes, all of them showing adverse clinical course with partial or null response to different chemotherapeutic agents.

Nine out of these 16 chromosome rearrangements described in this work had been previously reported by us as new alterations in CLL/SLL [48, 108, 110]. All of them are still novel rearrangements, and except for t(2;17), not recurrent in CLL/SLL. Two out of 9 were pseudo dicentric chromosomes, two deletions and the remaining five were translocations. In addition, two new chromosome rearrangements that have not been reported yet, and five infrequent chromosome abnormalities are shown. Chromosomes 17, 2, 12 and 14 were the most commonly involved.

Among these novel anomalies, a psu dic (17;2)(p11.2;p21) was found as a sole cytogenetic anomaly meanwhile a psu dic (12;21)(q24;q10) was associated to dic (17;18)(p11.2;p11.2) in the same karyotype. Both chromosome 17 abnormalities resulted in 17p loss, with *TP53* monoallelic deletion.

In hematological malignancies, including lymphoid and myeloid disorders rare dicentric anomalies have been described: dic(7;12), dic(9;12) and dic(17;18) in the novo or therapy-related acute myeloid leukemia and myelodysplastic syndrome [111]. Of interest, not all are associated with adverse clinical outcome, such as dic(9;12) in childhood B-ALL, which has favorable prognosis [112, 113].

Telomere dysfunction has been associated with genomic instability and mitotic disturbances in different type of neoplasms such colorectal and intestinal carcinomas and carcinomas of the head and neck [114-116]. Abnormal shortening of terminal TTAGGG repeats causes different mitotic defects with anaphase bridges triggering both numerical and structural chromosome changes, and with mitotic multipolarity leading to more dramatic genomic changes that may prove deleterious to the cell. It is known that one of them or both mechanisms are associated with the formation of dicentric chromosomes in these carcinomas [114-117]. In contrast, the mechanisms that prevail behind dicentric chromosomes generation in CLL are not known.

In reference to our cases with dicentric chromosomes, it is interesting to point out that case 1 with a psu dic (17;2)(p11.2;p21) [108], develop a HD nineteen months after the diagnosis of CLL. It is known that patients with CLL/SLL have approximately an 8-fold increased risk to develop HD [118], with HD representing one of the most frequent secondary neoplasms. In a few cases, clonality analysis of CLL/SLL and RS cells resulted in the detection of the same clonal immunoglobulin heavy chain gene rearrangement, indicating that the RS cells derived from a common precursor B-cell. However, in a high percentage of cases, EBV positive RS cells were not genetically related to the SLL tumour clone, suggesting that EBV might play a role in the development of the HD component [119-121], situation that is suggested for our patient.

Loss of the short arm of chromosome 17, where the *TP53* tumour suppressor gene localizes, has been observed in 4% of cytogenetically evaluable CLL patients. However, FISH studies have shown that monoallelic deletions of this gene occur in 9-15% of cases [19, 122]. Different reports have shown that *TP53* gene deletions have strong implications in the clinical course of the disease, with significantly shorter median survival, advanced clinical stage and resistance to chemotherapy. In addition, *TP53* abnormalities occur in about 60% of cases with Richter's syndrome [19, 122]. Moreover, Döhner et al. [122] have done FISH and molecular analysis of this gene, revealing a biallelic inactivation of the *TP53* in this pathology.

As for psu dic(12;21), to our knowledge this is the first CLL case where this alteration has been reported. It was found as a part of a karyotype that also include a dic(17;18), which represents the most frequent recurrent dicentric chromosome in CLL/SLL [111]. This patient showed disease progression to Richter's transformation. It is important to remark that, in the first study of this case, we found only one cell with a karyotype 45,XY,der(17;18)(q10;q10) (not presented in table 2), suggesting it as the early event leading to CLL that was followed by stepwise clonal evolution in the leukemic cells leading to lymphoma.

In a substantial subset of patients, new, and often more dangerous, mutations accumulate over time. These new alterations of the original clone may be better suited to survive and thrive, and they become the dominant clones. These clones with enhanced survival capabilities are much harder to kill with distinct protocols and disease progression is precipitated by the emergence of new and more aggressive clones [44, 57, 123]. At first, karyotypic clonal evolution seemed to be rare in CLL, but in last ten years it was detected from 16 to 43% of patients by sequential studies. Alterations more frequently involved are deletions at 6q, 13q and 11q [44, 123, 124]. In addition, the introduction of FISH studies was an important step to identify genomic changes, undetectable by conventional cytogenetic, implicated in the development and evolution of the malignant clone. In our experience, patients with clonal evolution showed a poor prognosis with a treatment-free survival significantly shorter than those observed in cases with 13q14 deletion or normal karyotypes ($p<0.001$) (data not shown).

Balanced translocations t(2;17)(p21;q23) (cases 2 and 3) and t(5;12;19) (q15;p11;q13) (case 5) have not been yet found in any other type of neoplasm. Our study also showed that chromosome 2 with breakpoint in p21 was the partner most frequently involved in rearrangements with chromosome 17. Bands 2p21-22 were found frequently rearranged in leukemia patients, region where, interestingly, the double-stranded RNA activated protein kinase (*PKR*) tumor suppressor gene [125], and the mismatch repair gene *MSH2* were mapped [126].

Moreover, t(10;14)(q12;q32) was described in other oncohematological diseases with a very low frequency [111]. Der(17)t(12;17)(q13;q25), der(14)t(11;14)(q13;q24) and del(3)(p11) were observed in the same karyotype. The first two abnormalities have not been reported just as in the literature neither in CLL/SLL nor in other neoplasm [111] but a der(11)t(11;14)(q13;q24) has been recently described in this disorder [102].

We have two more translocations in our series. One of them, t(11;13)(p15;q22), is a new chromosome rearrangements in CLL/SLL. The literature show a similar alteration but with different breakpoints on 13q arm in a CLL patient [44]. The other one, t(2;14)(p11;q32) (case 6), became recurrent in these pathology from our data [127]. At the moment of the present study, this patient was the only one that did not show an unfavorable clinical outcome.

Translocations are primary events useful for lymphoma classification, whereas secondary alterations are mainly chromosomal imbalances, many of which are associated with clinical outcome. By contrast, it is important to note that reciprocal chromosomal translocations are not frequently observed in CLL/SLL [111]. According to the literature, we did not find a CLL/SLL specific translocation but found instead many different translocations, all of them associated with unfavorable clinical course [88, 102].

As concerns to deletions, del(3)(p11), del(5)(q15) and del(18)(p11) are novel rearrangements in this pathology but they are recurrent anomalies in other neoplasms [111]. It has been described a human receptor tyrosine kinase gene on chromosome region 3p11.2, that is expressed by some pre-B and thymic T-cell lines, but is not detectable on normal adult human tissues. Gene cloning studies established that this gene is a new member of the EPH family of receptor tyrosine kinases, and its inappropriate expression may contribute to oncogenesis [128]. Losses of 5q are observed in different types of neoplasms and it has been suggested the possibility of gene products involved in cancer progression [129]. Recently the putative tumor suppressor gene, *DAL1* has been mapped on chromosome band 18p11.3. Future studies, will determinate whether this or another tumor suppressor gene is the target of the high frequency of allelic deletion found in different types of neoplasms [130, 131].

In addition, case 5 and case 9 showed deletions of a part of the long arm of chromosome 7: del(7)(q11) and del(7)(q34), which are infrequent findings in CLL. The former was previously described in only three cases [47, 132, 133]. Deletions 7q have been reported in myeloid diseases like acute myeloid leukemia and myelodisplastic syndrome but seem to be rare in lymphoproliferative diseases. Particularly, loss of 7q22-qter is common in hairy cell leukemia and splenic marginal zone lymphoma but not in CLL/SLL [134].

Among the remaining alterations, we had one case with dup(12)(q13q24) as a sole cytogenetic anomaly and the coexistence of 13q deletion identified by FISH. Duplication 12q is an uncommon chromosome abnormality not only in CLL but also in other hematological neoplasms [111]. Structural aberrations of the long arm of chromosome 12 have been observed in about 5% of CLL cases [88] and in approximately 15% of patients with Richter's syndrome [135]. Our case showed atypical morphology, high proliferation rate and progression of the disease, indicating that the duplication of this region may be similar to the complete trisomy 12 in CLL patients, in which this association is well known. This region contains several genes directly or indirectly associated with cancer but none of them seems affected in CLL. Some authors proposed that *MDM2* or *P2RX7* genes could be involved, but results are controversial [136-139]. More recently, a novel gene named CLLU1 (located at 12q22), was observed as uniquely overexpressed in CLL, suggesting their participation in CLL development [81].

Breakpoint analysis among these novel or infrequent chromosomal anomalies identified eight recurrent breaks. Each of these eight bands has previously been reported to be involved in lymphoid tumour [111] and they cannot be considered specific of CLL.

In reference to chromosome imbalances, most common gains involved chromosome region 12q13-q24, whereas 17p11.2 was the most frequently deleted regions. It is important to point out that one patient of our series (case 4) showed loss of 14q24-qter chromosome region and had an aggressive clinical course. Our finding and data from other reports suggest that the loss of 14q24-qter region is indicative of adverse clinical outcome and may be a new marker for prognostic stratification [140].

In summary, we have identified 16 novel or infrequent chromosome alterations, two of them reciprocal translocations, that might represent new recurrent abnormalities in CLL/SLL, and we described the salient clinical features of our patients. All these rare chromosome abnormalities, except for case 6, were observed in patients with adverse clinical outcome and short event free survival time. Our findings may facilitate the understanding of genetic events associated with development and progression of CLL/SLL and may have implications in the clinical management of patients with this disease.

Acknowledgements

This work was supported by grants from our National Research Council (CONICET) and the National Agency of Scientific and Technical Promotion (ANPCyT).

References

[1] Stilgenbauer S, Lichter P, Döhner H. Genetics of B-cell chronic lymphocytic leukemia. In: Chronic Lymphocytic Leukemia. *Molecular Genetics, Biology, Diagnosis, and Management*. Ed. Faguet GB 2004.

[2] Rozman C, Montserrat E. Chronic lymphocytic leukemia. *N. Engl. J. Med.* 1995;1333:1052-1057.

[3] Zwiebel JA, Cheson BD. Chronic lymphocytic leukemia: staging and prognostic factors. *Semis. Oncol.* 1998;25:42-59.

[4] SEER Cancer Statistics Review, 1975-2000, http://seer.cancer.gov/cgi-bin/csr/search.pl#results.

[5] American Cancer Society Facts and Figures for professionals, http://www.cancer.org/docroot/STT/stt_o.asp.

[6] Boggs DR, Chen SC, Zhang ZN, Zhang A. Chronic lymphocytic leukemia in China. *Am. J. Hematol.* 1987;25:349-354.

[7] Chang YH, Park J, Kim HC, Chun HK, Kim YR, Kim M, Han K, Lee J-H, Lee K-H, Cho HI, Lee, YS, Lee DS. Korean patients with chronic lymphocytic leukemia show the similar types of chromosomal aberrations as those in Europe and North America. *Leukemia Res.* 2006;30:695-699.

[8] Yamahihara ET, Blaisdell RK, Hayashi T, Lukes RJ. Malignant lymphoma in Hawaii-Japanese: a retrospective morphologic survey. *Hematol. Oncol.* 1989;7:219-232.

[9] Gale RP, Cozen W, Goodman MT, Wang FF, Bernstein L. Decrease chronic lymphocytic leukemia incidence in Asians in Los Angeles County. *Leukemia Res.* 2000;24:665-669.

[10] Tamura K, Sawada H, Izumi Y, Fukuda T, Utsunomiya A, Ikeda S, Uike N, Tsukada J, Kawano F, Shibuya T, Gondo H, Okamura S, Suzumiya J; Kyushu Hematology Organization for Treatment (K-HOT) Study Group. Chronic lymphocytic leukemia (CLL) is rare, but the proportion of T-CLL is high in Japan. *Eur. J. Haematol.* 2001;67:152-157,

[11] Mauro FR, Foa R, Giannarelli D Cordone I, Crescenzi S, Pescarmona E, Sala R, Cerretti R, Mandelli F. Clinical characteristics and outcome of young chronic lymphocytic leukemia patients: a single institution study of 204 cases. *Blood.* 1999;94:448-454.

[12] Dighiero G. CLL Biology and Prognosis. *Hematology Am. Soc. Hematol. Educ. Program.* 2005:278-284.

[13] Byrd JC, Stilgenbauer S, Flinn I. Chronic Lymphocytic Leukemia. *Hematology Am. Soc. Hematol. Educ. Program.* 2004:163-183.

[14] Rozman C, Montserrat E, Rodriguez-Fernandez JM, Ayats R, Vallespi T, Parody R, Rios A, Prados D, Morey M, Gomis F. Bone marrow histologic pattern--the best single prognostic parameter in chronic lymphocytic leukemia: a multivariate survival analysis of 329 cases. *Blood.* 1984;64:642-648.

[15] Montserrat E, Sanchez-Bisono J, Vinolas N, Rozman C. Lymphocyte doubling time in chronic lymphocytic leukaemia: analysis of its prognostic significance. *Br. J. Haematol.* 1986;62:567-575.

[16] Kröber A, Seiler T, Benner A, Bullinger L, Bruckle E, Lichter P, Dohner H, Stilgenbauer S. V(H) mutation status, CD38 expression level, genomic aberrations, and survival in chronic lymphocytic leukemia. *Blood.* 2002;100:1410-1416.

[17] Di Giovanni S, Valentini G, Carducci P, Giallonardo P. Beta-2-microglobulin is a reliable tumor marker in chronic lymphocytic leukemia. *Acta Haematol.* 1989;81:181-185.

[18] Hallek M, Langenmayer I, Nerl C, Knauf W, Dietzfelbinger H, Adorf D, Ostwald M, Busch R, Kuhn-Hallek I, Thiel E, Emmerich B. Elevated serum thymidine kinase levels identify a subgroup at high risk of disease progression in early, nonsmoldering chronic lymphocytic leukemia. *Blood.* 1999;93:1732-1737.

[19] Döhner H, Stilgenbauer S, Benner A, Leupolt E, Krober A, Bullinger L, Dohner K, Bentz M, Lichter P. Genomic aberrations and survival in chronic lymphocytic leukemia. *N. Engl. J. Med.* 2000,343:1910-1916.

[20] Lin K, Sherrington PD, Dennis M, Matrai Z, Cawley JC, Pettitt AR. Relationship between p53 dysfunction, CD38 expression, and IgV(H) mutation in chronic lymphocytic leukemia. *Blood.* 2002,100:1404-1409.

[21] Crespo M, Bosch F, Villamor N, Bellosillo B, Colomer D, Rozman M, Marce S, Lopez-Guillermo A, Campo E, Montserrat E. ZAP-70 expression as a surrogate for immunoglobulin-variable-region mutations in chronic lymphocytic leukemia. *N. Engl. J. Med.* 2003,348:1764-1775.

[22] Dewald GW, Brockman SR, Paternoster SF, Bone ND, O'Fallon JR, Allmer C, James CD, Jelinek DF, Tschumper RC, Hanson CA, Pruthi RK, Witzig TE, Call TG, Kay NE. Chromosome anomalies detected by interphase fluorescence in situ hybridization: correlation with significant biological features of B-cell chronic lymphocytic leukaemia. *Br. J. Haematol.* 2003,121:287-295.

[23] Raval A, Byrd JC, Plass C. Epigenetics in chronic lymphocytic leukemia. *Semin. Oncol.* 2006,33:157-166.

[24] Damle RN, Wasil T, Fais F, Ghiotto F, Valetto A, Allen SL, Buchbinder A, Budman D, Dittmar K, Kolitz J, Lichtman SM, Schulman P, Vinciguerra VP, Rai KR, Ferrarini M, Chiorazzi N. IgV gene mutation status and CD38 expression as novel prognostic indicators in chronic lymphocytic leukemia. *Blood.* 1999;94:1840-1847.

[25] Hamblin TJ, Davis Z, Gardiner A, Oscier DG, Stevenson FK. Unmutated IgV(H) genes are associated with a more aggressive form of chronic lymphocytic leukemia. *Blood.* 1999;94:1848-1854.

[26] Klein U, Tu Y, Stolovitzky GA, Mattioli M, Cattoretti G, Husson H, Freedman A, Inghirami G, Cro L, Baldini L, Neri A, Califano A, Dalla-Favera R. Gene expression profiling of B cell chronic lymphocytic leukemia reveals a homogeneous phenotype related to memory B cells. *J. Exp. Med.* 2001;194:1625-1638.

[27] Rosenwald A, Alizadeh AA, Widhopf G, Simon R, Davis RE, Yu X, Yang L, Pickeral OK, Rassenti LZ, Powell J, Botstein D, Byrd JC, Grever MR, Cheson BD, Chiorazzi N, Wilson WH, Kipps TJ, Brown PO, Staudt LM. Relation of gene expression phenotype to immunoglobulin mutation genotype in B cell chronic lymphocytic leukemia. *J. Exp. Med.* 2001;194:1639-1647.

[28] Wiestner A, Rosenwald A, Barry TS, Wright G, Davis RE, Henrickson SE, Zhao H, Ibbotson RE, Orchard JA, Davis Z, Stetler-Stevenson M, Raffeld M, Arthur DC, Marti GE, Wilson WH, Hamblin TJ, Oscier DG, Staudt LM. ZAP-70 expression identifies a chronic lymphocytic leukemia subtype with unmutated immunoglobulin genes, inferior clinical outcome, and distinct gene expression profile. *Blood.* 2003;101:4944-4951.

[29] Vasconcelos Y, De Vos J, Vallat L, Reme T, Lalanne AI, Wanherdrick K, Michel A, Nguyen-Khac F, Oppezzo P, Magnac C, Maloum K, Ajchenbaum-Cymbalista F, Troussard X, Leporrier M, Klein B, Dighiero G, Davi F; French Cooperative Group on CLL. Gene expression profiling of chronic lymphocytic leukemia can discriminate cases with stable disease and mutated Ig genes from those with progressive disease and unmutated Ig genes. *Leukemia.* 2005,19:2002-2005.

[30] Nolz JC, Tschumper RC, Pittner BT, Darce JR, Kay NE, Jelinek DF. ZAP-70 is expressed by a subset of normal human B-lymphocytes displaying an activated phenotype. *Leukemia.* 2005;6:1018-1024.

[31] Catherwood MA, Matthews C, Niblock R, Dobbin E, Morris TC, Alexander HD. ZAP-70 mRNA quantification in B-cell chronic lymphocytic leukaemia. *Eur. J. Haematol.* 2006;76:294-298.

[32] Corcoran M, Parker A, Orchard J, Davis Z, Wirtz M, Schmitz OJ, Oscier D. ZAP-70 methylation status is associated with ZAP-70 expression status in chronic lymphocytic leukemia. *Haematologica.* 2005;90:1078-1088.

[33] Shanafelt TD, Geyer SM, Kay NE. Prognosis at diagnosis integrating molecular biologic insights into clinical practice for patients with CLL. *Blood.* 2004;103:1202-1210.

[34] Scroers R, Griesinger F, Trümper I, Haase B, Klein-Hitpass L, Sellmann I, Dührsen U, Dürig L. Combined analysis of ZAP-70 and CD38 expression as a predictor of disease progression in B-cell chronic lymphocytic leukemia. *Leukemia.* 2005;19:750-758.

[35] Autio K, Elonen E, Teerenhovi L, Knuutila S. Cytogenetic and immunologic characterization of mitotic cells in chronic lymphocytic leukaemia. *Eur. J. Haematol.* 1987,39:289-298.

[36] Larramendy ML, Siitonen SM, Zhu Y, Hurme M, Vilpo L, Vilpo JA, Knuutila S. Optimized mitogen stimulation induces proliferation of neoplastic B cells in chronic lymphocytic leukemia: significance for cytogenetic analysis.The Tampere Chronic Lympocytic Leukemia group. *Cytogenet. Cell Genet.* 1998;82:215-221.

[37] Ross FM, Stockdill G. Clonal chromosome abnormalities in chronic lymphocytic leukemia patients revealed by TPA stimulation of whole blood cultures. *Cancer Genet. Cytogenet.* 1987, 25: 109-121.

[38] Nayak BN, Sokal J, Ray M. Clonal chromosomal changes in chronic lymphocytic leukemia. *Cancer Lett.* 1990 Feb;49(2):99-105.

[39] Stephenson CF, Desai ZR, Bridges JM. The proliferative activity of B-chronic lymphocytic leukaemia lymphocytes prior to and after stimulation with TPA and PHA. *Leuk. Res.* 1991;15(11):1005-12.

[40] Juliusson G, Gahrton G. Chromosome aberrations in B-cell chronic lymphocytic leukemia. Pathogenetic and clinical implications. *Cancer Genet. Cytogenet.* 1990,45:143-160.

[41] Stockero KJ, Fink SR, Smoley SA, Paternoster SF, Shanafelt TD, Call TG, Zent CS, Van Dyke DL, Kay NE, Dewald GW. Metaphase cells with normal G-bands have cryptic interstitial deletions in 13q14 detectable by fluorescence in situ hybridization in B-cell chronic lymphocytic leukemia. *Cancer Genet. Cytogenet.* 2006;166:152-156.

[42] Fink SR, Smoley SA, Stockero KJ, Paternoster SF, Thorland EC, Van Dyke DL, Shanafelt TD, Zent CS, Call TG, Kay NE, Dewald GW. Loss of TP53 is due to rearrangements involving chromosome region 17p10 approximately p12 in chronic lymphocytic leukemia. *Cancer Genet. Cytogenet.* 2006;167:177-181

[43] Han T, Sadamori N, Ozer H, Gajera R, Gomez GA, Henderson ES, Bhargava A, Fitzpatrick J, Minowada J, Bloom ML, et al. Cytogenetic studies in 77 patients with chronic lymphocytic leukemia: correlations with clinical, immunologic, and phenotypic data. *J. Clin. Oncol.* 1984,2:1121-1132.

[44] Finn WG, Kay NE, Kroft SH, Church S, Peterson LC. Secondary abnormalities of chromosome 6q in B-cell chronic lymphocytic leukemia: a sequential study of karyotypic instability in 51 patients. *Am. J. Hematol.* 1998,59:223-229.

[45] Buhmann R, Kurzeder C, Rehklau J, Westhaus D, Bursch S, Hiddemann W, Haferlach T, Hallek M, Schoch C. CD40L stimulation enhances the ability of conventional metaphase cytogenetics to detect chromosome aberrations in B-cell chronic lymphocytic leukaemia cells. *Br. J. Haematol.* 2002,118:968-975.

[46] Stilgenbauer S, Döhner H. Molecular genetics and its clinical relevance. *Hematol. Oncol. Clin. North. Am.* 2004,18:827-848.

[47] Reddy KS. Chronic lymphocytic leukaemia profiled for prognosis using a fluorescence *in situ* hybridization panel. *Br. J. Haematol.* 2006,132:705-722.

[48] Chena C, Arrossagaray G, Scolnik M, Palacios MF, Slavutsky I. Interphase cytogenetic analysis in Argentinean B-cell chronic lymphocytic leukemia patients: association of trisomy 12 and 13q14 deletions. *Cancer Genet. Cytogenet.* 2003;146: 154-160.

[49] Liu Y, Szekely L, Grander D, Soderhall S, Juliusson G, Gahrton G, Linder S, Einhorn S. Chronic lymphocytic leukemia cells with allelic deletions at 13q14 commonly have one intact RB1 gene: evidence for a role of an adjacent locus. *Proc. Natl. Acad. Sci. USA.* 1993;90:8697-8701.

[50] Stilgenbauer S, Dohner H, Bulgay-Morschel M, Weitz S, Bentz M, Lichter P. High frequency of monoallelic retinoblastoma gene deletion in B-cell chronic lymphoid leukemia shown by interphase cytogenetics. *Blood.* 1993;81:2118-2124.

[51] Liu Y, Hermanson M, Grander D, Merup M, Wu X, Heyman M, Rasool O, Juliusson G, Gahrton G, Detlofsson R, Nikiforova N, Buys C, Söderhäll S, Yankovsky N, Zabarovsky E, Einhorn S. 13q deletions in lymphoid malignancies. *Blood.* 1995;86:1911-1915.

[52] Stilgenbauer S, Nickolenko J, Wilhelm J, Wolf S, Weitz S, Dohner K, Boehm T, Dohner H, Lichter P. Expressed sequences as candidates for a novel tumor suppressor gene at band 13q14 in B-cell chronic lymphocytic leukemia and mantle cell lymphoma. *Oncogene.* 1998,16:1891-1897.

[53] Migliazza A, Bosch F, Komatsu H, Cayanis E, Martinotti S, Toniato E, Guccione E, Qu X, Chien M, Murty VV, Gaidano G, Inghirami G, Zhang P, Fischer S, Kalachikov SM, Russo J, Edelman I, Efstratiadis A, Dalla-Favera R. Nucleotide sequence,

transcription map, and mutation analysis of the 13q14 chromosomal region deleted in B-cell chronic lymphocytic leukemia. *Blood.* 2001;97:2098-2104.

[54] Calin GA, Dumitru CD, Shimizu M, Bichi R, Zupo S, Noch E, Aldler H, Rattan S, Keating M, Rai K, Rassenti L, Kipps T, Negrini M, Bullrich F, Croce CM. Frequent deletions and down-regulation of micro-RNA genes miR15 and miR16 at 13q14 in chronic lymphocytic leukemia. *Proc. Natl. Acad. Sci. USA.* 2002;99:15524-15529.

[55] Calin GA, Ferracin M, Cimmino A, Di Leva G, Shimizu M, Wojcik SE, Iorio MV, Visone R, Sever NI, Fabbri M, Iuliano R, Palumbo T, Picchiorri F, Roldo C, Garzon R, Sevignani C, Rassenti L, Alder H, Volinis S, Liu CG, Kipps TJ, Negrini M, Croce CM. A MicroRNA signature associated with prognosis and progression in chronic lymphocytic leukemia. N Engl J Med 2005;353:1793-1801. *Erratum N. Engl. J. Med.* 2006;355:533.

[56] Döhner H, Stilgenbauer S, James MR, Benner A, Weilguni T, Bentz M, Fischer K, Hunstein W, Lichter P. 11q deletions identify a new subset of B-cell chronic lymphocytic leukemia characterized by extensive nodal involvement and inferior prognosis. *Blood.* 1997;89:2516-2522.

[57] Cuneo A, Bigoni R, Rigolin GM, Roberti MG, Bardi A, Cavazzini F, Milani R, Minotto C, Tieghi A, Della Porta M, Agostini P, Tammiso E, Negrini M, Castoldi G. Late appearance of the 11q22.3-23.1 deletion involving the ATM locus in B-cell chronic lymphocytic leukemia and related disorders. Clinico-biological significance. *Haematologica.* 2002,87:44-51.

[58] Stilgenbauer S, Liebisch P, James MR, Schroder M, Schlegelberger B, Fischer K, Bentz M, Lichter P, Dohner H. Molecular cytogenetic delineation of a novel critical genomic region in chromosome bands 11q22.3-923.1 in lymphoproliferative disorders. *Proc. Natl. Acad. Sci. USA.* 1996,93:11837-11841.

[59] Monni O, Zhu Y, Franssila K, Oinonen R, Hoglund P, Elonen E, Joensuu H, Knuutila S. Molecular characterization of deletion at 11q22.1-23.3 in mantle cell lymphoma. *Br. J. Haematol.* 1999,104:665-671.

[60] Schaffner C, Stilgenbauer S, Rappold GA, Dohner H, Lichter P. Somatic ATM mutations indicate a pathogenic role of ATM in B-cell chronic lymphocytic leukemia. *Blood.* 1999;94:748-753.

[61] Stilgenbauer S, Winkler D, Ott G, Schaffner C, Leupolt E, Bentz M, Moller P, Muller-Hermelink HK, James MR, Lichter P, Dohner H. Molecular characterization of 11q deletions points to a pathogenic role of the ATM gene in mantle cell lymphoma. *Blood.* 1999;94:3262-3264.

[62] Schaffner C, Idler I, Stilgenbauer S, Dohner H, Lichter P. Mantle cell lymphoma is characterized by inactivation of the ATM gene. *Proc. Natl. Acad. Sci. USA.* 2000;97:2773-2278.

[63] Khanna KK, Jackson SP. DNA double-strand breaks: signaling, repair and the cancer connection. *Nat. Genet.* 2001,27:247-254.

[64] Eclache V, Caulet-Maugendre S, Poirel HA, Djemai M, Robert J, Lejeune F, Raphael M. Cryptic deletion involving the ATM locus at 11q22.3 approximately q23.1 in B-cell chronic lymphocytic leukemia and related disorders. *Cancer Genet. Cytogenet.* 2004;152:72-76.

[65] Stankovic T, Stewart GS, Fegan C, Biggs P, Last J, Byrd PJ, Keenan RD, Moss PA, Taylor AM. Ataxia telangiectasia mutated-deficient B-cell chronic lymphocytic leukemia occurs in pregerminal center cells and results in defective damage response and unrepaired chromosome damage. *Blood.* 2002,99:300-309.

[66] Canman CE, Lim DS. The role of ATM in DNA damage responses and cancer. *Oncogene.* 1998;17:3301-3308.

[67] Juliusson G, Robert KH, Ost A, Friberg K, Biberfeld P, Nilsson B, Zech L, Gahrton G. Prognostic information from cytogenetic analysis in chronic B-lymphocytic leukemia and leukemic immunocytoma. *Blood.* 1985;65:134-141.

[68] Bird ML, Ueshima Y, Rowley JD, Haren JM, Vardiman JW. Chromosome abnormalities in B cell chronic lymphocytic leukemia and their clinical correlations. *Leukemia.* 1989;3:182-191.

[69] Anastasi J, Le Beau M, Vardiman J, Fernald AA, Larson RA, Rowley JD. Detection of trisomy 12 in chronic lymphocytic leukemia by fluorescence in situ hybridization to interphase cells: A simple and sensitive method. *Blood.* 1992;79:1796-1801.

[70] Escudier SM, Pereira-Leahy JM, Drach JW, Weier HU, Goodacre AM. Fluorescent in situ hybridization and cytogenetic studies of trisomy 12 in chronic lymphocytic leukemia. *Blood.* 1993;81:2702-2707.

[71] Matutes E, Oscier D, Garcia-Marco J, Ellis J, Copplestone A, Gillingham R, Hamblin T, Lens D, Swansbury GJ, Catovsky D. Trisomy 12 defines a group of CLL with atypical morphology: correlation between cytogenetic, clinical and laboratory features in 544 patients. *Br. J. Haematol.* 1996;92:382-388.

[72] Crossen PE, Horn HL. Origin of trisomy 12 in B-cell chronic lymphocytic leukemia. *Cancer Genet. Cytogenet.* 1987;28:185-186.

[73] Que TH, Garcia-Marco J, Ellis J, Matutes E, Brito-Babapulle V, Boyle S, Catovsky. Trisomy 12 in chronic lymphocytic leukemia detected by fluorescence in situ hybridization: Analysis by stage, immunophenotype, and morphology. *Blood.* 1993;82:571-575.

[74] Athanasiadou A, Stamatopoulos K, Tsompanakou A, Gaitatzi M, Kalogiannidis P, Anagnostopoulos A, Fassas A, Tsezou A. Clinical, immunophenotypic, and molecular profiling of trisomy 12 in chronic lymphocytic leukemia and comparison with other karyotypic subgroups defined by cytogenetic analysis. *Cancer Genet. Cytogenet.* 2006;168:109-119.

[75] Dierlamm J, Wlodarska I, Michaux L, Vermeesch JR, Meeus P, Stul M, Criel A, Verhoef G, Thomas J, Delannoy A, Louwagie A, Cassiman JJ, Mecucci C, Hagemeijer A, Van den Berghe H. FISH identifies different types of duplications with 12q13-15 as the commonly involved segment in B-cell lymphoproliferative malignancies characterized by partial trisomy 12. *Genes Chrom. Cancer.* 1997;20:155-166.

[76] Merup M, Juliusson G, Wu X, Jansson M, Stellan B, Rasool O, Roijer E, Stenman G, Gahrton G, Einhorn S. Amplification of multiple regions of chromosome 12, including 12q13-15, in chronic lymphocytic leukaemia. *Eur. J. Haematol.* 1997;58:174-180.

[77] Wessendorf S, Schwaenen C, Kohlhammer H, Kienle D, Wrobel G, Barth TF, Nessling M, Moller P, Dohner H, Lichter P, Bentz M. Hidden gene amplifications in aggressive

B-cell non-Hodgkin lymphomas detected by microarray-based comparative genomic hybridization. *Oncogene.* 2003;22:1425-1429.

[78] Schwaenen C, Nessling M, Wessendorf S, Salvi T, Wrobel G, Radlwimmer B, Kestler HA, Haslinger C, Stilgenbauer S, Dohner H, Bentz M, Lichter P. Automated array-based genomic profiling in chronic lymphocytic leukemia: development of a clinical tool and discovery of recurrent genomic alterations. *Proc. Natl. Acad. Sci. USA.* 2004,101:1039-1044.

[79] Haslinger C, Schweifer N, Stilgenbauer S, Dohner H, Lichter P, Kraut N, Stratowa C, Abseher R. Microarray gene expression profiling of B-cell chronic lymphocytic leukemia subgroups defined by genomic aberrations and VH mutation status. *J. Clin. Oncol.* 2004;22:3937-3949.

[80] Kienle DL, Korz C, Hosch B, Benner A, Mertens D, Habermann A, Krober A, Jager U, Lichter P, Döhner H, Stilgenbauer S. Evidence for distinct pathomechanisms in genetic subgroups of chronic lymphocytic leukemia revealed by quantitative expression analysis of cell cycle, activation, and apoptosis-associated genes. *J. Clin. Oncol.* 2005;23:3780-3792.

[81] Buhl AM, Jurlander J, Jorgensen FS, Ottesen AM, Cowland JB, Gjerdrum LM, Hansen BV, Leffers H. Identification of a gene on chromosome 12q22 uniquely overexpressed in chronic lymphocytic leukemia. *Blood.* 2006;107:2904-2911.

[82] Offit K, Louie DC, Parsa NZ, Filippa D, Gangi M, Siebert R, Chaganti R. Clinical and morphologic features of B-cell small lymphocytic lymphoma with del(6)(q21q23). *Blood.* 1994; 83:2611-2618.

[83] Bigoni R, Cuneo A, Roberti MG, Bardi A, Rigolin GM, Piva N, Scapoli G, Spanedda R, Negrini M, Bullrich F, Veronese ML, Croce CM, Castoldi GL. Chromosome aberrations in atypical chronic lymphocytic leukemia: a cytogenetic and interphase cytogenetic study. *Leukemia.* 1997; 11:1933-1940.

[84] Gaidano, RS Hauptschein, NZ Parsa, K Offit, PH Rao, G Lenoir, DM Knowles, RS Chaganti and R Dalla-Favera. Deletions involving two distinct regions of 6q in B-cell non-Hodgkin lymphoma. *Blood.* 1992; 80:1781-1787.

[85] Menasce LP, Orphanos V, Santibanez-Koref M, Boyle JM, Harrison CJ. Common region of deletion on the long arm of chromosome 6 in non-Hodgkin's lymphoma and acute lymphoblastic leukaemia. *Genes Chromosomes Cancer.* 1994;10 :286-288.

[86] Zhang Y, Matthiesen P, Harder S, Siebert R, Castoldi G, Calasanz MJ, Wong KF, Rosenwald A, Ott G, Atkin NB, Schlegelberger B. A 3-cM commonly deleted region in 6q21 in leukemias and lymphomas delineated by fluorescence in situ hybridization. *Genes Chrom. Cancer.* 2000; 27:52-58.

[87] Merup M, Moreno TC, Heyman M, Ronnberg K, Grander D, Detlofsson R, Rasool O, Liu Y, Soderhall S, Juliusson G, Gahrton G, Einhorn. 6q deletions in acute lymphoblastic leukemia and non-Hodgkin's lymphomas. *Blood.* 1998; 91:3397-3400.

[88] Juliusson G, Oscier DG, Fitchett M, Ross FM, Stockdill G, Mackie MJ, Parker AC, Castoldi GL, Guneo A, Knuutila S, Elonen E, Gahrton G. Prognostic subgroups in B-cell chronic lymphocytic leukemia defined by specific chromosomal abnormalities. *N. Engl. J. Med.* 1990;323:720-724.

[89] Barbouti A, Stankiewicz P, Nusbaum C, Cuomo C, Cook A, Hoglund M, Johansson B, Hagemeijer A, Park SS, Mitelman F, Lupski JR, Fioretos T. The breakpoint region of the most common isochromosome, i(17q), in human neoplasia is characterized by a complex genomic architecture with large, palindromic, low-copy repeats. *Am. J. Hum. Genet.* 2004;74:1-10.

[90] Michaux L, Dierlamm J, Wlodarska I, Bours V, Van Den Berbhe H, Hagemeijer A. t(14;19)/BCL3 rearrangements in lymphoproliferative disorders: A review of 23 cases: *Cancer Genet. Cytogenet.* 1997;94:36-43.

[91] Dyer MJ, Zani VJ, Lu WZ, O'Byrne A, Mould S, Chapman R, Heward JM, Kayano H, Jadayel D, Matutes E. BCL2 translocations in leukemias of mature B cells. *Blood.* 1994;83:3682-3688.

[92] Chena C, Cerretini R, Noriega MF, Narbaitz M, Scolnk M, Palacios MF, Neme D, Bruno S, Slavutsky I. Cytogenetic, FISH and molecular studies in a case of B-cell chronic lymphocitic leukemia with karyotypic evolution. *Eur. J. Haematol.* 2002; 69:309-314.

[93] Seiler T, Dohner H, Stilgenbauer S. Risk stratification in chronic lymphocytic leukemia. *Semin. Oncol.* 2006; 33:186-194.

[94] Dohner H, Stilgenbauer S, Dohner K, Bentz M, Lichter P. Chromosome aberrations in B-cell chronic lymphocytic leukemia: reassessment based on molecular cytogenetic analysis. *J. Mol. Med.* 1999;77:266-81.

[95] Pittman S, Catovsky D. Prognostic significance of chromosome abnormalities in chronic lymphocytic leukaemia. *Br. J. Haematol.* 1984;58:649-660.

[96] Oscier DG, Stevens J, Hamblin TJ, Pickering RM, Lambert R, Fitchett M. Correlation of chromosome abnormalities with laboratory features and clinical course in B-cell chronic lymphocytic leukaemia. *Br. J. Haematol.* 1990;76:352-358.

[97] Gahrton G, Juliusson G. Clinical implication of chromosomal aberrations in chronic B-lymphocytic leukaemia cells. *Nouv. Rev. Fr. Hematol.* 1988;30:389-392.

[98] Glassman AB, Harper-Allen EA, Hayes KJ, Hopwood VL, Gutterman EE, Zagryn SP. Chromosome 6 abnormalities associated with prolymphocytic acceleration in chronic lymphocytic leukemia. *Ann. Clin. Lab. Sci.* 1998;28:24-29.

[99] Cuneo A, Rigolin GM, Bigoni R, De Angeli C, Veronese A, Cavazzini F, Bardi A, Roberti MG, Tammiso E, Agostini P, Ciccone M, Della Porta M, Tieghi A, Cavazzini L, Negrini M, Castoldi G. Chronic lymphocytic leukemia with 6q- shows distinct hematological features and intermediate prognosis. *Leukemia.* 2004; 18:476-483.

[100] Chevallier P, Penther D, Avet-Loiseau H, Robillard N, Ifrah N, Mahe B, Hamidou M, Maisonneuve H, Moreau P, Jardel H, Harousseau JL, Bataille R, Garand R. CD38 expression and secondary 17p deletion are important prognostic factors in chronic lymphocytic leukaemia. *Br. J. Haematol.* 2002;116:142-150.

[101] Jahrsdörfer B, Wooldridge JE, Blackwell SE, Taylor CM, Link BK, Weiner GJ. Good prognosis cytogenetics in B-cell chronic lymphocytic leukemia is associated in vitro with low susceptibility to apoptosis and enhanced immunogenicity. *Leucemia.* 2005;19:759-766.

[102] Mayr C, Speicher MR, Kofler DM, Buhmann R, Strehl J, Busch R, Hallek M, Wendtner CM. Chromosomal translocations are associated with poor prognosis in chronic lymphocytic leukemia. *Blood.* 2006;107:742-751.

[103] Harris NL, Jaffe ES, Stein H, Banks PM, Chan JK, Cleary ML, Delsol G, De Wolf-Peeters C, Falini B, Gatter KC, Grogan TM, Isaacson PG, Knowles DM, Manson DY, Muller-Hermelink HK, Pileri SA, Piris MA, Ralfkier E, Warnke RA. A revised European-American classification of lymphoid neoplasms: a proposal from the International Lymphoma Study Group. *Blood.* 1994;84:1361-1392.

[104] Harris NL, Jaffe ES, Diebold J, Flandrin G, Muller-Hermelink HK, Vardiman J, Lister TA, Bloomfield CD. The World Health Organization classification of neoplastic diseases of the hematopoietic and lymphoid tissues. Report of the Clinical Advisory Committee meeting, Airlie House, Virginia, November, 1997. *Ann. Oncol;* 1999;10:14191432.

[105] Cheson BD, Bennett JM, Grever M, Kay N, Keating MJ, O'Brien S, Rai KR. National Cancer Institute-sponsored Working Group guidelines for chronic lymphocytic leukemia: revised guidelines for diagnosis and treatment. *Blood.* 1996;87:4990-4997.

[106] Carbone PP, Kaplan HS, Musshoff K, Smithers DW, Tubiana M. Report of the Committee on Hodgkin's Disease Staging Classification. *Cancer Res.* 1971;31:1860-1861.

[107] Rai KR, Sawitsky A, Cronkite EP, Chanana AD, Levy RN, Pasternack BS. Clinical staging of chronic lymphocytic leukemia. *Blood.* 1975;46:219-234.

[108] Cerretini R, Chena C, Giere I, Sarmiento M, Arrossagaray G, Rodriguez A, Perez Bianco R, de Dios Soler M, Narbaitz M, Slavutsky I. Structural aberrations of chromosomes 17 and 12 in chronic B-cell disorders. *Eur. J. Haematol.* 2003;71:433-438.

[109] ISCN. An International System for Human Cytogenetic Nomenclature. In Mitelman F, ed. Basel; Karger, 1995.

[110] Chena C, Sarmiento M, Palacios MF, Scolnik M, Slavutsky I. Dup(12)(q13-q22) and 13q14 deletion in a case of B-cell chronic lymphocytic leukemia. *Acta Haematol.* 2000;104:197-201.

[111] Mitelman Database of Chromosome Aberrations in Cancer. Mitelman I, Johansson B, Mertens I; eds. Available at http//cgap.nci.nih.gov/Chromosomes/Mitelman. 2006.

[112] Carroll JA, Raimondi SC, Williams DL, Behm FG, Botowite M, Castelbery RP, Harris MB, Patterson D, Pullen DJ, Crist WM. tdic(9;12): a non random chromosome abnormality in childhood B-cell precursor acute lymphoblastic leukemia: a Pediatric Oncology Study group. *Blood.* 1987;6:1962-1965.

[113] Callet-Bauchu E, Salles G, Gazzo S, Poncet C, Morel D, Pages J, Coiffier B, Coeur P, Felman P. Translocations involving the short arm of chromosome 17 in chronic B-lymphoid disorders: frequent occurrence of dicentric rearrangements and possible association with adverse outcome. *Leukemia.* 1999;13:460-468.

[114] Rudolph KL, Millard M, Bosenberg MW, DePinho RA. Telomere dysfunction and evolution of intestinal carcinoma in mice and humans. *Nat. Genet.* 2001;28:155-159.

[115] Gisselsson D, Jonson T, Yu C, Martins C, Mandahl N, Wiegant J, Jin Y, Mertens F, Jin C. Centrosomal abnormalities, multipolar mitoses, and chromosomal instability in head and neck tumors with dysfunctional telomeres. *Br. J. Cancer.* 2002;87:202-207.

[116] Stewenius Y, Gorunova L, Jonson T, Larsson N, Hoglund M, Mandahl N, Mertens F, Mitelman F, Gisselsson D. Structural and numerical chromosome changes in colon cancer develop through telomere-mediated anaphase bridges, not through mitotic multipolarity. *Proc. Natl. Acad. Sci. USA.* 2005;102:5541-5546.

[117] Gisselsson D, Jonson T, Petersen A, Strombeck B, Dal Cin P, Hoglund M, Mitelman F, Mertens F, Mandahl N. Telomere dysfunction triggers extensive DNA fragmentation and evolution of complex chromosome abnormalities in human malignant tumors. *Proc. Natl. Acad. Sci. USA.* 2001;98:12683-12688.

[118] Travis LB, Curtis RE, Hankey BF, Fraumeni JF. Second cancers in patients with chronic lymphocytic leukemia. *J. Natl. Can. Inst.* 1992;84:1422-1427.

[119] Ohno T, Smir BN, Weisenburger DD, Gascoyne RD, Hinrichs SD, Chan WC. Origin of the Hodgkin/Reed-Sternberg cells in chronic lymphocytic leukemia with "Hodgkin's transformation". *Blood.* 1998;91:1757-1761.

[120] Kanzler H, Küppers R, Helmes S, Wacker H-H, Chott A, Hansmann M-L, Rajewsky K. Hodgkin and Reed-Sternberg-like cells in B-cell chronic lymphocytic leukemia represent the outgrowth of single germinal–centre B-cell-derived clones: potential precursors of Hodgkin and Reed-Sternberg cells in Hodgkin's disease. *Blood.* 2000; 95:1023-1031.

[121] Küppers R, Sousa AB, Baur AS, Strickler JG, Rajewsky K, Hansmann ML. Common germinal-centre B-cell origin of the malignant cells in two composite lymphomas, involving classical Hodgkin's disease and either follicular lymphoma of B-CLL. *Mol. Med.* 2001;7:285-292.

[122] Döhner H, Fisher K, Bentz M, Hansen K, Benner A, Cabot G, Diehl D, Schlenk R, Coy J, Stilgenbauer S, Volkmann M, Galle PR, Poustka A, Hunstein W, Lichter P. p53 gene deletion predicts for poor survival and non-response to therapy with a purine analogs in chronic B-cell leukemias. *Blood.* 1995;85:1580-1589.

[123] Fegan C, Robinson H, Thompson P, Whittaker JA, White D. Karyotypic evolution in CLL: identification of a new sub-group of patients with deletions of 11q and advanced or progressive disease. *Leukemia.* 1995;9:2003-2008.

[124] Oscier D, Fitchett M, Herbert T, Lambert R. Karyotypic evolution in B-cell chronic lymphocytic leukaemia. *Genes Chrom. Cancer.* 1991;3:16-20.

[125] Haus O. The genes of interferons and interferon-related factors: localization and relationships with chromosome aberrations in cancer. *Arch. Immunol. Ther. Exp.* 2000;48:95-100.

[126] Kotoula V, Hytiroglou P, Kaloutsi V, Barbanis S, Kouidou S, Papadimitriou CS. Mismatch repair gene expression in malignant lymphoproliferative disorders of B-cell origin. *Leuk. Lymphoma.* 2002;43:393-399.

[127] Geisler CH, Philip P, Christensen BE, Hou-Jensen K, Pedersen NT, Jensen OM, Thorling K, Andersen E, Birgens HS, Drivsholm A, Ellegaard J, Larsen JK, Plesner T, Brown P, Andersen PK, Hansen MM. In B-cell chronic lymphocytic leukaemia chromosome 17 abnormalities and not trisomy 12 are the single most important

cytogenetic abnormalities for the prognosis: a cytogenetic and immunophenotypic study of 480 unselected newly diagnosed patients. *Leuk. Res.* 1997;21:1011-1023.

[128] Wicks IP, Lapsys NM, Baker E, Campbell LJ, Boyd AW, Sutherland GR. Localization of a human receptor tyrosine kinase (ETK1) to chromosome region 3p11.2. *Genomics.* 1994;19;38-41.

[129] Jain AN, Chin K, Borresen-Dale AL, Erikstein BK, Eynstein Lonning P, Kaaresen R, Gray JW. Quantitative analysis of chromosomal CGH in human breast tumors associates copy number abnormalities with p53 status and patient survival. *Proc. Natl. Acad. Sci. USA.* 2001; 98:7952-7957.

[130] Tran YK, Bogler O, Gorse KM, Wieland I, Green MR, Newsham IF. A novel member of the NF2/ERM/4.1 superfamily with growth suppressing properties in lung cancer. *Cancer Res.* 1999;59:35-43.

[131] Karkera JD, Ayache S, Ransome RJ Jr, Jackson MA, Elsayem AF, Sridhar R, Detera-Wadleigh SD, Wadleigh RG. Refinement of regions with allelic loss on chromosome 18p11.2 and 18q12.2 in esophageal squamous cell carcinoma. *Clin. Cancer Res.* 2000; 6:3565-3569.

[132] Offit K, Louie DC, Parsa NZ, Noy A, Chaganti RSK. Del(7)(q32) is associated with a subset of small lymphocytic lymphoma with plasmacytoid features. *Blood.* 1995;86:2365-2370.

[133] Oscier DG, Gardiner A, Mould S. Structural abnormalities of chromosome 7q in chronic lymphoproliferative disorders . *Cancer Genet. Cytogenet.* 1996;92:24-27.

[134] Andersen CL, Gruszka-Westwood A, Ostergaard M, Koch J, Jacobsen E, Kjeldsen E, Nielsen B. A narrow deletion of 7q is common to HCL, and SMZL, but not CLL. *Eur. J. Haematol.* 2004;72:390-402.

[135] Hebert J, Jonveaux P, Dágay M-F, Berger R. Cytogenetic studies in patients with Richter's syndrome. *Cancer Genet. Cytogenet.* 1994;73:65-68.

[136] Watanabe T, Hotta T, Ichikawa A, Kinoshita T, Nagai H, Uchida T, Murate T, Saito H. The MDM2 oncogene overexpression in chronic lymphocytic leukemia and low-grade lymphoma of B-cell origin. *Blood.* 1994;84:3158-3165.

[137] Huang YQ, Raphael B, Buchbinder A, Li, JJ, Zhang WG, Friedman-Kien AE. Rearrangement and expression of MDM2 oncogene in chronic lymphocytic leukemia. *Am. J. Hematol.* 1994;47:139-141.

[138] Wiley JS, Dao-Ung LP, Gu BJ, Sluyter R, Shemon AN, Li C, Taper J, Gallo J, Manoharan A. A loss-of-function polymorphic mutation in the cytolytic P2X7 receptor gene and chronic lymphocytic leukaemia: a molecular study. *Lancet.* 2002; 359:1114-1119.

[139] Nuckel H, Frey UH, Durig J, Duhrsen U, Siffert W.1513A/C polymorphism in the P2X7 receptor gene in chronic lymphocytic leukemia: absence of correlation with clinical outcome. *Eur. J. Haematol.* 2004;72:259-263.

[140] La Starza R, Barba G, Matteucci C, Crescenzi B, Romoli S, Pierini V, Beacci D, Cantaffa R, Martelli MF, Mecucci C. Chronic lymphocytic leukaemia Is terminal del(14)(q24) a new marker for prognostic stratification? *Leuk. Res.* 2006; Apr 29. Epub ahead of print.

Index

A

B

C

D

E

F

G

H

I

J

K

L

M

N

O

P

Q

R

S

T

U

V

W

X

Y

Z